BIOTELEMETRY III

ACADEMIC PRESS RAPID MANUSCRIPT REPRODUCTION

Proceedings of the
THIRD INTERNATIONAL SYMPOSIUM
ON BIOTELEMETRY
Asilomar, Pacific Grove, California, May 1976

BIOTELEMETRY III

Edited by

THOMAS B. FRYER
NASA-Ames Research Center
Moffett Field, California

HARRY A. MILLER
Stanford University School of Medicine
Palo Alto, California

HAROLD SANDLER, M.D.
NASA-Ames Research Center
Moffett Field, California

Associate Editors
MANLEY J. HOOD
PAUL E. PURSER

ACADEMIC PRESS, INC. New York San Francisco London 1976
A Subsidiary of Harcourt Brace Jovanovich, Publishers

ACADEMIC PRESS, INC.
111 Fifth Avenue, New York, New York 10003

United Kingdom Edition published by
ACADEMIC PRESS, INC. (LONDON) LTD.
24/28 Oval Road, London NW1

Library of Congress Cataloging in Publication Data

International Symposium on Biotelemetry, 3d, Pacific Grove, Calif., 1976.
Biotelemetry III.

Includes bibliographical references and index.
1. Biotelemetry–Congresses. I. Fryer, Thomas B. II. Miller, Harry A. III. Sandler, Harold. IV. Title.
QH324.9.B5I57 1976 574′.028 76-49826
ISBN 0-12-269250-0

PRINTED IN THE UNITED STATES OF AMERICA

CONTENTS

Session II
IMPLANTABLE TELEMETRY SYSTEMS
Chairmen: J. D. Meindl and J. Bojsen

Session III
PATIENT MONITORING, CLINICAL TELEMETRY
Chairman: E. L. Nagel

Session IV
NEUROLOGICAL APPLICATIONS
Chairmen: J. A. J. Klijn and A. M. Dymond

Session VII
BIOTELEMETRY OF CARDIOVASCULAR PARAMETERS
Chairmen: E. P. McCutcheon and R. D. Rader

Session VIII
BIOMEDICAL TELEMETRIC EQUIPMENT
Chairmen: T. B. Fryer and G. Matsumoto

PREFACE

In the Third International Symposium on Biotelemetry, which forms the basis for this book, the scope of the meaning of "biotelemetry" varied to meet the needs of each author. "When I use a word . . . it means just what I choose it to mean—neither more nor less."[1] It's that way with "biotelemetry" in this book. "Bio-" always means life or life processes in the broadest sense. However, "-telemetry" means whatever each author chooses it to mean. Sandler in this volume defines it as "measuring from afar," a broad definition including even voice communication. The term even includes telestimulation and telecontrol as described in some papers. As will be clear by context, most authors in this book restrict "biotelemetry" to transmission of biologic information by radio.

Telemetry of biomedical information, especially by radio transmission, offers many advantages in medical diagnosis and monitoring and in research. Patients are monitored by physicians or nurses as the patients move about freely, even while they are in ambulances, unencumbered by trailing wires. Electronic monitors sound alarms when irregularities occur in the patient's status, thus freeing nurses from constant individual attention. Some functions which are not measurable by other means, such as local temperatures throughout the gastrointestinal tract, are telemetered from ingested, capsule-like radiosondes. Physiological and neurological systems of man and animals are monitored as they function in their natural environments. Implanted instruments telemeter information about internal bodily functions without the problems of leads through the skin. Some use is being made of telemetrically controlled bioelectric stimulation for prosthetic purposes.

The emergence of radio and then of transistors and other miniature, low-power, electronic elements has made possible the blossoming of biotelemetry during the last few years (see the chapters by Sandler and by Holter in this volume). As "necessity is the mother of invention," biotelemetry was advanced rapidly by the National Aeronautics and Space Administration to monitor the well-being of test pilots and astronauts as they performed unusual tasks while under extraordinary stresses in unusual environments many miles from attending physicians.

To promote further development and use of biotelemetry for improved patient care and for research, the Third International Symposium on Biotelemetry was held at Asilomar Conference Grounds, located in Pacific Grove, California, May 17–20,

[1]Carroll, Lewis: *Alice Through the Looking Glass.*

1976. One hundred forty-five physicians, engineers, physicists, and businessmen from 13 countries gathered and exchanged information on the latest ideas and developments, and the as yet unsolved problems in biotelemetry. This volume contains the papers prepared for that symposium.

There was extensive and valuable interchange of information through formal discussion during the sessions, and probably even more so during informal discussions outside of the sessions. Only a small portion of the formal discussion which the Editors consider to be of the broadest interest is included in condensed, paraphrased form. The discussion sections of this publication reflect the collective impressions of the editors and are a compilation of Chairmen's remarks, tape transcriptions of questions and answers, and audience interest.

As an added attraction, the symposium was favored with a keynote address by Harold Sandler and a special lecture by Norman J. Holter, a pioneer in biotelemetry. These lectures describe different historical aspects of biotelemetry. Sandler spoke from the viewpoint of a "recent" eminent researcher. He looked primarily at the written historical record and supplemented the recent record with a few personal experiences. Holter spoke from the viewpoint of the "pioneer" eminent researcher. He looked primarily at his personal experiences and supplemented the record with a few references to others' publications. These two complementary lectures are included as introductions to this book and add an interesting perspective of the field.

The symposium was inspired and promoted by the International Society on Biotelemetry, which contributed especially to the international participation. Further sponsorship, management of the symposium, and editing of this book have been provided by the Cardiology Division of the Stanford University School of Medicine. Cooperating in promoting the symposium were the Institute of Electrical and Electronics Engineers, the American Heart Association, the Association for the Advancement of Medical Instrumentation, and the Stanford Electronics Laboratory of Stanford University.

Special thanks are due to the National Aeronautics and Space Administration for financial support of the Symposium under a grant from the Technology Utilization Division to the Stanford University School of Medicine.[2] NASA has also supported a number of the technological developments reported at the symposium and within this publication.

The organization and smooth running of the symposium as well as coordination of this publication were due to the dedicated efforts of Mrs. Jane Merrill Strauss. The Editors gratefully acknowledge her contributions and thank her publicly for her emotional and professional support. Special thanks are due to the Staff consisting of Carla Dunham, Linda Gunzel, Dr. Gene Schmidt, Sam Davidsen, and Cecil Profitt who each contributed in a major way to the success of the Symposium and assistance with this publication. In addition, the Editors wish to acknowledge the contribution of Dr. Peter Kimmich, President of the International Society of Biotelemetry, for his assistance in planning this Conference and promotion of Biotelemetry as a professional entity.

[2]This work was supported in part by NASA Grant NGR 05-020-634 to the Stanford University School of Medicine, Cardiology Division, Palo Alto, California.

LIST OF SYMPOSIUM PARTICIPANTS

Allen, Henry V., Research Associate, Integrated Circuits Laboratory, Stanford University, Stanford, California 94305

Ames, James G., Director, Biomedical Systems Engineering, Samaritan Health Service, Good Samaritan Hospital, 1033 E. McDowell Road, Phoenix, Arizona 85006

Andersson, Tomas, EEG Laboratory, Vipeholms Sjukhus, 221 01 Lund, Sweden

Andersson, Torsten, EEG Laboratory, Vipeholms Sjukhus, 221 01 Lund, Sweden

Barr, R. E., Ph.D., Dalton Research Center, University of Missouri, Columbia, Missouri 65201

Becker, Winfried M. A., Dr.-Ing., Euratom-CCR, I-21020 Ispra (VA), Italy

Black-Cleworth, Patricia Ann, Assistant Research Anatomist, NPI, UCLA, 405 Hilgard Avenue, Los Angeles, California 90024

Bojsen, Jørgen, The Finsen Institute, Strandboulevarden 49, Copenhagen, DK-2100, Denmark

Bourne, Geoffrey H., Director, Yerkes Primate Center, Emory University, Atlanta, Georgia 30322

Brastad, Brian A., Project Engineer, Medtronic, Inc., 6970 Central Avenue NE, Minneapolis, Minnesota 55432

Brockelsby, William Keith, Gallagher Electronics Ltd., P. O. Box 5324, Frankton, Hamilton, New Zealand

Burchard, Detlef A., Dipl-Ing., Engineering Office Burchard, Tiroler Weg 2, Ebringen 7801, West Germany

Butler, Patrick J., Dr., Lecturer, Department of Zoology, University of Birmingham, Birmingham B15 2TT, England

Carlson, Charles E., Sc.D., Research Associate, Massachusetts Institute of Technology, 3-455, Cambridge, Massachusetts 02139

Carlson, Edwin L., M.D., Research Associate, NASA-Ames Research Center; and Cardiology Division, Stanford University School of Medicine, Stanford, California 94305

Catalano, Genard T., Assistant Professor, Electronics Systems, Division of Engineering, San Francisco State University, San Francisco, California 94132

Cathignol, D., Chargé de Recherche, Institute National de la Santé et Recherche Medicine, Chirurgie Vasculaire et Transplantation D'Organes, 18 Avénue du Doyen Lépine, Bron, 69, France

Chapman, Loring F., Professor, University of California, Davis, California 95616

Christiansen, Robert A., Sales Engineer, Motorola C & E, Redding, California 96001

Collins, John J., 2037 West 177 Street, Torrance, California 90504

Conti, Luis A., Senior Design Engineer, SPACELABS, 20550 Prairie Street, Chatsworth, California 91311

Corbin, Steven D., Biomedical Engineer, Department of Cardiovascular Surgery, Stanford University School of Medicine, A206, Stanford, California 94305

Cormey, Walter P., Staff Engineer, Pacific Communications, Inc., 2041-A South Grand Avenue, Santa Ana, California 92705

Couvillon, L. A., Director of Engineering, Utah Biomedical Test Laboratory, 520 Wakara Way, Salt Lake City, Utah 84108

Cupal, Jerry J., Ph.D., Biologist, Rocky Mountain Forest and Range Experiment Station, Box 3313 University Station, Laramie, Wyoming 82071

Davidsen, J. S. W., Conference Staff, Cardiology Division, Stanford University School of Medicine, Stanford, California 94305

Davies, Roderick D., Research Assistant, Integrated Circuits Laboratory, Stanford University, Stanford, California 94305

Devaney, Michael J., Associate Investigator, Dalton Research Center, University of Missouri, Columbia, Missouri 65201

Dunham, Carla, Conference Staff, Cardiology Division, Stanford University School of Medicine, Stanford, California 94305

Dwight, Arthur D., Biomedical Engineering Technician, Veterans Hospital, Palo Alto, California 94306

Dymond, Anthony M., Ph.D., Veterans Hospital (Brentwood) MS 530/151D, Los Angeles, California 90073

Engstrom, Everis R., Biomedical Engineer, Biomedical Applications Team, University of Wisconsin, 1500 Johnson Drive, Madison, Wisconsin 53706

Fletcher, Bruce, Konigsberg Instruments, 2000 East Foothill Boulevard, Pasadena, California 91107

Ford, Leta Frances, Research Physical Therapist, Children's Hospital at Stanford, 520 Willow Road, Palo Alto, California 94304

Freund, William R., Research Engineer, Applied Electronics Laboratory, Stanford University, Stanford, California 94305

Frumin, M. Jack, M.D., Clinical Professor of Anesthesia, Stanford University School of Medicine, Stanford, California 94305

Fryer, Thomas B., Chairman, Third International Symposium on Biotelemetry; Assistant Chief, Electronic Instruments Development Branch, NASA-Ames Research Center, Mail Stop 213-3, Moffett Field, California 94035

Garay, Gabriel L., Ph.D., Institute of Biological Sciences, Syntex Corporation, 3401 Hillview Avenue, Palo Alto, California 94304

Geating, John A., Laboratory Director, General Electric Company, Box 8555, Philadelphia, Pennsylvania 19101

Gessaman, James A., Ph.D., Department of Biology, UMC 53, Utah State University, College of Science, Logan, Utah 84322

Gottlieb, Philip, Chairman, Biomedical Technology, Texas State Technical Institute, Waco, Texas 76705

Graystone, Peter, Ph.D., Biomedical Engineer, University of British Columbia, I.R.C. Building, Room 350, Vancouver, V6T 1W5 British Columbia, Canada

Guckel, H., Professor, Department of Electrical Engineering, University of Wisconsin, Madison, Wisconsin 53706

Gunzel, Linda, Conference Staff, Cardiology Division, Stanford University School of Medicine, Stanford, California 94305

Hahn, Allen W., DVM, Ph.D., Dalton Research Center, University of Missouri, Columbia, Missouri 65201

Hall, Robert John, Chief Executive Officer/Chief Engineer, Medical Telectronics Ltd., P.O. Box 4323, 4 Quay Street, Auckland, New Zealand

Hannibal, Steen, Copenhagen County Hospital, Niels Andersensvej 65, Copenhagen 2900, Denmark

Harrison, Donald D., M.D., Chief, Cardiology Division, Department of Medicine, Stanford University School of Medicine, Stanford, California 94305

Hartley, J., Dalton Research Center, University of Missouri, Columbia, Missouri 65201

Hess, Orvan W., M.D., Clinical Professor, Obstetrics-Gynecology, Yale School of Medicine, 333 Cedar Street, New Haven, Connecticut 06510

Hobbs, Wesley D., Section Head, Health Service, Los Angeles County Department of Communications, 500 West Temple Street, Los Angeles, California 90012

Holter, Norman J., D.Sc., 7688 Caminito Coro Mandel, La Jolla, California 92037

Hood, Manley J., Conference Staff, Cardiology Division, Stanford University School of Medicine, Stanford, California 94305

Hughes, Roy, Dr./Pharmacologist, Wellcome Research Laboratories, Beckenham BR3 3BS, Kent, England

Hull, Maury L., Ph.D., Postdoctoral Fellow, University of California, 5144 Etcheverry Hall, Berkeley, California 94720

Hyndman, Bruce H., Manager, Biomedical Engineering, Room E006, Stanford University Medical Center, Stanford, California 94305

Indyk, Leonard, Ph.D., Assistant Professor of Medical Physics, Department of Pediatrics, College of Physicians and Surgeons, 630 West 168 Street, New York, New York 10032

Ingram, Mike, AVM Instrument Company, 810 Dennison Drive, Champaign, Illinois 61820

Ives, John R., Biomedical Engineer, Montreal Neurological Institute, 3801 University Avenue, Montreal H3A 2B4, Canada

James, Gordon W., Staff Engineer, Motorola, Inc., 8000 W. Sunrise, Plantation, Florida 33317

Jennings, Robert G., Biomedical Engineering Technician, Upstate Medical Center, 117 Monroe Street, Syracuse, New York 13210

Jilek, Jiri, Biomedical Engineer, USC Medical Center, Women's Hospital, 5K40, 1240 N. Mission Road, Los Angeles, California 90033

Johnson, Paul, M.D., Nuffield Institute for Medical Research, Oxford, England

Kapp, Hermann, Biomedical Engineer, Bio–Ing. Experimental Lab. Neurological Hospital for Neurophysiology, Hansastr. 9a, Frieburg D7800, Germany

Karlson, Douglas A., Research Engineer, Stanford Electronics Laboratory, AEL 121, Stanford, California 94305

Kimmich, H. P., Ph.D., President, International Society on Biotelemetry; Vice Chairman, Third International Symposium on Biotelemetry; Director, Department of Physiology, University of Nijmegen, Geert Grooteplein, Noord 21a, Nijmegen, The Netherlands

Klein, Fritz, Ph.D., Department of Anesthesiology, Duke Medical Center, Box 3094, Durham, North Carolina 27710

Klewe, Hans J., Dipl.–Ing., Group Leader, Institute for Flight Mechanics, DFVLR Flughafen, Braunschweig 3300, West Germany

Klijn, J. A. J., Ph.D., Department of Neurophysiology, University of Nijmegen, Geert Grooteplein Noord 21, Nijmegen, The Netherlands

Knutti, James W., Research Associate, Integrated Circuits Laboratory, Stanford University, Stanford, California 94305

Ko, Wen H., M.D., Director, Engineering Design Center, Case Western Reserve University, Cleveland, Ohio 44106

Koelle, Alfred R., Staff Member, Los Alamos Scientific Laboratory, Box 1663, MS 428, Los Alamos, New Mexico 87545

Kojima, Gilbert K., Electronics Engineer, NASA-Ames Research Center, MS 213-3, Moffett Field, California 94035

Konigsberg, Eph, President, Konigsberg Instruments, 2000 East Foothill Boulevard, Pasadena, California 91107

Kravtsov, Valeriy I., Consultant to International Department of the USSR Central Council of Trade Unions, Moscow, USSR

Küchler, Georg, Professor, Deutsches Zentralinstitut fur Arbeitsmedizin, Noldnerstr. 40-42, DDR-1134 Berlin-Lichtenberg, East Germany

Laird, Gregory D., Biomedical Engineer, Good Samaritan Hospital, 1120 N.W. 20th, Portland, Oregon 97209

Leung, Albert, Graduate Student, Engineering Design Center, Case Western Reserve University, Cleveland, Ohio 44106

Lieb, Jeffrey P., Assistant Research Neurophysiologist, Reed Neurological Research Center, UCLA School of Medicine, Los Angeles, California 90024

Lund, Gordon F., Research Associate, NASA-Ames Research Center, Moffett Field, California 94035

MacQuivey, Donald, Stanford Electronics Laboratory, Stanford University, Stanford, California 94305

Madsen, Jack, M.D., Division of Pediatric Neurology, University of Utah, College of Medicine, 50 North Medical Drive, Salt Lake City, Utah 84132

Matsumoto, Goro, Prof./Dr., Research Institute of Applied Electricity, Hokkaido University, Sapporo, Japan

Matsuo, Fred, M.D., Department of Neurology, University of Utah, 50 North Medical Drive, Salt Lake City, Utah 84132

McClure, Derek, Engineer, Alberta Systems Development Group, 10025–108 Street, Edmonton, Alberta T5J 1K9, Canada

McCutcheon, Ernest P., M.D., Research Medical Officer, NASA-Ames Research Center, Moffett Field, California 94035

McEwen, G.N., Jr., Ph.D., Research Scientist, NASA-Ames Research Center, Moffett Field, California 94035

McHenry, Robert B., Product Support Manager, Hewlett-Packard Co., 175 Wyman Street, Waltham, Massachusetts 02154

Meindl, James D., Ph.D., Professor of Electrical Engineering, Stanford University, Stanford, California 94305

Miller, Harry, Coordinator, Third International Symposium on Biotelemetry; Administrator, Cardiology Division and Deputy Director, Biomedical Technology Transfer, Stanford University School of Medicine, Stanford, California 94305

Miranda, Rafael, Electrical Engineer, NASA-Ames Research Center, Moffett Field, California 94035

Montgomery, Judy, Vice President, AVM Instrument Company, 810 Dennison Drive, Champaign, Illinois 61820

Morgan, Wayne, Electronics Engineer, Pacesetter Systems, Inc., 12741 San Fernando Road, Sylmar, California 91342

Murphy, John M., Assistant Manager, Clinical Chemistry, Norden Laboratories, Inc., 601 W. Cornhusker, Lincoln, Nebraska 68501

Nagel, Eugene L., M.D., Department of Anesthesia, Harbor General Hospital (UCLA), Torrance, California 90509

Naik, Kamal, Ohio Medical Products, 3030 Airco Drive, Madison, Wisconsin 53701

Neukomm, Peter A., Dipl. El. Ing. ETH, Laboratory of Biomechanics, ETH-Z, Gloriastr. 35, Zurich 8006, Switzerland

Neuman, Michael R., Associate Professor, Engineering Design Center, Case Western Reserve University, Cleveland, Ohio 44106

O'Handley, Douglas A., Ph.D., Assistant Manager, Biomedical Systems Office, Jet Propulsion Laboratory, 4800 Oak Grove Drive, Pasadena, California 91103

Ordon, Vladimir, Ph.D., Consultant, 312 Loma Avenue, Long Beach, California 90814

Pauley, James D., Ph.D., Assistant Professor, University of Colorado Medical Center, 4200 East Ninth Avenue, Denver, Colorado 80220

Paulsen, A. William, Instructor, Department of Anesthesiology, Emory University, 656 Woodruff Memorial Building, Atlanta, Georgia 30322

Perachio, Adrian A., Ph.D., Head, Laboratory Neurophysiology, Yerkes Regional Primate Research Center, Emory University, Atlanta, Georgia 30322

Pettengill, Robert C., Research Assistant, Stanford Electronics Laboratory, Stanford University, Stanford, California 94305

Phillips, Robert L., ECG Laboratory Supervisor, Cardiology Division, Stanford University School of Medicine, Stanford, California 94305

Portner, Peer M., Ph.D., Vice President, Andros Inc., 2332 Fourth Street, Berkeley, California 94710

Profitt, Cecil L., Engineer, Cardiology Division, Stanford University School of Medicine, Stanford, California 94305

Purser, Paul E., Conference Staff, Cardiology Division, Stanford University School of Medicine, Stanford, California 94305

Pykhtin, Victor S., Institute of Physical Culture, Moscow, USSR

Quanbury, Arthur O., Acting Head, Biomedical Engineering, Shriners Hospital, 633 Wellington Crescent, Winnipeg, Manitoba, Canada

Rader, Roland D., Ph.D., Laboratory Director, University of Southern California, Centrifuge Building, 815 W. 37th Street, Los Angeles, California 90007

Ratov, Igor P., Dr. ped., Professor, All-Union Scientific Research Institute of Physical Culture, Head of Biomechanics Laboratory, Kazakova str. 18, Moscow K-64, USSR

Reischl, Peter, Ph.D., Biomedical Engineering Department, University of Southern California, Los Angeles, California 90007

Reite, Martin L., M.D., Associate Professor of Psychiatry, University of Colorado Medical Center, 4200 E. 9th Avenue, Denver, Colorado 80220

Rice, Dale H., Research Staff, Yerkes Regional Primate Research Center, Emory University, Atlanta, Georgia 30322

Ruddle, Jerry C., Biomedical Engineer, Research Triangle Institute, P.O. Box 12194, Research Triangle Park, North Carolina 27709

Salmons, Stanley, Ph.D., Lecturer, Department of Anatomy, University of Birmingham Medical School, Birmingham B15 2TJ, England

Sander, Craig Steven, Research Assistant, Stanford Integrated Circuits Laboratory, AEL 123, Stanford University, Stanford, California 94305

Sandler, Harold, M.D., Vice Chairman, Third International Symposium on Biotelemetry; Chief, Biomedical Research Division, NASA-Ames Research Center, Moffett Field, California 94035

Schmidt, Gene, M.D., Biomedical Technology Transfer, Cardiology Division, Stanford University School of Medicine, Stanford, California 94305

Schulman, Joseph H., Ph.D., Vice President, Research and Development, Pacesetter Systems, Inc., 12740 San Fernando Road, Sylmar, California 91342

Scidmore, Allan K., Ph.D., Professor, Electrical and Computer Engineering, University of Wisconsin, Madison, Wisconsin 53706

Seawright, Gary L., D.V.M., Veterinary Medical Officer, Veterinary Services Laboratories, P.O. Box 70, Ames, Iowa 50010

Silverberg, Gerald David, M.D., Assistant Professor of Neurosurgery, Stanford University School of Medicine, Stanford, California 94305

Smith, Lee M., Materials Engineer, College of Medicine, College of Engineering, University of Utah, Building 518, Salt Lake City, Utah 84112

Steinhagen, Wolfgang, Dipl.–Ing., Institut für Theoretische Elektrotechnik, Kopernikusstr. 16, 51 Aachen, West Germany

Stone, H. Lowell, Ph.D., Professor and Chief, Cardiovascular Control Section, Marine Biomedical Institute, 200 University Boulevard, University of Texas Medical Branch, Galveston, Texas 77550

Strauss, Jane M., Conference Staff, Cardiology Division, Stanford University School of Medicine, Stanford, California 94305

Strosberg, Arthur M., Ph.D., Principal Scientist, Section Head, Cardiovascular Pharmacology, Syntex, 3401 Hillview Avenue, Palo Alto, California 94304

Stuart, Jerry L., Consultant, Systems Design, J. Stuart Enterprises, P. O. Box 2219, Grass Valley, California 95945

Stupin, Yuriy V., Chairman, Kislovodsk Territorial Council on Trade Union Health Resorts Management, Kislovodsk, USSR

Tatarenko, Leonid I., Director, Research Biotelemetry Laboratory, P.B. 41, Kislovodsk 357700, USSR

Updike, Otis L., Professor, Biomedical Engineering, Thornton Hall, University of Virginia, Charlottesville, Virginia 22901

Verberne, Gerda, Ph.D., Laboratory of Animal Physiology, Kruislaan 320, Amsterdam, The Netherlands

Vreeland, Robert W., Associate Development Engineer, Research and Development Laboratory, University of California Medical Center, San Francisco, California 94143

Wagner, Lektor Gorm, Med Fys. Institute B, Juliana Mariesvy 30, 2100 Copenhagen, Denmark

Weller, Clive, Scientist, Bioengineering Division, Medical Research Council, Clinical Research Centre, Watford Road, Harrow, Hal 3UJ, England

Westbrook, Richard M., Electronics Engineer, NASA-Ames Research Center, N-213-3, Moffett Field, California 94035

Williams, Bill A., Research Physiologist, NASA-Ames Research Center, Moffett Field, California 94035

Wilson, Daniel, Ph.D., Director, Biomedical Systems Science, Samaritan Health Service, Good Samaritan Hospital, 1033 E. McDowell Road, Phoenix, Arizona 85006

Woltring, Herman J., Ir.Msc., Laboratory of Psychology, University of Nijmegen, Erasmuslaan 16, Nijmegen 6804, The Netherlands

Yarborough, John M., Jr., Senior Research Engineer, Stanford Research Institute, 333 Ravenswood Avenue, Menlo Park, California 94025

Young, Donald R., Ph.D., Chief, Environmental Physiology Branch, NASA-Ames Research Center, N-239-7, Moffett Field, California 94035

Zerzawy, Reinhard, M.D., Med. Poliklinik, Oestl. Stadtmauerstr. 29, D 8520 Erlangen, West Germany

Zimmerman, Robert, Biomedical Research Division, NASA-Ames Research Center, Moffett Field, California 94035

BIOTELEMETRY: ITS FIRST 50 YEARS

Harold Sandler

Biomedical Research Division
National Aeronautics and Space Administration
Ames Research Center
Moffett Field, California

".... Nothing is complicated, its our understanding of it that makes it so ...."

Alfred Whitehead

Biotelemetry or obtaining biologic information from afar has made incredible advances over the past 50 years. It is for the reason of assessing these changes and progress over the past few years that the present meeting is being held. Technologic advances involving the increasing use of integrated circuit techniques, improved packaging and advanced construction methods have allowed for an increasing number of measurements to be made from animal or man without encumberances by wires or interferences with physiologic function. The task set for these introductory remarks is to briefly review the history of this progress, to detail the various classes of operation available and to survey resultant useages. As continued improvement of these systems is realized, it is fully expected that biotelemetry systems will be increasingly used for monitoring animals during research operations and for human applications at rest and during various environmental stresses.

Medical telecommunication is not a new phenomenon. In its broadest form, human language is a form of telemetry -- a way of delivering information from a distance. At sea, flags and lamps are still used today as ancient man used smoke signals to pass on information, or summon help. In the context of its modern use, biotelemetry has come to mean the detection and transmission of biologic information with the least possible interference to the phenomenon being examined, doing so from near or far away. An additional objective is to deliver this information in as unambiguous or as noise-free a form as possible. Most recently, investigations are coming to realize that the system is not a one-way street, but interactive. This has allowed for communication from the observer to the animal, or human being, being observed (e.g., nerve stimulation or demand cardiac pacing) as well as reception of the resultant information.

The first reported successful transmission of information from living unanesthetized animals was accomplished by Marey in France in 1869 (1). Flapping vanes (sensor or transducer) were attached to the wing tips of a pigeon. These served to close electrical contacts which passed current through trailing wires (transmission channel) to an electromagnet (receiver) which drove the stylus of a smoked drum (display). Almost all modern systems continue to follow this model. In 1903 Einthoven (2), was the first to record and transmit the human electrocardiogram. The subjects' hands and feet were placed in buckets of saline and signals transmitted over the wires of the Leiden Telephone System where they were recorded a mile away by a string galvanometer system. Similar hard wire telemetry to connect various wards of a hospital to a heart station within a hospital or to a physiology laboratory was presented by Barker in 1910 (3).

The modern history of biotelemetry begins with the transmission of heart sounds over a marine radio link in 1921 by Winters (4). In this case, the U.S. Army Signal Corps developed a system for transmission of heart sounds from ships without on-board physicians to medical facilities on shore. Existing radio devices were used, but were applied for a biomedical purpose.

By 1948 Fuller and Gordon (5), had developed a system for transmission of pulse wave forms and respiration rate from freely moving animals for a distance of 100 feet. Over the period 1947 to 1953 Holter (6), and later Parker and co-workers (7), accomplished the first reliable radio transmission of human electrocardiograms and electroencephalograms. In 1952 a team of investigators lead by James Henry (8), accomplished the first transmission of biologic data during rocket flight as shown in Figure 1.

Figure 1. Transmission of biological data during rocket flight

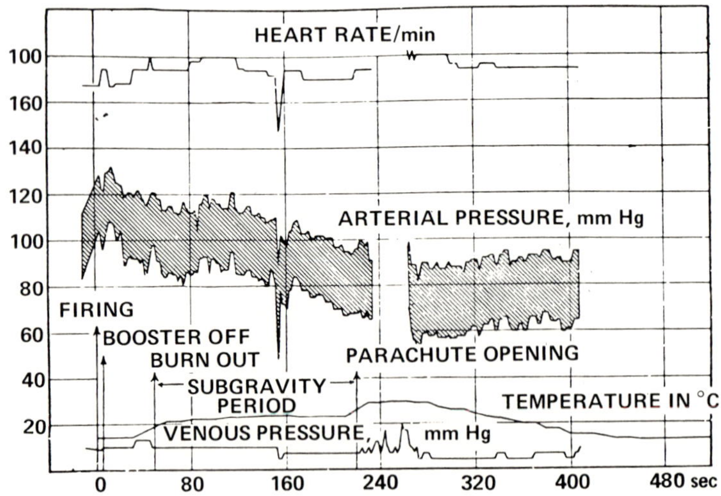

As can be seen, blood pressure, heart rate and venous pressure were all recorded from intravascular catheters from the time of rocket firing to parachute opening. Also starting in 1952, aircraft to ground recordings of ECG were accomplished (9). The first telemetry recordings of brain potentials of unanesthetized animals were reported by Gold and Perkins (10), in 1957. Telemetric studies of pressure, temperature and motility of the gastrointestional tract were reported in the same year by Farrar and co-workers (11). Finally in 1957, the Russians accomplished the first transmission of biologic data from space on November 3, with the historic flight of the dog, Laika, in Sputnik II (12), and the modern era of radiotelemetry had now come into its own.

It is now clear that zoologists, physiologists and medical practioners have used telemetry systems to measure almost every parameter of interest in animals or man using invasive and/or noninvasive means. Much of this has been made possible during the past two decades due to miniaturization and battery operation made possible by the invention of the transistor. To detail each and every system and its use is beyond the scope of this review. However, their uses do fall into several broad categories which can be grouped. First, with respect to class of operation, then later with respect to criteria for engineering design and, finally, for their areas of use. The various classes of operation are shown in Figure 2 and describe ways in which the transmitter may be arranged in relation to the subjects (13). The transmitter may be fixed or move with (on or within) the subject.

Figure 2. Classes of transmitter operation denoting its relationship on the subject with respect to receiver

CLASS I	SUBJECT DISTANT TO TRANSMITTER CONNECTED BY WIRE TRANSMITTER-RECEIVER FIXED
CLASS II	SUBJECT DISTANT TO RECEIVER CONNECTED BY WIRE TRANSMITTER-RECEIVER FIXED BUT PORTABLE
CLASS III	TRANSMITTER ON SUBJECT NO WIRES TRANSMITTER MOVES WITH RESPECT TO RECEIVER
CLASS IV	TRANSMITTER INSIDE SUBJECT = IMPLANTED NO WIRES TRANSMITTER MOVES WITH RESPECT TO RECEIVER
CLASS V	TRANSMITTER SWALLOWED BY SUBJECT NO WIRES TRANSMITTER MOVES RELATIVE TO SUBJECT AND RECEIVER

Figure 3 gives an example of these differences in practical terms. The great bulk of present work is being conducted under Class III operation. Class IV and V uses have received increasing attention over the past decade.

Figure 3. Examples of Class II and Class III biotelemetry transmitter operation

CLASS II MAN IN SPACE	COMPARED	CLASS III BRAIN POTENTIAL FROM FREELY MOVING CAT
TRANSMITTER IN SPACE CAPSULE		TRANSMITTER ON CAT
SIZE NO PROBLEM		SIZE SMALL
POWER NO PROBLEM		BATTERY OPERATION
ELECTRODES EXTERNAL		ELECTRODES INTERNAL

All telemetry systems are composed of the building blocks shown in Figure 4: sensor, communications system, and user or recipient.

Figure 4. Schematic of components of typical biotelemetry system

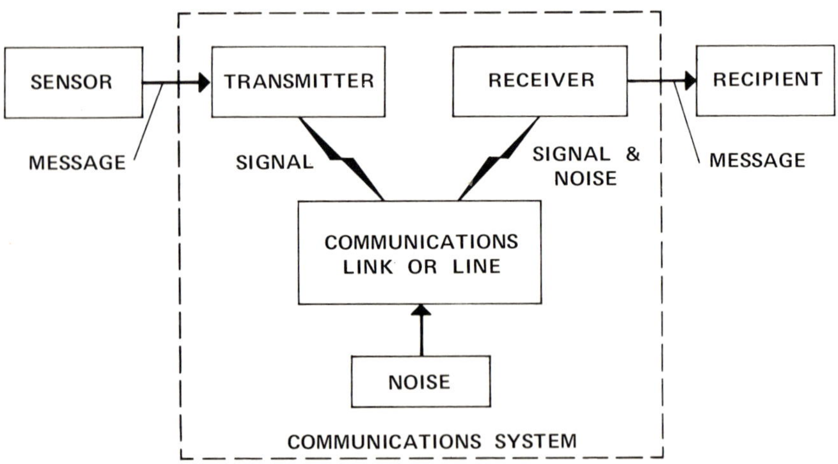

Design objectives are to provide a communication link which relays the message provided by the sensor as reliably and accurately as possible under the circumstances. Noise is always present, as shown, and system design must be directed towards eliminating or minimizing this feature.

An analysis of the factors which must be considered when designing a telemetry system are summarized under the 11 categories shown in Figure 5 (13).

Figure 5. Categories of design considerations for biotelemetry systems

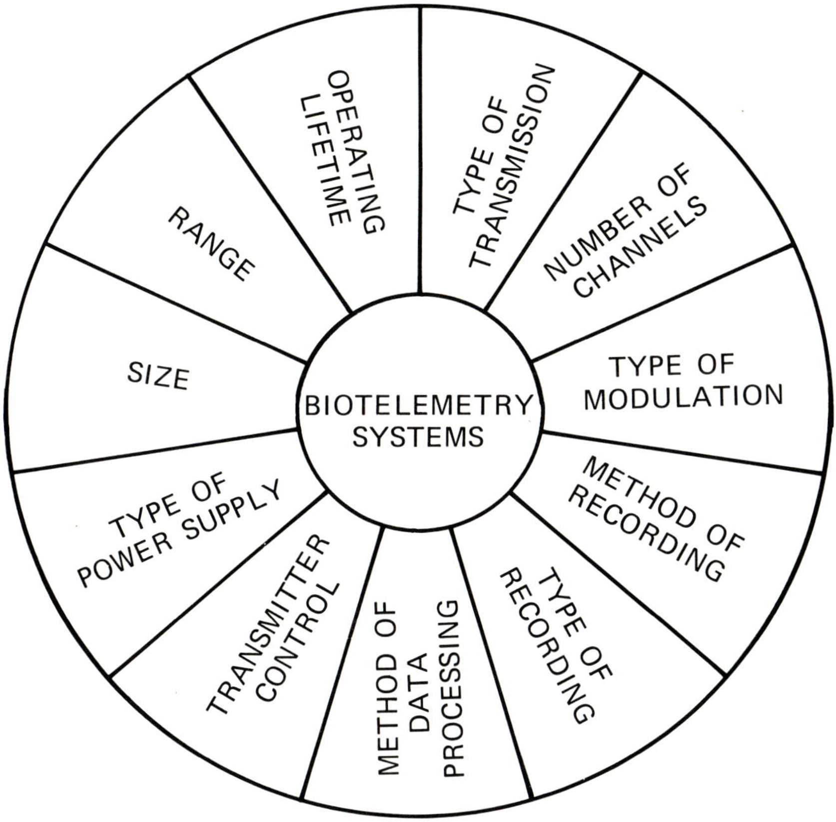

These involve concern for type of transmission, number of channels, type of modulation, method of recording, type of recording, method of data processing transmitter control, type of power supply, size, range and operating life time of units. These criteria have become most important as need and use of Class IV and Class V transmitting systems have become more wide spread. Almost every portion of the electromagnetic spectrum has been used for transmission which include radiowaves, microwaves, ultrasound, optical waves and lasers. For purposes of these discussions, only radiowave useage will be considered. Single, multiple, single or multichannel systems are now available for use. In the latter case, at least 8 to 10 channels of information can be transmitted on a single carrier from the body surface or from inside the body (14,15). Power supplies have changed drastically over the past several years, and now include nuclear batteries (16), piezoelectric crystals (17), and body electrodes (18-20). Simple carbon-zinc (flashlight) or mercury cell batteries can now be replaced by rechargeable nickel-cadmium cells (21,22). More recently, radio frequency (passive) induction of power has been reused to entirely replace the implanted battery (23). Size has also constantly decreased, especially for implant use, as discrete parts construction has been replaced by microminiature elements, thin film (24), and finally, solid state parts (25).

Biotelemetry systems may be used for two purposes: for transmission of physiological information or for remote ecological monitoring. The manner of system useage in either instance is shown in Figure 6.

Figure 6. System diagram for use of biotelemetry units for human or animal monitoring

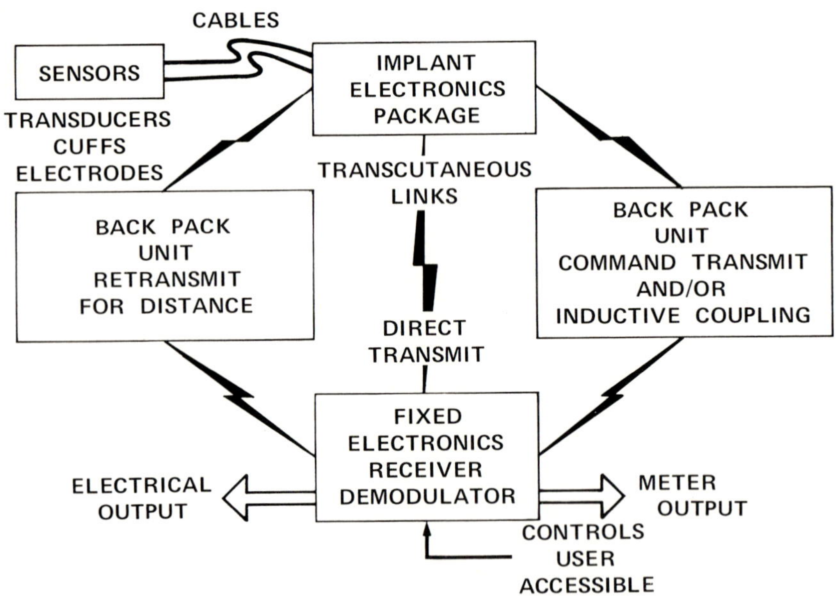

Sensors are connected by cable to a transmitter and relayed for subsequent observation or storage. The most common and economical unit is a back-pack system which uses body surface sensors (nothing is implanted inside the body). Implanted units have operated as shown by direct transmission to a receiver or through a back-pack relay (transponder) which may then re-radiate a received signal for several hundred feet to several miles. The back-pack may also be used to power the implant or recharge its power supply (22,23). Figure 7 summarizes the broad categories of telemetry use for ecologic purposes which provide information relating to determination of location (26), or social interaction (27,28).

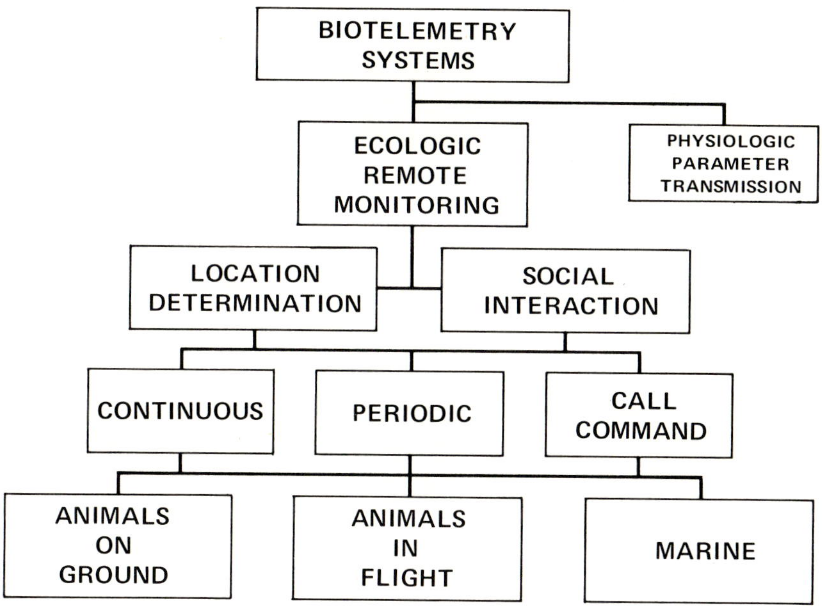

Figure 7. Categories for use of biotelemetry systems for ecologic monitoring

Figure 8 details the broad categories of useage for animal or human physiological parameter transmission.

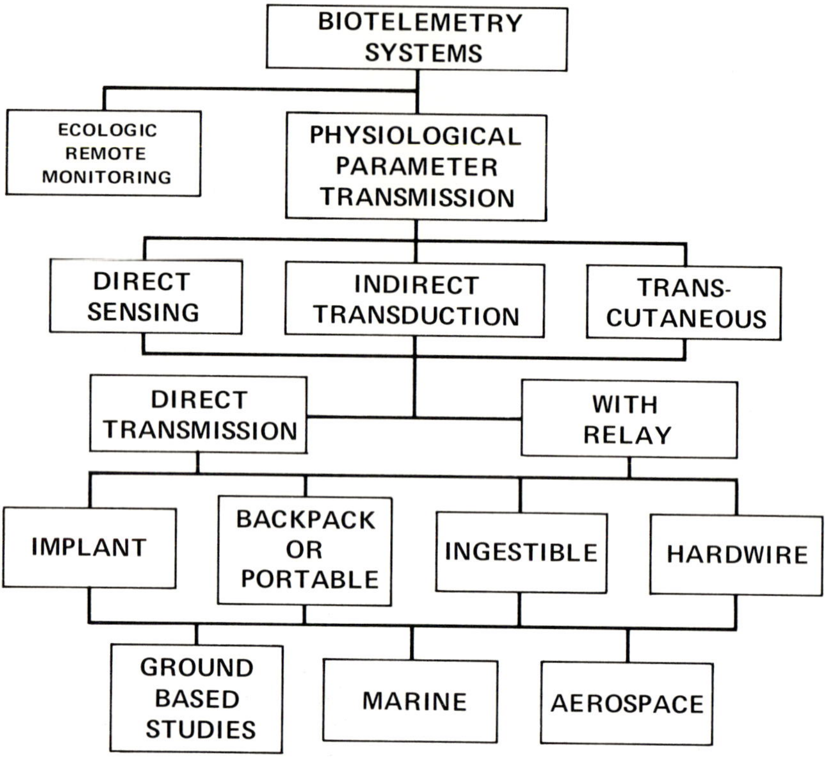

Figure 8. Categories of use of biotelemetry systems for monitoring of physiologic parameters

By far the largest activity has involved direct transmission of data from units worn external to the body providing physiological recordings during sports and work (29,30), normal activity (31), or sleep (32); monitoring of astronauts' health (33); surveillance of critically ill patients (34-36). Recent activity and developments have been directed at design of implant or swallowable devices to provide additional data or information not readily available by surface monitoring methods (37). Medical diagnosis using a telemetry link has now been proven useful and feasible as well (38,39).

Progress over this half century has relied on the production of continuously smaller devices with improved circuit design. Further developments will depend on improved reliability, decrease in cost and standardization of procedures for use and manufacture. These approaches will result in an absolute increase in the number of units available for study of man under normal daily activity and in research animals in an awake unanesthetized state. The future looks exceedingly bright. Implant telemetry remains as a field open to rapid development and increased application as various systems become used for drug testing or study of the progress of induced disease states. With continued advances, small units will eventually be used in human subjects as parts or controllers of prosthetics such as limbs (40), or artificial hearts (41), or left at the time of surgery to measure various variables such as ECG, blood pressure or brain potentials (37).

REFERENCES

1. Marey M: Phenomenon of flight in the Animal Kingdom. Smithsonian Annual Report, Washington, D.C., U.S. Government Printing Office, 1869, pp 226-285
2. Einthoven W: Galvanometrische Registrirung des Menchlichen Electrokardiograme. Arch Ges Physiol 99: 472, 1903
3. Barker LF: Electrocardiography and phonocardiography, A Collective Review. Johns Hopkins Hosp Bull 20: 358, 1910
4. Winters SR: Diagnosis by wireless. Sci Am 124: 465, 1921
5. Fuller SL and Gordon TM Jr: The radio inductograph, a device for recording physiological activity in unrestrained animals. Sci 108: 287, 1948
6. Holter NJ and Generelli JA: Remote recording of physiologic data by radio. Rocky Mountain Med J 46: 747, 1949
7. Parker CS, Breakell CC and Christopherson F: The radio-electrophysiologogram; radio transmission of electrophysiological data from the ambulant and active patient. Lancel 1: 1285, 1953
8. Henry JP, Augerson WS, Belleville RE et al.: Effects of weightlessness in ballistic and orbital flight: A Progress Report. Aerospace Med 33: 1056, 1962
9. Smith GB and Lamb LE: Vectorcardiography in aerospace flight.

Applications and rationale. Amer J Cardiol 6: 62, 1960
10. Gold DC and Perkins WJ: A miniature electroencephalograph telemeter system. Electronic Eng 31: 337, 1959
11. Farrar JT, Zworykin VK and Baum J: Pressure-sensitive telemetering capsule for study of gastrointestinal motility. Sci 126: 975, 1957
12. Chernov VN and Yakolev VI: Research on the flight of a living creature in an artificial earth satellite. Ars J 29: 736, 1959
13. Parin VV, Bayevskiy RM and Geller YeS: Structure and classification of biotelemetric systems. In Biologischeskaya Telemetriya (ed V Parin), Moscow, Meditsina Press, 1971, pp 33-56
14. Olsen DE, Moise SL and Huston SW: An eight channel semi-implantable telemetry system for animal research. In Biotelemetry II (ed P Neukomm), 2nd Int Symp, Karger, Basel, 1974, pp 173-175
15. Fryer TB, Sandler H and Datnow B: A multichannel telemetry system. Med Res Eng 8: 9, 1969
16. Greatback W and Bustard TS: A $Pu^{238}O_2$ nuclear power source for implantable cardiac pacemakers. IEEE Trans Biomed Eng BME 20: 336, 1973
17. Enger CC and Kennedy JH: Piezoelectric power sources utilizing the mechanical energy of the human heart. Proc Conf Eng Med Biol 5: 60, 1963
18. Roy OZ: Biological energy sources: A Review. Biomed Eng 6: 250, 1971
19. Cooper, JB and Hahn AW: Hybrid biological power cells for cardiac pacemakers: Material Evaluation. IEEE Trans Biomed Eng BME 20: 336, 1973
20. Fester K and Doty RL: Solid state batteries for cardiac pacemakers. Excerpta Med, Med Instrum 7: 172, 1973
21. Love JW, Lewis KB and Fischell RE: The Johns Hopkins rechargeable pacemaker. Historical Aspects. Spec Comm, JAMA 234: 64, 1975
22. Fryer TB, Sandler H, Freund W et al.: A multichannel implantable telemetry system for flow, pressure and ECG measurements. J Appl Physiol 39: 318, 1975
23. Fryer TB: Power sources for implanted telemetry systems. Biotelemetry 1: 31, 1974
24. Ko WH and Neuman MR: Implant biotelemetry and microelectronics. Report on developments in implant telemetry associated problems and the potential of microelectronics. Sci 156: 351, 1967.
25. Gill RW and Meindl JP: An integrated circuit implantable pulsed doppler ultrasonic blood flowmeter. In Biotelemetry II (ed P Neukomm), 2nd Int Symp, Karger, Basel, 1974, pp 37-39
26. Cole LC: Radiotelemetry of environmental conditions for ecological field studies. In Biotelemetry (ed LE Slater), MacMillan, New York, 1963, pp 203-213

27. Delgado JMR and Bracchatta H: Free and instrumental behavior in monkeys during radio stimulation of the caudate nucleus. Int J Psychobiol 2: 233, 1972
28. Reite M, Pauley JD, Walker S et al.: A systems approach to studying physiology and behavior in infant monkeys. J Appl Physiol 37: 417, 1974
29. Nevison TO Jr: Telemetry and monitoring of man in unusual environments. Telemetry J 2: 15 1967
30. Skutt HR, Kertzer R and Fell RB: The use of telemetry to obtain physiological data during exercise. In Biotelemetry (eds HP Kimmich and JA Vos), Leiden, The Netherlands: Meander, 1972, pp 21-29
31. Harrison DC, Fitzgerald JW and Winkle RA: Ambulatory ECG's to Diagnose and Treat Cardiac Arrhythmias. New Eng J Med 294: 373, 1976
32. Hanley J, Hahn PM and Adey NR: Electrode systems for recording the EEG in active subjects. In Biomedical Electrode Technology (eds HA Miller and DC Harrison), Academic Press, New York, 1974, pp 283-313
33. Pool SL: Physiological measurement systems for advanced manned space missions. In Advanced in Biomedical Engineering (eds JHU Brown and JF Dickson, III), Academic Press, New York, 1975, pp 151-215
34. Updike SJ: Patient monitoring by radio telemetry. Reliability on a medical war. J Ass Adv Med Instrum 6: 240, 1972
35. Johnstone M: Radio telemetry-simple apparatus for cardiovascular monitoring during surgery. Med and Biol Eng 6: 327, 1968
36. Nagel EL, Hirschman JC, Nussfled SR et al.: Telemetry-medical command in coronary and other mobile energy care systems. J Amer Med Assn 214: 332, 1970
37. Fryer TB and Sandler H: A review of implant telemetry systems. Biotelemetry J 1: 351, 1974
38. Hudson H and Parker E: Medical communication in Alaska by satellite. N Eng J Med 289: 1351, 1973
39. Murphy RLH and Bird KT: Telediagnosis: A new community health resource: Observations on the feasibility of telediagnosis-based on 1,000 patient transactions. Am J Pub Health 64: 113, 1974
40. Childress DS: Powered limb prostheses, their clinical significance. IEEE Trans Biomed Eng BME 20: 200, 1973
41. Olsen DB, Smith L, Sandler H et al.: Monitoring the total artificial heart by telemetry under treadmill exercise. Circ 52: 223, 1975

THE GENESIS OF BIOTELEMETRY

Norman J. Holter

President, The Holter Research Foundation, Inc.
Helena, Montana and La Jolla, California

This discussion is intended to illustrate how ideas progress in unanticipated ways in the pursuit of non-goal-directed scientific research. One doesn't just say "I will now invent biotelemetry". The process of the formation of ideas follows a quite circuitous path and often leads to results never originally visualized or planned. A new idea can either succeed or fail and often still lead to successes unrelated to the original effort. This is more likely in so called non-goal-directed research as I will show. On the other hand, goal-directed work such as that of an engineer in industry assigned to build a better ice-cube maker is not likely to result in a better paint brush. Examples in several fields will be described, with emphasis on the sequence of events leading to the concept of biotelemetry.

Legend has it that there were three princes of Serendip, whatever that is or was, and that they set out in the world to see specific places and find specific things. They did not get to these places or find these things but got to other places and found other things. Hence the word "serendipity", which plays such a part in probing the unknown. This, together with the non-rigid goals in the work of our laboratory, played a large part in the development of biotelemetry. Years ago the concept was thought by most to be just a laboratory curiosity and too radical to be useful in the study of living systems. If Newton did indeed see an apple fall and developed the law of gravitation as a result, it would be a case of serendipity. However if Galileo did indeed drop two different masses from the Pisa tower, it would not involve serendipity because he planned what he was going to seek without it being the result of a chance observation.

I will give some more examples of goal versus non-goal directed research and of where serendipity did or did not play a part, and then outline the series of events that led to biotelemetry. War research is of necessity goal-directed, but it still leads to often unexpected results. I was involved in measuring the size of ships' wakes (which reflect sound and could be measured acoustically) to see if enemy submarines could hide in them as the ship enters a harbor. The measurement of width and length was pretty straightforward, but the thickness posed a problem because sound could not penetrate from the surface. So a submarine was used with sound and timing equipment aimed upwards and the signals from the bottom of the wake were reflected back downwards and timed. Thus the goal was achieved without realizing that this would later make it possible for the submarine Nautilus to cross the North Pole under the ice, keeping an eye on the bottom of the ice above. Thus a new method of specialized navigation came out an entirely unrelated original goal.

Very early in the World War, Dr. Roger Revelle and I did some swimming off a ship some distance from shore with only the goal of finding what noises, if any, were present under water, and how far we could hear the old obsolete World War I sound projector. At a certain point we heard what sounded like bacon frying. This was later traced by marine biologists to a species of what were called snapping shrimp. Since these shrimp interfered with underwater sound equipment used to detect enemies, we had to place our defensive equipment at points where they did not exist. Conversely, they were also dropped into enemy harbors to confuse their efforts to detect us. Here was practically non-goal "playing around" that led to a useful and unusual weapon.

Serendipity also turned a war-type experiment into an art form. At Bikini in 1946 I was working on the measurement of ocean waves generated by atomic explosions. One of our instruments was crushed in a manner that made it look like modern art. I made a note to try some steel sculpturing with lesser explosives and a few years ago got around to it. After much experimenting I produced some pieces of deformed plates, cubes, cones, pyramids, cylinders, etc. which I entitled a "Geometry in Steel" series and held an art show at one of the Poindexter Galleries of Contemporary Art. Serendipity again, I believe.

In the early 1950's, one experiment done in giving a course to physicians on handling radioactive drugs consisted of showing how various radiation, alpha particles, beta rays, and gamma rays were absorbed by various thicknesses of plastic. In measuring out extremely small volumes of a highly radioactive fluid, it was necessary to get close enough to see the syringe and this was not safe. On thinking about the absorption experiment I turned a sheet of 1/4"-thick plastic, 8 inches square, on its side, polished the edges and made a sandwich with the plastic in the middle and thick sheets of lead on either side. Thus a valuable and simple instrument was conceived from a minor experiment (1).

Now about a project that failed and at the same time led to successful discoveries by my not being rigidly bound to the original goal. I thought a lazy man's way to hunt for uranium back when this was a national craze would be to record the normal atmospheric radioactivity which results from the gas radon produced by the decay of uranium in the ground and to record wind direction at the same time. If we found significant correlation between wind direction and radioactivity level, we thought it would give us a better idea of the direction in which to hunt. I never did think much of the idea and when I tried it, it didn't work. However, one day the normal fluctuations of activity showed a higher peak than ever before and I thought I would verify that it was caused by normally expected radon (or by thoron, another radioactive gas originating from thorium in the ground). By measuring the rate of decay, the gradual loss of radioactivity by all such substances, it was found to be neither one; but it fit the decay curve for fission products, the many different kinds of radioactivity mixed together resulting from an atomic explosion. This was verified from rain and snow for several days and nights, and the conclusion was inescapable that the Russians were making atomic tests and the prevailing westerly winds were bringing the results over Montana. President Truman had announced and then denied such tests and I never have known why the vacillation. I went to Washington at my own expense and donated my findings to the Atomic Energy Commission. After deciding that I wasn't a kook, they listened and I published the results (2,3). The mathematics of the decay curves also enabled us to tell the date of each explosion and this led to another discovery in Germany. (A Navy scientist had used the same method independently from air samples collected by the Air Force and he received the Navy's highest civilian award for his work: I am still awaiting a postcard of simple thanks.) But it was exciting and that is more important.

By knowing the date a given large air mass became contaminated it was, in effect, fingerprinted and could be detected elsewhere and known to be the identical air. Scientists at the University of Frieburg did that and found air masses over Germany that had come from Helena, Montana. They thus learned much about large-scale meterology, including data on the jet stream (4).

When melting snow and evaporating water for the bomb-detection work, I noticed that a drop of water splashing out of a beaker looked square. Later I pursued this odd observation together with my assistant physicist, W.R. Glasscock, who was of major assistance in all our work. After much effort we found water drops to be vibrating polygons under certain conditions and were able to photograph triangles, squares, polygons, and right up the line to a drop with eighteen straight sides (5).

So much for this series of examples. We started to find uranium and ended up finding square water drops. Serendipity indeed! The non-scientist asks "so what" since the water-drop work has no application. Just as the climber goes up a dangerous mountain "because it is there", we put time and effort in studying drops "because they were there, and besides they looked funny" a basic discovery of no possible use to anybody--so far as we know now.

Now, on to the series of projects which led (we followed-we didn't lead, as some scientists think they are doing) to biotelemetry. It started with some orange juice and a frog nerve in 1936. I was assisting Dr. Lawrence Detrick at UCLA in studying the effect of vitamin C on frog muscle fatigue and later on wound healing in rats. I was concerned mostly with instrument design but became interested in the well-known frog muscle preparation (this was used by Galvani in 1771 in his historic work on animal electricity). Then, a little before World War II, I worked with Dr. J.A. Gengerelli on the idea of stimulating a frog nerve and hence its attached muscle without touching it directly by mechanical or electric contacts. We found it possible to do so by an electric field at some distance from the nerve (6). We reasoned that if we could do something to the nerve, what might we get from the nerve at a distance. Although a nerve impulse differs from an electric current, we thought it possible that a magnetic field might emanate from an active nerve. It wasn't until after the war that we found this to be true by using the better equipment then available (7,8). The field of biomagnetics is just now blooming and has made great strides.

We then worked on stimulating a rat brain from a distance by imbedding electrodes in the brain through the skull and screwing a miniature radio receiver into the skull. Dr. Gengerelli continued for some years to work on this since the first work in 1948 and 1949 and has been able to modify rat behavior using the proper signals. In 1946 we discussed the general idea of how to study organisms without making the mistake of early zoologists who studied caged animals and concluded that their behavior was the same as it would be in their natural environment. In 1947 I formed the Holter Research Foundation, Inc. for the primary purpose of trying to broadcast by radio the more obvious electrophysiological phenomena occurring in humans so they could be free to do something besides lie quietly on a couch. The electronics of those days were crude, transistors had not been invented, and it was far from clear how we would eliminate muscle interference and other unwanted signals.

We realized that, if such a thing was possible, it would not be practical unless it could be made small. However, the first goal was to see if transmission of electroencephalograms or electrocardiograms was possible, so we did not consider size until later. Our first success was in broadcasting electroencephalograms a short distance from a subject riding a bicycle. Fidelity was improved so that a wired record taken at the time of radiotelemetering was achieved (9). I then concentrated on the ECG because it provided about a ten-fold increase in electrical signal and because it would be more important in terms of numbers in case of future clinical applications, heart deaths being of epidemic proportions. The equipment weighed 85 pounds and is shown in several publications (10). The work was presented at the Annual Meeting, Montana Medical Association, Butte, Montana, 1949.

After several more models, the unit was reduced to two units fastened to the chest with straps. At that time we had trouble with interference between amplifier and transmitter and had to separate them (11). This was 1952 and our final radio unit was designed for one container and could fit into a man's shirt pocket (12,13).

With any of these units the subject was required to remain within radio range so we developed a receiving and tape-recording system a little larger than a brief case to be carried by the subject wherever he went. This gave him more freedom of action and location (14). Finally with the advent of adequate transistors we "uninvented" the radio link by putting amplifier, tape recorder, temperature-control circuits, motor speed control circuits, and batteries into a single unit small enough for a man's coat pocket or woman's purse. This eliminated transmitter, receiver, and the need for the "ECG brief case". The unit was oval with a maximum length and width of 19.5 and 9.8 cm, respectively. The thickness was 4.6 cm, and the weight was 1 kilogram. A ten-hour record was standard for this unit. This is shown in Ref. 14. We later built a still smaller multichannel unit. Both the radio telemetry and the magnetic-tape units have been improved by industry although the first tape unit manufactured had some undesirable features thought necessary by the manufacturer to speed production. These deficiencies have since been corrected and units are made by many companies. I am not involved financially, having contributed our results to the public at large as a non-profit institution.

When we found it feasible to make tape records, it was apparent that very voluminous data would result and the tapes could not be examined as one does with an office-writing ECG. Development was started on a method to read tapes by speeding them up and presenting them electronically so that unusual heart beats could be located by sight and sound in a practical time. Thus a ten-hour record could be analyzed in 10 minutes. Details of this have been published elsewhere (12). We originally dubbed this analyzer by the acronym AVSEP standing for "Audio Visual Superimposed Electrocardiographic Presentation". Manufacturers now give it a variety of names--whatever their sales department thinks is catchy.

Incidentally, I tend to date the beginning of the bioengineering industry from the meeting (14) in New York in 1961 because that is the first time I had encountered almost as many manufacturers representatives as scientists. They had been sent to locate gadgets that might be profitable to manufacture. This is, of course, desirable as it does no good to develop a helpful product and not have it available to those who can be helped by it. After my paper I was swamped by these scouts but won't go into the many headaches involved in trying to select a company to do the job of getting the equipment to the medical profession.

We have studied many normal hearts since our final tape model was developed. One original motivation was to learn more about the normal heart as well as to apply the Holter ECG, as it is now correctly called, to the study of disease progress, drug evaluation, and detection of aberrant heart behavior when the patient is out living daily routine life. The unit is often called the Holter Monitor and this might stick although I don't think it is correct since the data are studied retrospectively. We have encountered sudden death in a subjectively healthy individual (10), and it is my hope that enough can be learned to prevent many of the deaths so common today.

We are now in 1976, and the participants at this Symposium are more qualified to know what is going on in this field with their many new applications and ideas. One current project that I find interesting is a study of coronary patients who have their attack in the Hawaiian Islands. One out of four people admitted to the CCU unit at Queens Medical Center there is a tourist. Thus, the interesting problem arises of when he can be flown home (where he would naturally wish to be) as soon as he can travel safely. So a study is being made using the Holter ECG on those patients whose first activity out of the hospital will probably be a long plane flight when ready; but the question is "when are they ready?" I am myself a guinea pig for a group of cardiologists. I had a Holter ECG before, shortly after, and at intervals since I had a coronary about two years ago. I will continue under planned protocols and have arranged for an autopsy in case of heart death. My only complaint is in not being able to read the final reports.

Thus endeth the trip from orange juice to sudden death, hardly a direct and planned trip like one has when setting forth on a vacation.

I will close on a sad but significant note by quoting a letter received recently from Dr. Paul Dudley White. In the early days I found practically no encouragement from the medical profession except from Dr. White who could see a future for the idea.

"Dear Norman Holter:

I'll never forget that day in Helena when we took electrocardiograms on one of your friends in the street below (note by N.J.H.-this was 1951). That for me was a real red-letter day, and I cherish the memories of it. You have been very important in the evolution of tele-electrocardiography since those pioneer days. I recently had an accident, a fall, and a cranial operation but am now convalescing nicely. Thank you so much for being in contact with me.

Sincerely yours,

Paul D. White

P.S. I'll be delighted to enscribe my new book for you.
P.S.S. Dr. White had another stroke since writing this letter and was unable to sign it. Unfortunately, he will also be unable to enscribe the book."

This letter was written October 12th and he died October 31st. I wish here to publicly credit and thank Dr. White for preventing me from giving up on a project beset with far more problems than are realized by today's users.

REFERENCES

1. Holter NJ, Glasscock WR: A Close Shield for Use In Micropipetting Radioisotope Stock Solutions. Rev Sci Instr 23: 3, March 1952
2. Holter NJ, Glasscock WR: Tracing Nuclear Explosions. Nucleonics 10:8, August 1952
3. Arnold JR: Tracing Nuclear Explosions. Bull Atomic Sci 9:2, March 1953
4. Herbst VW, Neuwirth R, Philipp K: Betrachtungen uber die Eignung radioaktiver atomtechnischer Aerosole als Markierungsmittel bei Arbeiten auf dem Gebiete der meteorologischen Stromungsforschung. Die Naturwissenschaften Heft 7:156,1954
5. Holter NJ, Glasscock WR: Vibrations of Evaporating Liquid Drops. J Acoust Soc Amer 24:6, p 682, Nov 1952
6. Gengerelli JA, Holter NJ: Experiments on Stimulation of Nerves by Alternating Electrical Fields. Proc Soc Exp Biol and Med 46:532, 1941
7. Gengerelli JA, Holter NJ, Glasscock WR: Magnetic Fields Accompanying Transmission of Nerve Impulses in the Frog's Sciatic. J Psychol 52:317, 1961
8. Gengerelli JA, Holter NJ, Glasscock WR: Further Observations on Magnetic Fields Accompanying Nerve Transmission and Tetanus. J Psychol 57:201, 1964
9. Holter NJ, Gengerelli JA: Remote Recording of Physiological Data by Radio. Rocky Mtn Med J September 1949
10. Holter NJ: Storage Telemetry and Sudden Death. In Biotelemetry. Univ Nijmegen, 1972, Chapter 3
11. Glasscock WR, Holter NJ: Radioelectroencephalograph for Medical Research. Electronics 1952
12. Holter NJ: New Method for Heart Studies. Science 134:3486, p 1214, Oct 20, 1961
13. Holter NJ: Radioelectrocardiography: A New Technique for Cardiovascular Studies. Ann New York Acad Sci,p 913, 1957
14. Holter NJ, Glasscock WR: The Electrocardiocorder and Related Instruments-Applications in Clinical Cardiology. Digest of the 1961 International Conference on Medical Electronics, July 1961

SESSION I

HUMAN RESEARCH APPLICATIONS

Chairmen

H.P. KIMMICH, Ph.D.

H. SANDLER, M.D.

CHAIRMEN'S OVERVIEW

Combining the techniques from the presentations in this session should make it possible to follow completely the movement of a subject. At the same time one could monitor a large number of physiological parameters such as ECG, EEG, temperature, respiration and movement of different body elements, as well as the ambient parameters of temperature and humidity. In addition it might be possible to estimate the probability of the subjects's surviving the experimental situation without trauma. We should be more concerned about possible damage to man. To that extent, we would prefer not to use invasive methods as has been done frequently in the past for measuring blood pressure, blood flow and other internal processes in humans. Possible hazards of radio frequencies from instrumentation should be better understood.

TELEMETRY OF RESPIRATORY FLOW WITH A NEW MAGNETIC TYPE OF FLOWMETER

H. Peter Kimmich

Dept. of Physiology, University of Nijmegen
Nijmegen, Netherlands

The continuous telemetry of respiratory gas flow poses many problems, especially during work and exercise, where often a single transducer is required from rest (0 to 20 l/min) up to maximum exercise (0 to 500 l/min). Even for a linear relationship between gas flow and electrical signal an accuracy of 0.04 % would be required at maximum flow to ensure an accuracy of 1 % for flow values as measured at rest. This is especially true for the zero drift which generally is only slightly dependent on the maximum flow sensitivity. For the special application of the flow meter in dynamic oxygen uptake ($\dot{V}O_2$) measurements (Kimmich and Kreuzer, 1972) drift of the zero point influences the final value to a much larger extent than for example inaccuracies of the calibration line and drift of its slope.

The force (F) generated by the air flow on a round plate (area equal to 2A) inserted perpendicular into the air stream may be calculated to be $F = c_r \rho v^2 A$. The resistance coefficient c_r is a function of the Reynold's number, and is constant for Re greater than 1000. In other words with turbulent flow the force F is independent of the viscosity of the gas and only dependent on the density (ρ).

Method.

The principle of this flowmeter is based on an earlier design (Kimmich and Kreuzer, 1972) and has been described (Kimmich and Spaan, 1975). The respiratory flow is measured by detection of the position of a flow sensing flag (fig. 1). This plate is connected to the middle magnet M2 of a series of three magnets and can rotate around an axis. The firm magnets M1 and M3 (fig. 2) are poled such that there is a stable equilibrium of the forces between magnets M1 and M2, and magnets M2 and M3 in the middle of the space between M1 and M3. Such an arrangement guarantees an exact and reproducible position of M2 with zero flow and thus a stable zero point since the magnetic forces determine the exact position of M2 and not the position of the axis. The actual inaccuracy of the zero positioning is in the order of 0.05 to 0.1 μm. The force on the flow sensing plate (induced by the air flow) is proportional to the deflection of M2 for small deflections and quadratic for "large" deflections. The weight of the magnet M2 is such that it counterbalances that of the flow sensing plate in order to negate artefacts due to accelerations.

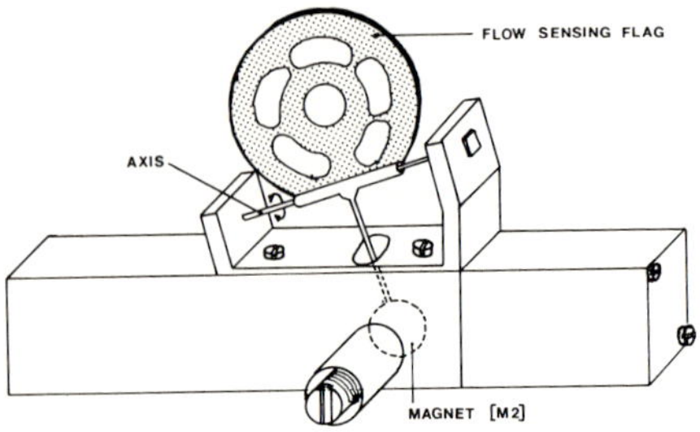

Figure 1. Air flow transducer as it is inserted into the respiratory tube of an "open" mask. The air flow is measured by detecting the force enervated on the flow sensing plate with the aid of a magnetical system (positioning the flag) and an optical position detecting system (see fig. 2).

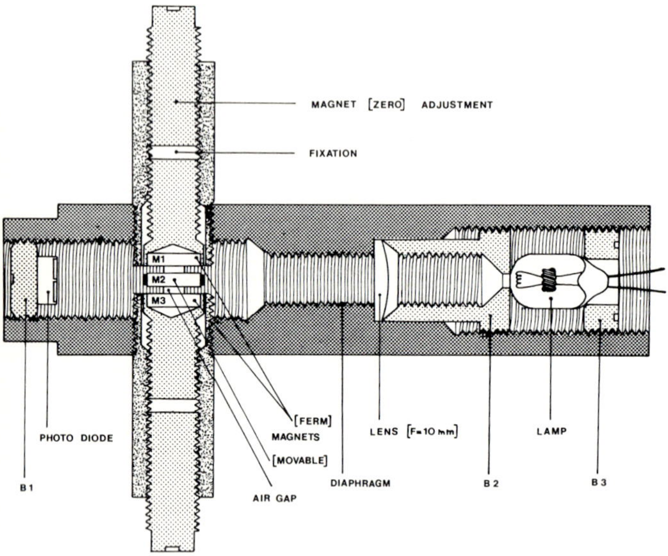

Figure 2. Cross section of the lower part of figure 1 viewed from the bottom. It reveals the optical system for detection of the movable magnet M2 and thus the flow sensing plate. For explanation of its working see text. The parts B1, B2, and B3 fix the photo diodes, the lens and the lamp to its correct position.

The position of the magnet M2 is detected by the optical-flag-detection system (fig. 2). Parallel light (originating from an infrared-light source and a lens system) is detected by two light-sensitive diodes of rectangular shape which are mounted on a single chip. Part of the light sensitive areas of both diodes is screened by the movable magnet. The difference in light detected by the diodes is proportionally related to the deflection of the movable magnet M2. It may be calculated that the maximum possible deflection (for proportionality) corresponds to an air flow of approximately 560 l/min which is sufficient for application of the flow meter under all sorts of conditions. The optical system has been constructed as simply as possible. The firm connection to the flow sensing system, and the symmetrical construction allow temperature expansion of the respiratory tube and the parts of the transducer itself, without markedly changing the characteristics of the flow sensor (Calibration line better than 1 %, and zero drift = 0.005 %/°C of maximum deflection).

Characteristics.

The calibration line of the flowmeter (inserted into a resp. tube with an inner diameter of 36 mm) shows a quadratic relationship between flow and transducer output signal from 0 to 300 l/min and a linear relationship from 300 to approx. 530 l/min (fig. 3: V_{IN}). It can be shown that this result is only possible if the flow in the respiratory tube is turbulent down to practically zero flow, which has been achieved by the mechanical layout.

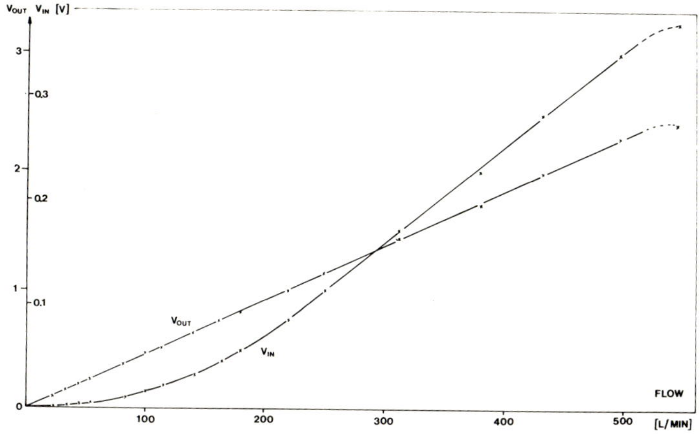

Figure 3. Calibration line of the respiratory gas flow transducer (V_{IN}) and the electronically linearized signal (V_{OUT}). The calibration line for neg. flow is symmetrical to the origin.

Sensitivity and calibration curve: see fig. 3, linearity ± 1 %,
stability over several weeks: 0.1 % of max. deflection,
transient response to a step function in flow: better than 10 ms,
sensitivity to temperature (0 to 50 °C): drift of the zero point = 0.02 mV/°C, calibration line = $T_o/T_o + T_F$ (T_o = 273 °K, T_F = temperature of the flowmeter),
sensitivity to acceleration (2g parallel to flow): 0.25 mV per g,
sensitivity to viscosity of the gas: zero, resonance freq.: 180 Hz

Electronic linearization.

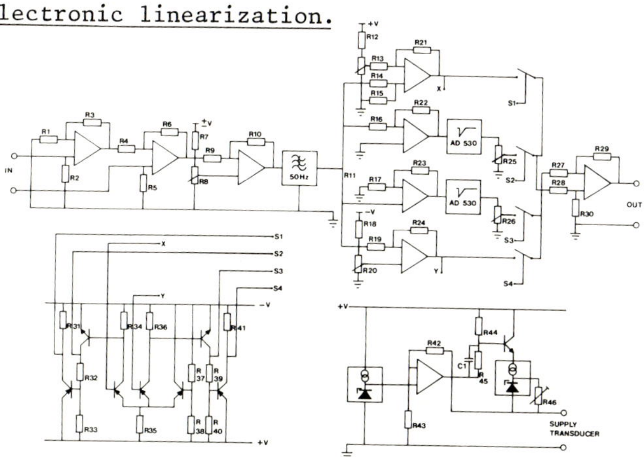

Figure 4. Electronic circuit for linearization of the signal of the flow sensor. The great zero stability of the transducer requires careful circuit design in order to guarantee the required accuracy of better than 0.05 %.

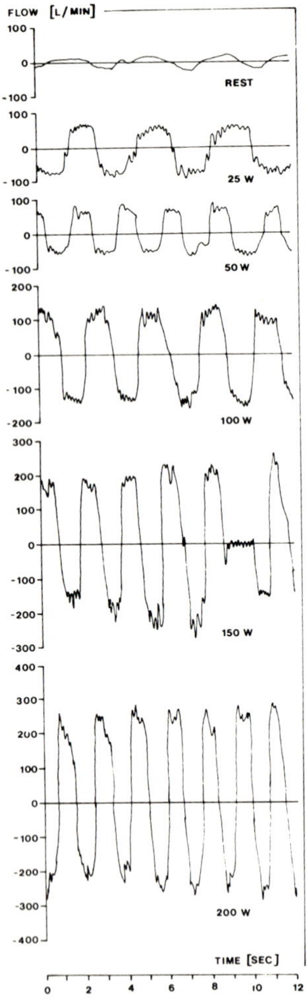

Figure 5. Examples of telemetered respiratory flow at different work loads.

The principle of the electronic linearization (fig. 4) is separate processing of the positive and negative flow signals as well as of the quadratic and linear portions, and its correct sampling by the FET switches S1 to S4. The actual linearization of the quadratic portions is achieved by IC square rooters (AD 530) which are at the same time used for accurate detection of the zero line. The 50 Hz low-pass filter is necessary for application to telemetry to reduce possible oscillations at the resonance frequency (180 Hz) of the flowmeter. The current supply to the light emitting diode of the transducer has a stability of 0.01 %.

Results and conclusions.

Examples of telemetered flow curves at rest and during exercise with different work loads during a single experiment are shown in fig. 5. Note the curve measured at rest. The peak flows of this subject were relatively small if measured at constant work load. In other subjects, as well as in experiments without fixed work load, peak flow values greater than 400 l/min have been measured frequently.

Under normal conditions the drift of the zero line is less than 1/4 $^{o}/oo$ of the maximum signal over several hours which is good for accurate measurements in the range of 0 to 40 l/min (very light exercise) up to 0 to 530 l/min (heavy exercise). With careful calibration of the zero line in situ accurate pneumotachograms can also be measured at rest. It is easy to design a similar flowmeter of larger sensitivity, thus useful for measurement at rest up to moderate exercise. It may be concluded that this telemetry system is able to measure accurately the respiratory flow over a wide range of activity with a single transducer.

References:

Kimmich, H. P.: Telemetry of respiratory air flow. In: Biotelemetry, Meander NV, Leiden, 111-120 (1972)

Kimmich, H. P. and F. Kreuzer: Telemetric determination of oxygen uptake by man during exercise. Int. J. Biomed. Engng. 1/1, 27-36 (1972)

Kimmich, H. P. and J. G. Spaan: Magnetic respiratory flow transducer. Biocapt 75, 383-388 (1975)

RADIOTELEMETRY SYSTEM FOR OCCUPATIONAL HEALTH AND INDUSTRIAL HYGIENE RESEARCH

Peter Reischl, Department of Biomedical Engineering
University of Southern California, Los Angeles, California 90007

Uwe Reischl, Program in Social Ecology
University of California, Irvine, California 92717

A prototype environmental monitoring system capable of measuring thermal stress and physiological responses of persons working or resting in occupationally stressful environments has been developed at the University of California at Irvine. The parameters measured for assessing heat stress include ambient air temperature, heat radiation and relative humidity while physiological responses monitored include body skin temperature, rectal temperature and cardiac pulse rate.

In the past, industrial hygiene methods for monitoring heat stress were slow, inaccurate and cumbersome. Large black-globe thermometers and sling-psychrometers were used. Measurements were limited to station samples. No versatile personal environmental stress monitoring system had been available for industrial hygiene work although the need for such equipment has long been recognized.

Several investigators designed and constructed radiotelemetry systems useful for monitoring test animals or human subjects exposed to limited environmental stress settings (2,3,4,8,9,10); however, no one adapted a system to industrial hygiene needs.

It has been shown that a radiotelemetry system using a bridge type transducer network with pulse width modulation can maintain a high degree of accuracy (3,11) in transmitting slowly-varying parameters. Also, such a system has a low power drain. The variations in the environmental and physiological parameters encountered in occupational health work are also relatively slow, therefore needed information can easily be accommodated by such a system. The radio frequency section of the system consists of a multichannel pulse width modulated, time division multiplexed transmitter-receiver system which is produced for the remote control industry by Kraft Systems, Inc.[θ] Pulse time stability for the system is guaranteed to be ±1% between -15°C. and 50°C.

The complete 7-channel Environmental Monitoring Telemetry System is diagramatically illustrated in Figure 1. Temperatures are sensed with precision thermistors,* while humidity is detected with a Xeritron Humidity Sensor.** Time constants for these transducers are in the order of a few seconds. Cardiac pulse rate is detected by a pulsimeter[≠] attached to an ear lobe of the subject. For the reference channel either a resistor is used in place of a thermis-

---

[θ]Kraft Systems, Inc., 450 W. California Ave., Vista, California 92083

tor, or a thermistor is used to telemeter internal transmitter temperature. Any variation in the output level of the reference channel may be correlated to a drift in the electronics for compensation.

Transducer bridge networks couple the signals to the signal conditioner. The signal conditioners are low noise, broad band, high-gain differential integrated circuit amplifiers which provide signals between 0 and 2 volts to the multiplexer. The multiplexer combines the signals from the amplifiers into a time serial stream which modulates the transmitter. Commutation rate is 50 Hz.

The battery powered receiver reshapes and decommutates the pulse train which in turn branches to the servo amplifiers. The servo amplifiers control a meter or a strip chart interface. The entire receiver package measures 3 cm. x 24 cm. x 5 cm. and weighs 80g.

The transmitter system, multiplexer, signal conditioner and batteries are mounted into a chassis 15 cm. x 27 cm. x 5 cm. It weighs 1 kg. and has an operating time of 5 hours with a 9.6 V, 500 mahr battery pack. The power into the final radio frequency stage is presently limited to 100 mw in the 27 MHz frequency band to be in compliance with FCC regulations (7). The package can be conveniently mounted on a back pack or harness with transducer rack (Figure 2). Signal leads from transducers are connected to the transmitter by a 15-pin connector. Ambient air temperature and relative humidity are measured within the sensor chassis ventilated via a small electric fan. The entire system may also be incorporated into a vest.

Calibration is perhaps the most critical in any data collection system of this type. Examples of calibration data are given in Figure 3.

The concept of an Environmental Monitoring System has been realized with this system. The basic multichannel telemetry system can be adapted readily to transmit several other kinds of information, i.e., bioelectric signals, strain-gauge respirometry and plethysmography, direct or indirect blood pressure manometry, basal skin resistance, and galvanic skin response, to mention but a few.

Worldwide data acquisition and transmission will also be possible whenever telephone lines are available. By utilizing a data phone it will be possible to transmit data over the standard telephone system. This method is used in several biomedical applications (1,6). Real-time computer based data analysis will also be possible.

---

*Precision thermistor ±0.1°C. interchangeability, YSI #44034; General purpose esophageal or rectal temp. probe YSI #401; Attachable surface skin temp. probe, YSI #409; Direct reading/wavelength independent/radiometer probe YSI #6551; Yellow Springs, Ohio 45387

**Model XNAM 10205, Hygrometrix, Inc., 285 Fifth St., Oakland, CA 94607

†Item No. 015023801 OPT/MOD Ref. 06A00601, Tektronix Inc., P.O. Box 500, Beaverton, Oregon 97077

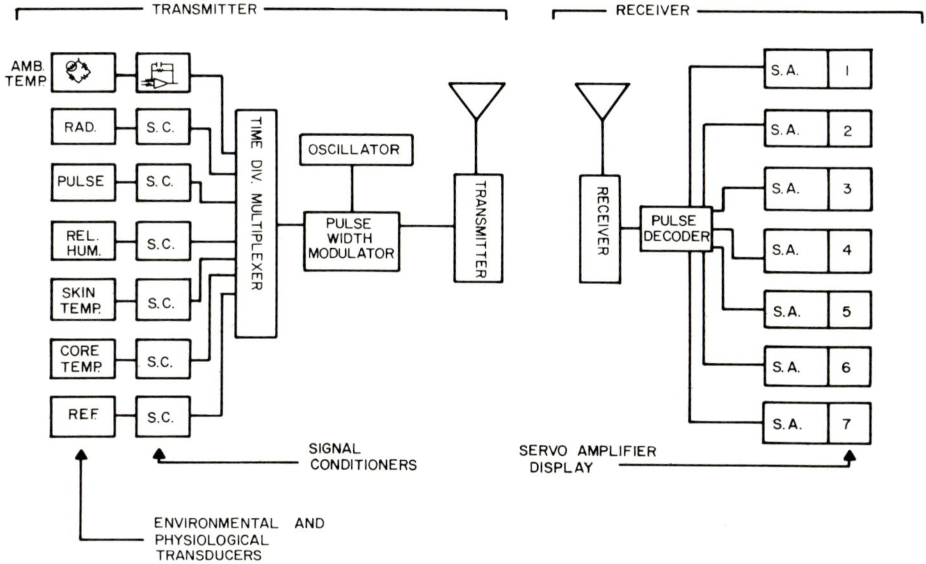

Figure 1. Schematic Diagram of Telemetry System

Figure 2. Details of Back-pack Transmitter Unit →

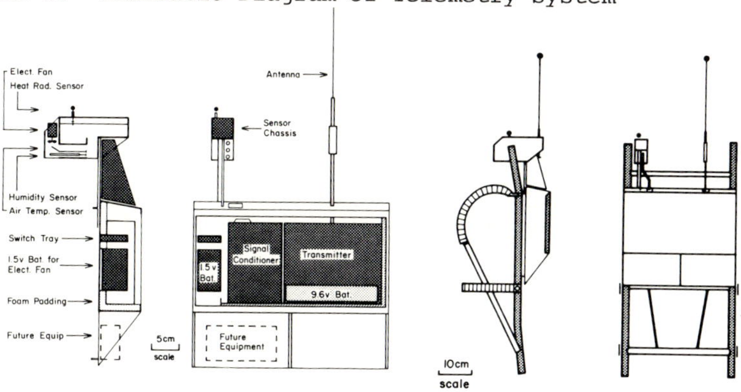

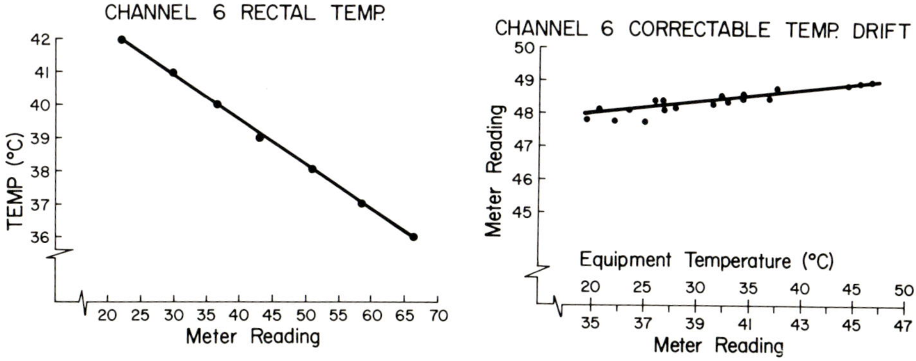

Figure 3. Example of Calibration Curves

References

1. Bennett, D.R., Gardner, R.M.: A Model for the Telephone Transmission of Six-Channel Electroencephalograms. Electroenceph. Clin. Neurophys., Vol. 29, 1970, pp. 404-408.

2. Bergey, G.E., Sipple, W.C., et al: Personal FM/AM Biotelemetry. Aerosp. Med., May, 1968, pp. 488-492.

3. Bodenlos, L.J.: Transmitter Back-Pack for Free-Roaming Animals. Lab. Animal Care, Vol. 16, No. 5, 1966, pp. 454-458.

4. Cupal, J.J., Ward, A.L., et al: A Repeater Type Biotelemetry System for Use on Wild Big Game Animals, ISA BM 74329, 1974, pp. 145-152.

5. Fryer, T.B., Deboo, G.J. et al: Miniature Long-Life Temperature Telemetry System. J. Appl. Phys., Vol. 21, Jan. 1966, pp. 295-298.

6. Hanley, J., Zweizig, J.R., et al: Combined Telephone and Radiotelemetry on the EEG. Electroenceph. Clin. Neurophys., Vol. 26, 1969, pp. 323-324.

7. Makay, R.S.: Bio-Medical Telemetry, New York, John Wiley, 1970.

8. Murray, R.H., Marko, A., et al: A New, Miniatured Multichannel Personal Radiotelemetry System, J. Appl. Phys. Vol. 24. No. 4, April 1968, pp. 588-592.

9. Nagasaka, T., Ando, S., et al: A Radio Telemetering System and the Changes of EKG and Heart Rate of...Mountaineering. Nagoya J. Med. Sc., Vol. 29, 1966, pp. 93-103.

10. Watson, N.W., Franklin, D.L., et al: Back-Pack for Free Ranging Primates. J. Appl. Phys., Vol. 24, No. 2, Feb. 1968, pp. 252-253.

11. Winget, C.M., Averkin, E.G., et al: Quantitative Measurement by Telemetry of Ovulation and Oviposition in the Fowl. Am. J. Physiol., Vol. 20, No. 9, Sept. 1965, pp. 853-858.

EEG TELEMETRY: METHODOLOGY FOR THE EXAMINATION OF THE MENTALLY RETARDED

Torsten Andersson

EEG Lab, Vipeholms Infirmary, Lund, Sweden

INTRODUCTION

Electric encephalographic examination (EEG) is now automatically included in the investigation of mental retardation. For a conventional EEG it is necessary for the patient to lie or sit absolutely motionless during circa 20-30 minutes. The patient is usually moved to the EEG lab, a milieu that is usually completely strange to him. Many times pre-medication has to be ordained so that the patient can be sufficiently immobile during the examination. In this way, however, the EEG register receives disturbances and medication effects, and often the patient's alertness is reduced.

Because of the difficulties of achieving a conventional EEG examination in some of our patients, we began some years ago to employ a new method--telemetric transmission.

PROCEDURE FOR EEG TELEMETRY

During the last few years a small number of companies and designers have given information on an international scale in medical journals and other professional literature about multichannel EEG telemetry systems. These systems have had a clear limitation with respect to the number of simultaneously distributed channels. The telemetry systems on the market today can distribute a maximum of eight channels.

The transmitting set usually carried by the patient during distribution and registration of the EEG, ECG or other medical data should naturally not be too big or too heavy.

In all EEG registering it is of great value to take readings from as many different points as possible, for it is by this means that any damage can most accurately be tracked to a specific area of the brain. To keep the different channels apart during telemetric distribution of the EEG, most designers now use a subcarrier modulation of the transmitter; in other words every distributed EEG channel has a carrier wave with its own frequency. This has one big disadvantage. With an increased number of channels comes an astronomical increase in the number of electrotechnical components in the transmitting set. Current consumption also increases, necessitating frequent battery changes which in turn means that round-the-clock registering is impossible.

EEG telemetry's ideal is to have a transmitting set fastened to the patient's head with the help of straps to make the wires between the electrodes and the transmitting set as short as possible. The necessary amplification of the EEG signals is of such a magnitude that the static electricity, which can be generated when the wires distributing the EEG signals from the electrodes touch each other, may cause interference which prevents accurate reading of the EEG. As well as the paste which holds the electrodes in position, a thin layer of collodium-soaked cotton wool is also needed for the same purpose. This prevents the electrodes from coming loose or moving as the patient is usually in motion during a telemetrically distributed EEG examination.

If, during an investigation, a patient's convulsive behaviour is suspected of having an epileptogenic character, direct registering in connection with an EEG recording can be of use.

The patient's behaviour is recorded with the aid of a TV camera. The graphical registering of the EEG telemetry by the writer is also recorded by a TV camera. Information from both cameras goes via a mixer to a video tape recorder. A split picture is then obtained when the tape is played back on a TV monitor. One-half of the screen shows the EEG registering and the other half shows the patient.

Sound is also recorded on a separate channel on the video tape. This method has made it possible to correlate the patient's behaviour with the EEG activity.

Distribution of EEG Signals in the Community Antenna System Intended for Radio and TV Transmission

If the patient's EEG is to be monitored but it is inconvenient to move the patient from the ward to the EEG laboratory, the community antenna system found in most hospitals can be used to distribute the patient's EEG.

Operation of the EEG in the Community Antenna System

A transceiver is placed in the room from which the EEG telemetry is to be transmitted. Signals from the EEG transmitter located on the patient are received via an antenna mounted on the transceiving set. This signal is amplified within the transmitter and then fed into an ordinary antenna outlet. From the antenna outlet the signal is distributed in reverse on the existing branch circuit system to the nearest distribution amplifier. At the distribution amplifier the EEG signal is passed through a filter, amplified and then distributed to the trunk-line grid. The signal is led via the trunk line to the antenna system's head amplifier. In the head amplifier the EEG signal is separated from other radio and TV signals present in the system, regulated by automatic high frequency level control, and then fed into a crystal-controlled frequency converter with conversion from 224.75 MHz to 591.75 MHz. The signal is redistributed through a power stage to the trunk-line grid on its new carrier frequency.

DISTRIBUTION OF EEG-TELEMETRI AND VIDEO IN COMMUNITY ANTENNA SYSTEMS

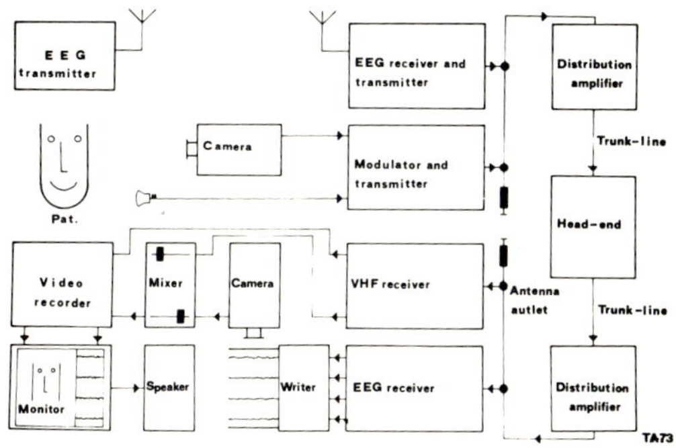

Block Diagram

Reception of the EEG signal can now take place either in the EEG laboratory or at any chosen place in the hospital. The frequency of the EEG signal received from the antenna outlet is converted to its original 224.75 MHz and then demodulated. Distribution of TV pictures and sound occurs simultaneously and in a similar fashion and together with the EEG charts are registered at the place of reception, either directly or by video taping.

8- to 16-channel Telemetry System for the Distribution of Medical Data with Pulse-width Modulation (PWM)

Digital computer technology has recently led to new types of integrated circuit solutions analogous to the previous types but with extremely low current consumption. With the aid of these components a practically unlimited number of channels can be distributed in the process of medical telemetry.

The design worked out in my private laboratory has none of the disadvantages ascribed to telemetry with subcarriers.

a)

b)

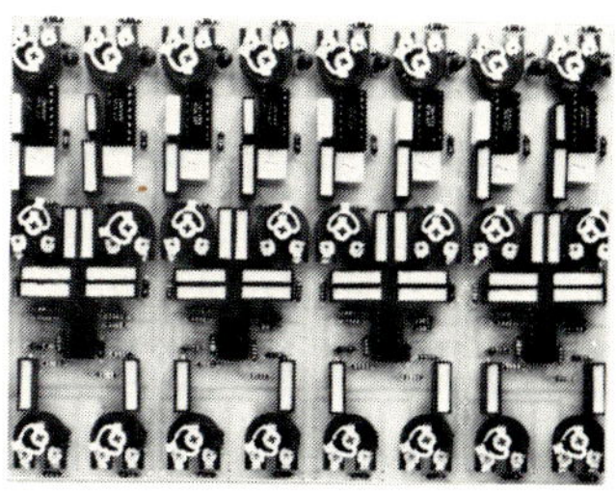

Picture a) shows the few components which make up a telemetry transmitter for pulse-width modulation. With respect to their function these components completely replace those shown in picture b) for telemetry with subcarriers.

From the EEG, ECG and EMG points of view the frequency range which is to be registered is easily covered by each telemetry channel even in a 16-channel system without making any excessive claims as to the frequency of the carrier wave.

Current consumption is taken care of in the transmitter with the aid of cheap standard batteries which can be bought at any radio shop.

CONCLUSION

We have found that telemetric EEG examination is a very useful method which supplements a conventional EEG in the investigation of mentally retarded patients. With the portable transmitter equipment the patient can be examined in his accustomed milieu, during normal activity and occupation. Where the usual EEG method has failed with restless patients telemetric examination has given a satisfactory result and prolonged recording has been made possible.

REFERENCES

1. Glasscock, W.R., and Holter, N.J.: Radioelectroencephalograph for medical research. Electronics, 25 (1952) 126-129.

2. Reneau, J.P.: Telemetric EEG audiometry instrumentation for use with the profoundly retarded. Amer. J. rent. Defic. 72 (1968) 506-511.

3. Stevens, J.R., Milstein, V.M., Dodds, S.: Prolonged recording of EEG by radiotelemetry: an aid to localisation and treatment of epilepsy. Electroenceph. clin. Neurophysiol., 27 (1969) 544-549.

4. Stålberg, E.: Telemetric long-term EEG recording. Electroenceph. clin. Neurophysiol., 26 (1969) 341.

5. Vidart, L., and Geier, S.: Enregistrements téléencéphalographiques chez des sujets épileptiques pendant leur travail. Rev. Neurol. 117 (1967) 475-480.

NINE CHANNEL ACCELEROMETER

S. Levy, G. Plank*, E. Konigsberg, and J. Gilbert

Konigsberg Instruments, Inc., Pasadena, California
and *U.S. Department of Transportation,
Transportation Systems Center, Cambridge, Massachusetts

INTRODUCTION

As part of a program to establish human tolerance levels to crash impact accelerations, we have undertaken the development of an instrumented boxer's mouthpiece. This system is designed to measure the linear and rotational accelerations for three orthogonal axes with a 3 dB response to 0.1 to 500 Hz. The accelerations are measured using nine linear accelerometers, and are transmitted from a multichannel radio telemetry unit contained in the mouthpiece to a portable receiving, decoding and processing system.

This paper presents a progress report on this system, including the results of tests on a breadboard of the end-to-end system.

METHODS

Accelerometers

The accelerometer sensor unit consists of nine linear accelerometers mounted at fixed relative spatial positions. From knowledge of the sensitivities and relative positions of the nine accelerometers, the three rotational components of acceleration can be derived from the nine linear acceleration channels. This calculation is accomplished in the receiver-decoder console.

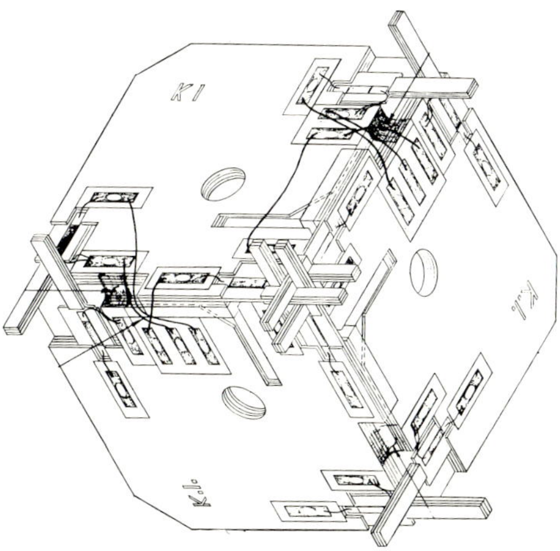

Figure 1
Accelerometer
Array Assembly

Figure 1 shows the geometric arrangement of the nine accelerometers, which are configured as an array of cantilever beams whose deflections are measured by bonded semiconductor strain gages. One pair of gages is mounted on each cantilever beam.

Design of the individual accelerometers was accomplished with the aid of a computer program. This program used the material constants and size constraints to generate tables of beam solutions, including data such as resonant frequencies, and static deflection. This accelerometer system, which is presently being fabricated, is predicted to have a full scale range of ±125 g, and a minimum resolution of 150 radians/sec^2. The resonant frequency of the cantilever beams is on the order of 5 kHz and the array is to be contained within the volume of a cube 0.58 inches on each side.

Transmitter and Electronics

The electronics circuits contained within the mouthpiece provide excitation power for the transducers, sample and hold circuits for individual accelerometer channels, channel amplification, multiplex coding, frequency shift modulation, FM transmission, and battery power. These circuits plus the accelerometer must fit within the confines of a custom-fitted athletic mouthpiece, and must not cause excessive discomfort or present a safety hazard to the test boxer.

The design of the mouthpiece electronics has been based on multichannel telemetry systems developed previously for cardiovascular instrumentation (1). Ten channels are provided, nine for the individual accelerometers, and one subcommutated channel which includes framing, a zero reference, two half scale references, and two full scale references. The multiplexer samples each channel once per 800 μs.

The nine accelerometer signals are amplified by nine FET common source stages, which are ac coupled in order to ameliorate the potential for long term drift. Trim resistors are used to achieve balance of all channels to be within a few mV of the same DC level. The amplified outputs are connected to an integrated circuit multiplexer, as are the synchronization and calibration channels. Synchronous counting logic accomplishes sequence control of the multiplexer. The multiplexer output is pulse-width modulated by a circuit which compares the multiplex signal to a linear ramp voltage. Small channel offsets caused by individual mismatches are calibrated and corrected by data processing in the receiver console.

The radio frequency portion of the mouthpiece electronics consists of a modulated oscillator and tripler. The transistors in this circuit are components of a single integrated circuit array (RCA CA 3027). The transmitter circuit diagram is shown in Figure 2.

Frequency modulation is accomplished by varying the collector to substrate capacitance of one of the transistors in the array, which is possible because all of the transistors are diffused into reversed biased wells to provide mutual isolation. These reverse

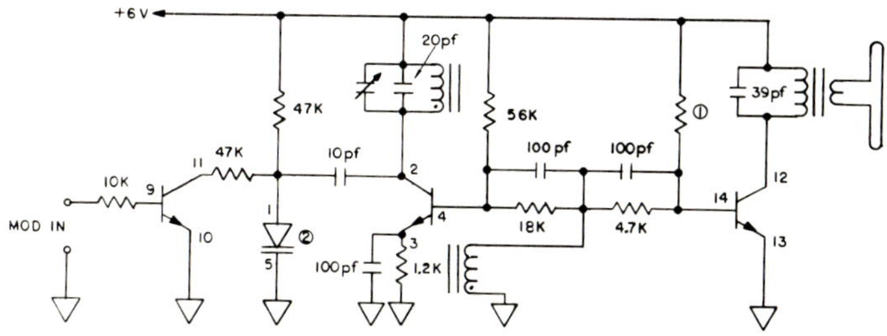

Figure 2
Transmitter Circuit Diagram

biased wells act as diodes of large areas, so that the capacitance voltage characteristic of the diode may be used to provide a voltage variable capacitance for modulation.

Other stages in the RF circuit provide buffer amplification and tripling in order to isolate the frequency determining elements from external loading effects in the antenna circuit. Unloaded, this circuit will deliver approximately one volt of RF voltage with a current drain of less than 2 mA DC.

Receiver-Decoder

A conventional entertainment band frequency modulation receier is used to recover the pulse-width modulated multiplex data stream from the RF carrier. A block diagram of this part of the system is shown in Figure 3.

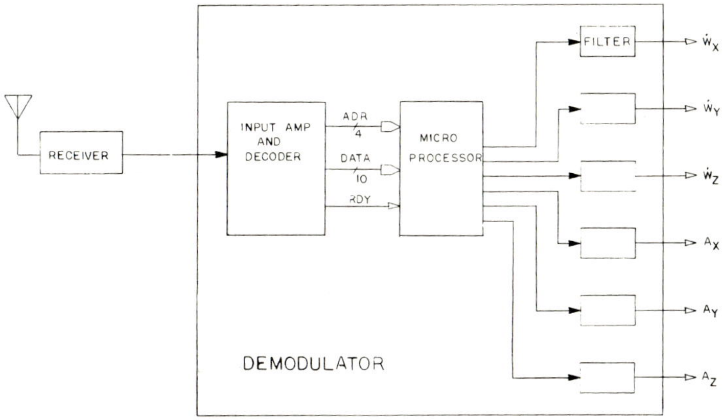

Figure 3
Block Diagram - Demodulator

The demodulator uses phase-lock technology to track the commutator clock in order to accommodate variations in the transmitter clock rate. Other digital logic circuitry quantitates the PWM data and provides it to a digital microprocessor. The microprocessor (Intel 3000 series) accomplishes correction for individual channel offset and gain variations, and performs calculations on the combination of individual channel outputs in order to extract the rotational components of acceleration. It is also available for the monitoring of data for quality control, and for re-formatting for access to the data by larger computer systems.

RESULTS

A breadboard of the mouthpiece electronic system has been completed and a series of subsystem tests has been run. It has been found that the Johnson noise of the sensor strain gages masks the internal noise of the processing and multiplexing circuitry. Cross channel leakage was determined to be less than 0.12%.

Tests on the transmitter circuitry have shown that unwanted spurious outputs near the output frequency were down at least 40 dB, and the second and fourth harmonics of the oscillator (at 60 and 120 MHz) were approximately 30 dB down. These levels of spurious outputs are considered to be acceptable.

Final mechanical package design has not yet been completed but a compilation of the total area of electronic component chips indicated the requirement to be approximately 250,000 square mils, or approximately 1.2 square inches at a nominal density of 20%. On the basis of these results, the electronic package size has been estimated at 1.1 x 0.55 x 0.2 inches. Mock-ups of these dimensions have been successfully fit into the appropriate athletic mouthpiece.

Proper operation of the phase-lock demodulator has been demonstrated. The mouthpiece transmitter clock can drift ±10% with no degradation in output data.

Laboratory tests of the mouthpiece-demodulator system indicate the system residual noise to be less than 5 mV pp in a full scale of 3 V, or 0.17%.

CONCLUSIONS

We have presented this report prior to the completion of an entire operating system but we are encouraged by the performance of elements of this system in subsystem tests to date.

ACKNOWLEDGEMENT

This work was carried out under the sponsorship of the U.S. Department of Transportation, Transportations Systems Center, Cambridge, Massachusetts, under Contract DOT-TSC-1095.

REFERENCES

1. Fryer T: Implantable telemetry systems. NASA, SP-5094.

HEALTH HAZARDS FROM TELEMETRY RF EXPOSURE ? A REVIEW ON THE INTERACTIONS BETWEEN ELECTROMAGNETIC FIELDS AND BIOSYSTEMS

Peter A. Neukomm

Laboratory of Biomechanics (Head: Prof.Dr.J.Wartenweiler)
Swiss Federal Institute of Technology, Zürich

INTRODUCTION

Telemetry transmitters are applied in the monitoring of physiological and other data, and great efforts are made to reduce the influences of the devices on the subject. Since the transmitting antenna is within or very close to the body of the subject, the question arises whether the RF radiation may affect the functions of biosystems or not. In many cases the recommended power density safety level of several countries may be exceeded.
Since a continuous monitoring of the subject is required, the body-mounted antenna should radiate omnidirectionally. Measurements of radiation patterns from man-mounted antennas have shown that the body acts like a reflector and absorber, if the frequency is higher than approx. 100 MHz. The large field intensity variations and the phase shifts may disturb the data transmission.

TELEMETRY FREQUENCIES AND POWER DENSITIES

As a result of practical considerations, most of the telemetry equipment operates with frequencies between 10 and 1000 MHz. The radiated power varies from a few µW (implants) up to approx. 2 W (modified walkie-talkies). The power density Pd in mW/cm^2 in a biological medium, produced by the near field of the antenna, can only be determined accurately by means of measurements. The formula $Pd = P/2\pi r \cdot L$ (P = radiated power, L = length of a whip antenna, r = distance from the antenna to the surface of the subject) gives only the first approximation. A standard UHF walkie-talkie delivers a Pd in the range of 10 mW/cm^2 when mounted on the chest of a man. The USA Standard (1) recommends the following safety level: 10 mW/cm^2, averaged over 0.1 hour periods, energy density 1 mWh/cm^2, valid for the frequency range 10 MHz to 100 GHz and for both partial and whole body irradiation. The safety levels of most of the eastern countries are much more stringent, e.g. USSR: 0.01 mW/cm^2, 6 hr/day, f > 300 MHz (2).

DIELECTRIC PROPERTIES OF BIOLOGICAL MEDIA

The biological media may be classified into two groups (3,4):

Muscle, Skin, and Tissues with High Water Content

The cell membranes act as insulating layers and as high capacities at low frequencies, so that current flows only in the extracellular medium. With increasing frequency the capacitive reactance of the cell decreases, resulting in increasing currents in the intracellular medium with an increase in the total conductivity of the tissue (4). Therefore the dielectric constant ε_{rm} decreases and the conductivity σ_m increases with increasing frequency: 40.7 MHz: $\varepsilon_{rm} = 97$, $\sigma_m = 0.7$ mho/m ; 1500 MHz: $\varepsilon_{rm} = 49$, $\sigma_m = 1.8$ mho/m (3,4).

Fat, Bone, and Tissues with Low Water Content

A similar trend is observed also in this group, but the values are an order of magnitude lower: 40.7 MHz: $\varepsilon_{rm} = 14.6$, $\sigma_m = 13-53$ mmho/m; 1500 MHz : $\varepsilon_{rm} = 5.6$, $\sigma_m = 71-171$ mmho/m (3,4). The conductivity depends on the ratio of free to bound water (4). Specifications of other tissues such as brain, lung, etc. lie between those of the two groups (4).

Wavelength within the Medium and Depth of Penetration

The wavelength λ_m in the medium is reduced by a factor of 2.3-14.3 (both groups, 40.7-1500 MHz) (4). The depth of penetration δ is defined as the distance required to reduce the power density by a factor of e^{-2}. In the first group, δ is 11 cm for 40.7 MHz and 2.4 cm for 1500 MHz. In the second group, δ is as much as ten times greater (4).

Power Absorption and Hot Spots related to the Radius of the Body

The Relative Absorption Cross Section RAC of a spherical object with characteristics similar to those of body tissues depends on the circumference $2\pi a$ of the object and on the free space wavelength λ_0 (5). RAC is defined as the ratio of absorbed to incident power, providing that the sphere is irradiated by a plane wave. Measurements (5) with phantoms ($\varepsilon_{rm} = 60$, $\sigma_m = 1$ mho/m) at the two frequencies 400 and 2880 MHz have shown that the RAC is small for small radii ($2\pi a/\lambda_0 < 0.2$: RAC < 0.1). All of the incident power (or more) is absorbed, if the circumference is approx. λ_0 ($2\pi a/\lambda_0 = 0.4-2$: RAC $= 0.7-1.6$). For larger radii, the RAC decreases slowly to the value 0.5. This means furthermore (5), that the body is transparent for low frequencies, absorbs tremendously if $2\pi a/\lambda_0$ is in the region of 1 and that the body absorbs the half of the incident power at higher frequencies. Hot spots, i.e. points inside a sphere with a higher E- or H-field intensity than at the irradiated surface, have been detected in the brain of animals (8). A theoretical investigation (6) revealed, that hot spots appear only inside spheres with radii 8 cm $> a >$ 0.1 cm and only in the frequency range $300 < f < 12000$ MHz. The hot spots have a heating potential which is 10-15 times higher than predicted by the RAC method and appear in depths which are up to 5 times larger than the predicted penetration depth δ of plane slabs (6).

HUMAN BODY EFFECTS ON RADIATION PATTERNS OF BODY-MOUNTED ANTENNAS

An antenna radiates waves which are not plane within the Fresnel region. However, one is able to roughly estimate the frequency range in which a small whip antenna will have a disturbed azimuthal radiation when mounted dorsally on the body. With a trunk radius of a = 15 cm, the most critical RAC is between $0.4 < 2\pi a/\lambda_0 < 2$, i.e. 130 to 640 MHz. In fact, our field measurements have shown that the human body has little influence below 50 MHz and acts like an absorber and reflector in the predicted region. Measurements in an anechoic chamber revealed field attenuation up to 30 dB on the axis transmitting antenna - heart - receiving antenna (231 MHz). Of great importance are the distances from the antenna to the body, the polarization of both antennas and the detuning of the antenna by the body tissue.

BIOLOGICAL EFFECTS OF RF RADIATION

Biological effects of strong microwave radiation in small animals have been treated by many researchers (2 et al), but unsufficient data are available on the human effects of weak RF radiation (7). RF radiation up to many GHz are nonionizing (2) and with low power densities the following effects may occur:

Thermal Effects

The heating results from both ionic conduction and vibration of the dipol molecules in water and proteins (4). The power absorbed by the tissue will produce a temperature rise, if the thermoregulatory capability of the tissue is exceeded (8). A power density of 10 mW/cm^2 produces less heat than the metabolic rate (1.3 W/kg for the total human body (8)) and is harmless from this point of view (8). The lowest level for irreversible damages seems to be indicated by the lens' opacity of the rabbit's eye: >100 mW/cm^2, > 1 hr, 2.4 GHz (2) and 10 mW/cm^2, 7 hr (9). A temperature rise of 4^0C may be destructive (2), but even some 0.001 ^{0}C/sec produces sensations (7,pp 3).

Non-Thermal Effects

Forces induced by alternating electrical fields (e.g. pearl chain effects, orientation of microorganisms) and dielectric saturation occuring in solutions of proteins and other biological macromolecules occur only with high field levels (4,5).

The teratogenic effects of irradiation with 9 GHz have been studied (12) using the pupae of the darkling beetle *Tenebrio molitor*. Statistically significant increase in teratogenesis (malformation in the adult insects) were observed at power levels down to 170 μW/cm^2. The pupation time increased monotonically with the power density (2 hr irradiation). The damages increased linearly with the logarithmus of the dosage, and the effects started at approx. 40 μW/cm^2 and 0.1 mWh. Western researchers stress destructive effects, in contrast to East Europeans (7) who place great emphasis on functional disturbances in the CNS, on physiological alterations and behavioral reactions, which occur by power density levels down to a few μW/cm^2 (7,9,10).

Various low level effects may be considered as selective absorption of radiation at the interfaces of heterogeneous biological systems (e.g. hypothalamic - hypophyseal - suprarenal system (9)). Electrophysical investigations of isolated nerves and muscle fibres in frogs (7,pp 31) at 5 $\mu W/cm^2$ have revealed slowed conduction of impulses, an increased synaptic delay, a lengthening of latent and refractionary periods and changes in action potentials. Alteration in EEG, in conditioned reflex activity (longer latent period, weakened reactions to positive stimuli) and in several metabolic processes, were observed (10) in rats and rabbits after irradiation with less than 10 $\mu W/cm^2$ (50 MHz, 12 hr/day). Vague effects are reported to occur in occupational exposure (2,11). The effects are manifested by weakness, fatigue, headache etc. (2), and dysfunctions in the vegetative system (11), which are apparently reversible.

CONCLUSION

The radiation of telemetry transmitters will not damage the subject, but artefacts in behavior and in EEG studies must be taken into account.

REFERENCES

1. ANSI: Safety Level of Electromagnetic Radiation with Respect to Personnel. ANSI C95.1-1974. New York, IEEE, 1974
2. Michaelson SM: Effects of exposure to microwaves: Problems and perspectives. Environ Health Persp 8: 133, 1974
3. Schwan HP, Piersol GM: The absorption of electromagnetic energy in body tissues. Am J Phys Med 33: 371, 1954
4. Johnson CC, Guy AW: Nonionizing electromagnetic wave effects in biological materials and systems. Proc IEEE 60: 692, 1972
5. Schwan HP: Microwave radiation : Biophysical considerations and standards criteria. IEEE Trans Biomed Eng BME 19: 304, 1972
6. Kritikos HN, Schwan HP: The distribution of heating potential inside lossy spheres. IEEE Trans Biomed Eng BME 22: 457. 1975
7. Biologic Effects and Health Hazards of Microwave Radiation. Proc Int Symp Warsaw,1973. Warsaw, Polish Med Publishers, 1974
8. Guy AW: Quantitation of induced electromagnetic field patterns in tissue and associated biologic effects. In Ref (7),pp 203-216
9. Gordon ZV, Roščin AV, Byčkov MS: Main directions and results of research conducted in the USSR on the biologic effects of microwaves. In Ref (7), pp 22-35
10. Dumanskij JD, Šandala MG: The biologic action and hygienic significance of electromagnetic fields of superhigh and ultrahigh frequencies in densely populated areas. In Ref (7),pp 289-293
11. Kalada TV , Fukalova PP,987čarova NN: Biologic effects of radiation in the 30-300 MHz range. In Ref (7), pp 52-57
12. Liu LM, Rosenbaum FJ, Pickard WF: The relation of teratogenesis in Tenebrio Molitor to the incidence of low level microwaves. IEEE Trans Microwave Theory Tech MTT 23: 929, 1975

TELEMETRIC CORRECTION OF MOVEMENTS WITH THE USE OF A COMPUTER AND AN ELECTROSTIMULATING DEVICE

I. Ratov, E. Boyko, O. Byvshev, V. Merkulov

All-Union Research Institute of Physical Education, Moscow, USSR

PURPOSE

The purpose of the present work was to study the possibilities of using two-way radiotelemetry in a "man-computer" functional system to provide better performance of a complicated motor task by means of telemetric correction of movements. Telemetric correction was achieved by the stimulation of appropriate muscles with electric pulses.

METHODS

In our study we used newly developed methods of radiotelemetric correction based on computerized evaluation of the movement during its development by comparison with the previously selected principal parameter. The use of these methods enabled us to determine the most favourable moment for muscle activation in the development of an exercise.

To meet the needs of our investigation a special telemetry system was designed. The first part of the system consists of the devices carried by the subject and includes a transducer, signal amplifier, modulator, transmitter, receiving set, stimulating unit, and power supply. The second part of the system consists of a radio-receiver, demodulator, computer, and control signal transmitter. During the series of experiments, this second part of the telemetry system was carried in a special automobile. A magnetic tape recorder and a multi-channel monitor were added to the telemetry system. The telemetry system is shown as a block diagram in Figure 1.

Utilization of this system and methods of telemetric correction allowed us to improve considerably both the technique and final results of the exercise.

Let us take the series of experiments involving telemetric correction of movements of shot putters as an example.

A piezoelectric accelerometer is fastened on the palm of a shot-putter's hand. The transducer signal generated in the process of shot putting is conducted to the amplifier fixed to the subject's belt. A radio-transmitting unit is also fixed to the belt of the subject. After amplification the transducer signal modulates, on the AM/FM principle, the transmitter operating

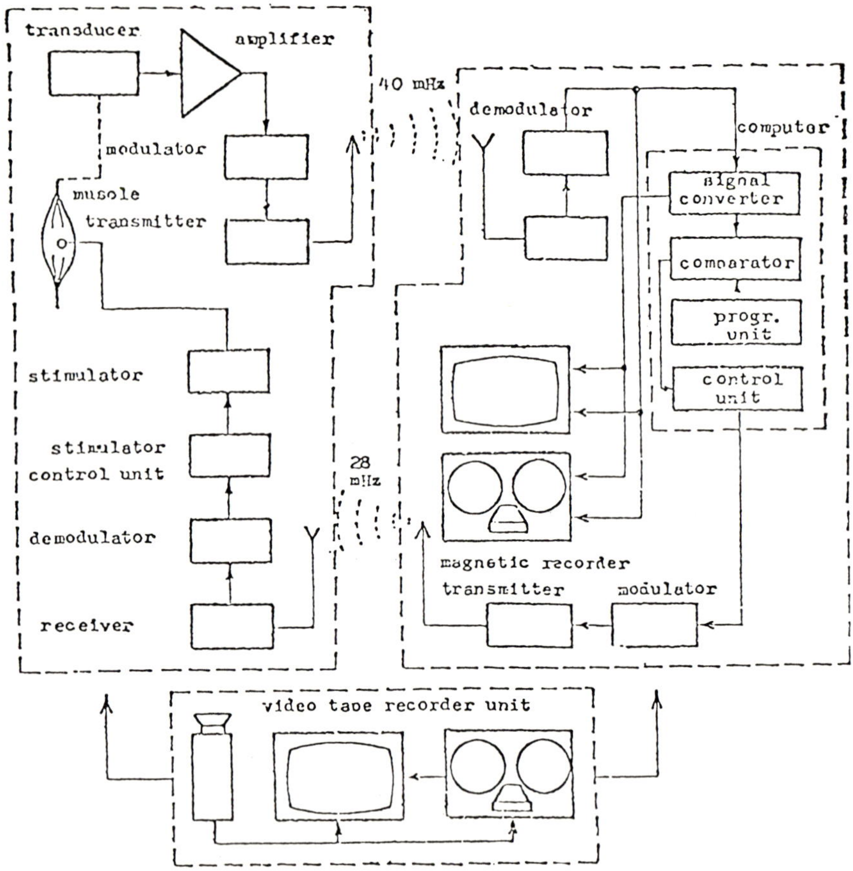

Figure 1. Block diagram of the telemetry system for telemetric correction of movements.

frequency. The signal is then emitted from the transmitting antenna.

The radio signals are received and demodulated by the"Sport-4" system and then are fed to the computer input. By integration the acceleration signal is converted to velocity signal and with the aid of a comparator is compared with the previously programmed value. This programmed value is the value to impart maximum momentum to the shot characteristic of high achievement. In case of coincidence of the received and programmed signal values, the computer produces a control pulse which causes the transmitter to send a command signal back to the subject.

The command signal is detected by the receiving set carried by the subject, and the electrostimulating device is switched on for a given period of time. With the aid of wires and electrodes the signal pulse produced by the stimulator is then conducted to the motor point of the muscle whose role in performing the motor action is at the present moment most important. In our example of using telemetric correction in shot putting, the leading muscle was musculus triceps brachii. The activation of this muscle with stimulating pulse signal in the final phase of the movement causes an increase in the velocity of the shot.

The computer supplies the muscle with control stimulating pulse only if the movement is being performed correctly. Under such conditions the use of telemetric correction makes it possible to convert any good attempt in shot putting into an excellent one, i.e. at the level of personal best achievement.

RESULTS

The results of our study make possible considerable increase in the quality of performance of physical exercises through telemetric correction of movements. Telemetric correction in the final phase of shot putting significantly increased the force applied to the shot. This resulted in increased shot

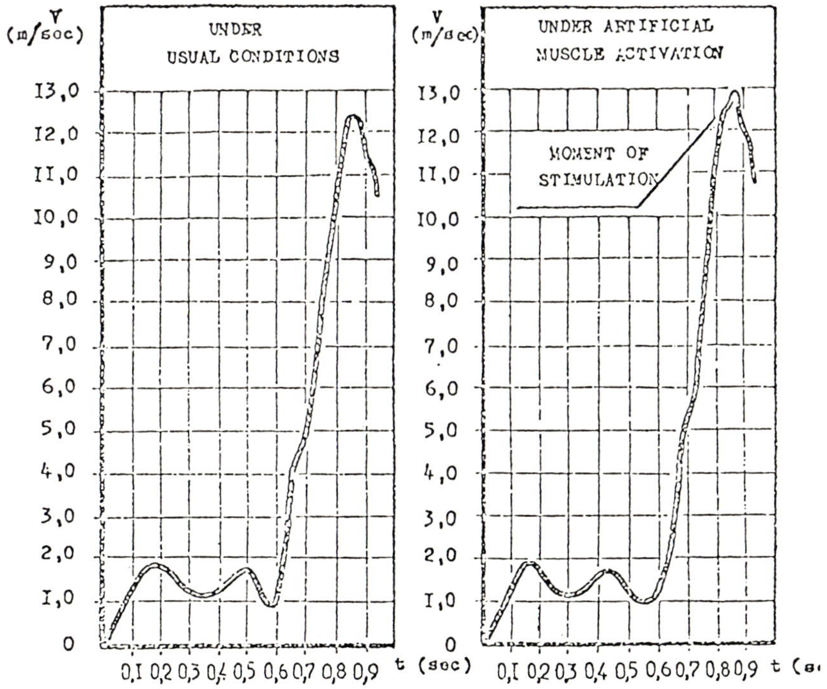

Figure 2. Velocities of the shot under usual conditions (left) and under artificial muscle stimulation (right).

velocity at release.

Comparison of the velocity curves obtained through oscillographic analysis shows a noticeable rise in the right-hand velocity curve resulting from muscle activation with the correcting pulse. This serves as evidence that the subject applies additional force to the shot (Figure 2). The comparison of the experimental data obtained in shot putting without and with telemetric correction demonstrated, in the latter case, a statistically significant increase in average velocity of 5.36% ($t = 5.4$; $p = 0.001$), and in average distances of 3.34% ($t = 8.2$; $p = 0.001$).

The experiments also revealed a positive after-effect which found its expression in the results of the subject's attempts in shot putting that followed the use of telemetric correction. In these attempts, velocity and distance were higher compared with the initial data. Thus, maximum velocities averaged 2.95% higher than the background data ($t = 6.5$; $p = 0.001$), and the distances were on an average 1.7% higher than the initial data ($t = 6.2$; $p = 0.001$).

It should be noticed that the positive qualitative changes in movements resulting from the use of stimulating correction were relatively stable. The use of telemetric stimulation for correcting movements typical of other physical exercises apart from shot putting also showed the stable character of these positive changes.

CONCLUSION

The results of the study show that the method of telemetric correction of movements has a great future in terms of improving the quality of fulfilling complicated motor tasks. The use of this method facilitates development of the most efficient techniques of athletic exercises. This is especially true when the object of the exercise demands maximum strength and speed from an athlete.

A COMBINED SENSOR SUITABLE FOR TELEMETERING RESPIRATORY FUNCTIONS

Georg Küchler, Rudolf Schneiderreit, Ilse Wolburg, and Wolfgang Wagner

Zentralinstitut für Arbeitsmedizin der DDR
FB Arbeitsphysiologie, 1134 Berlin

The combined sensor consists of an impeller system called "Wirbelrespirometer" which contains the pO_2-sensor and an additional thermistor. The construction of the "Wirbelrespirometer" is shown in fig. 1.

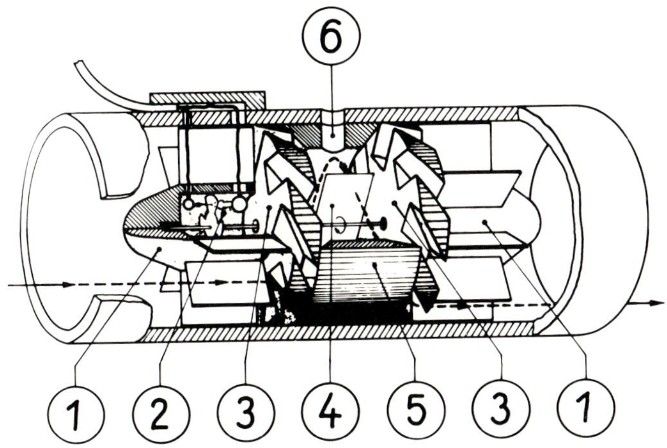

Fig. 1 Construction of the "Wirbelrespirometer": (1) guiding star to linizarize the whirling air after it has passed the second slotted disc (3); (2) arrangement (luminescent diode, interrupter disc, phototransistor) to transduce respiratory flow into electric pulses; (3) slotted disc rotates the air spirally around the axle; (4) impeller; (5) whirl intensifier: reduction of the tube diameter accelerates the rotation of the eddies; (6) borehole: position of the pO_2-electrode.

Two slotted discs which generate whirls are fixed in a plexiglass tube. The slots form acute angles with the axial direction. The impeller is supported between the discs. The flowing air passing the slots is rotated and drives the impeller. The number of impeller revolutions is measured by an opto-electronic cicuit. There are 8 pulses to one revolution, each pulse corresponding to an air volume of 1.5 ml. The relationship between airflow and output of the circuit is sufficiently linear (fig. 2).

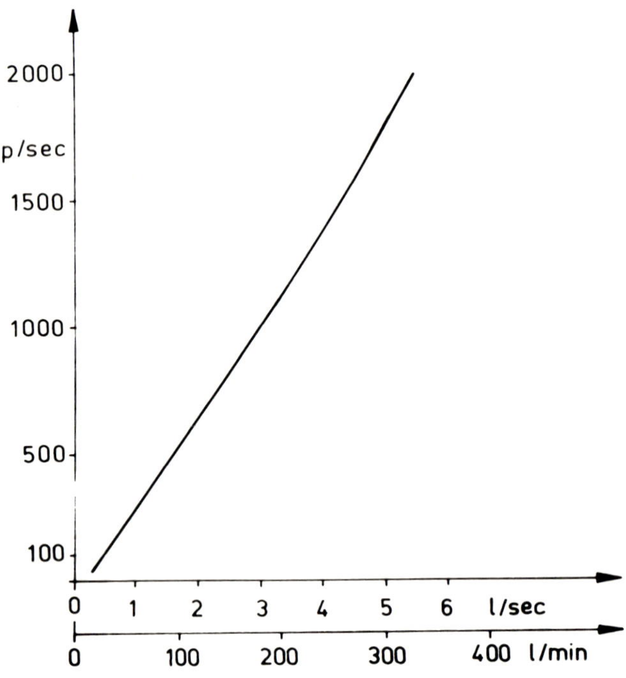

Fig. 2 Calibration curve: relationship between flow (abscissa) and the number of electric pulses (ordinate).

The opto-electronic circuit also detects the direction of impeller rotation, and hence, the direction of airflow. There is an asymmetry in the electric output related to the direction of airflow in the order of 1 to 3 p.c. This can be corrected electronically.

The respiratory resistance of the device is shown in fig. 3, revealing its lower resistance over a WRIGHT-respirometer.

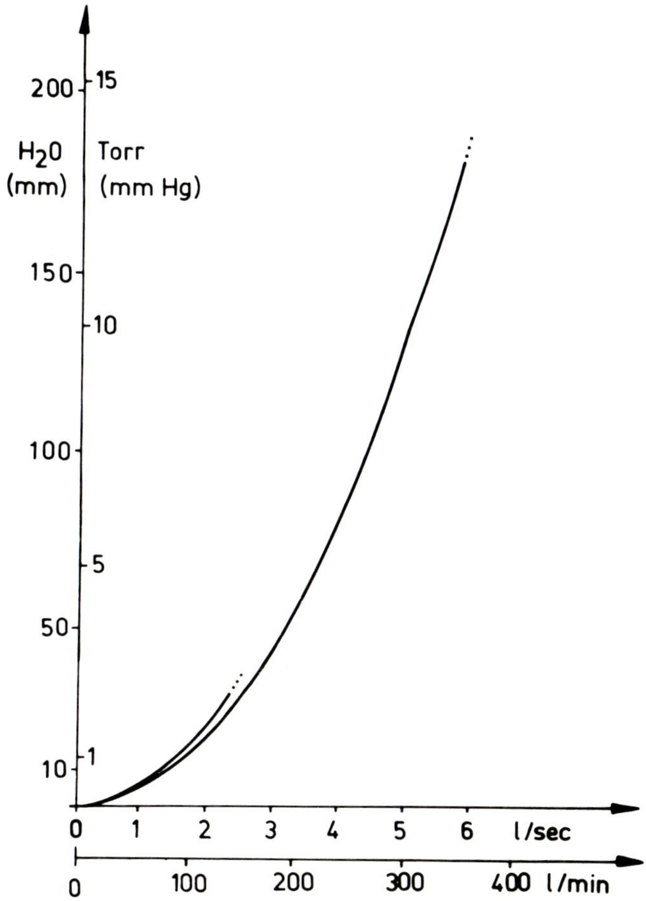

Fig. 3 Relationship between flow (abscissa) and respiratory resistance of the "Wirbelrespirometer" (ordinate); the short curve indicating higher resistance was produced by a WRIGHT-respirometer.

The influence of different temperatures (inspiration, expiration) on air volume, and consequently on flow, is compensated by an electronic circuit which includes the fast-reacting thermistor situated at the periphery of the impeller compartment.

The signal delivered by this fast-reacting thermistor is simultaneously used to correct the temperature effect on the pO_2-sensitivity of the polarographic

pO_2-sensor. The pO_2-sensitive surface of the sensor is likewise situated at the periphery of the impeller compartment. The whirling air at this point prevents water condensation on the teflon membrane of the pO_2-sensor.

The plotted courses of oxygen partial pressure, airflow, and temperature are given in fig. 4.

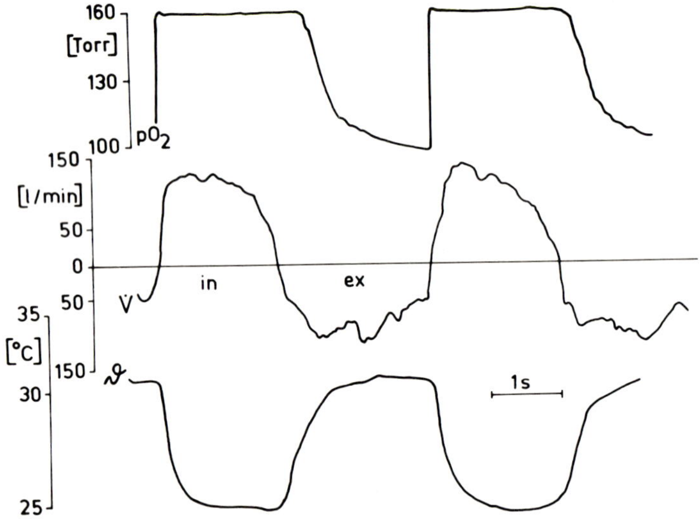

Fig. 4 Courses of partial pressure of oxygen (Torr), respiratory flow (l/min), and temperature (°C) within the "Wirbel-respirometer" during physical exercise on a bicycle ergometer.

The proband breathed through the combined "Wirbelre-spirometer" while working on a bicycle ergometer at 160 Watts. The temperature of the inspired air is near to 25 °C, of the expired air about 31 °C. The momentary values of maximal flow amount to 130 l/min; the endexpiratory partial pressure of oxygen drops to 100 Torr. The computed values give a tidal volume of about 2.5 l (BTPS), a ventilation minute volume of 40 l/min, and an oxygen consumption of about 1.1 l/min (STPD).

DISCUSSION

KIMMICH: This is a very difficult and important subject. It will be tremendously difficult to learn the effects of radiation on humans from experiments with animals.

NEUKOMM: (Reply to unidentified questioner) It is possible for the body to converge the incoming radiation and thus absorb more than falls on the body area in free space.

SCHULMAN: Radiation found normally in our environment, as from police transmitters and taxi transmitters, exceeds safety standards.

NEUKOMM: You are right, but field strength decreases rapidly with distance. Walkie-talkies radiate large power densities on the chests of the users. Safety standards are for the whole population rather than individual users of such devices.

UNIDENTIFIED: Have any behavioral effects been demonstrated?

NEUKOMM: The influence of radiation on animals has been investigated. On humans there is only some information on occupational exposure, and the causes of some changes are not clear.

VREELAND: The present standard of 10 milliwatts per square centimeter corresponds to 194 volts per meter field strength. We have found that as little as 0.6 volts per meter can stop some of the early pacemakers.

RADER: Some researchers in Southern California have found that calcium flow in the central nervous system is affected by modulated RF radiation at field strengths about a thousand below the cell voltage or approximately 7 millivolts across the cell. This effect has been correlated with behavioral action in, I believe, either baboons or chimpanzees.

SALMONS: The discussion has been concerned with deleterious effects of radiation. There is a firm which makes a device transmitting 27 MHz pulsed radiation for which they claim beneficial effects on tissue repair such as nerve repair. No one really knows how it works. We need a good system for looking at effects of radiation, perhaps with tissue cultures and observations of ion movements.

KUCHLER: We have investigated the effects of electric fields on membrane functions. All of the effects we have found are temperature effects.

SESSION II

IMPLANTABLE TELEMETRY SYSTEMS

Chairmen

J.D. MEINDL, Ph.D.

J. BOJSEN

CHAIRMEN'S OVERVIEW

The most significant technical development in biotelemetry was the invention of the high-frequency, diffused-base transistor, about 20 years ago. It gave us the capability in a very miniature device to telemeter radio-frequency signals. About 10 years ago we saw a second very significant development in the field of implantable biomedical telemetry. This was the commercial availability for the first time of high-speed, low-powered, digital, integrated circuits, particularly the CMOS variety which is described here. Today one can purchase commercially, integrated circuits containing 5,000 to 10,000 transistors in a single chip of silicon. These are classed as microprocessors and 4,000-bit memory chips. It has been predicted in the forseeable future that there will be integrated circuits with 50,000,000 to 100,000,000 transistors in a single chip of silicon. I think this is where the exciting future for implantable biomedical telemetry lies. These developments in semiconductor electronics make the future seem very promising and very exciting indeed.

MULTI-CHANNEL IMPLANTABLE BIOTELEMETRY: PROBLEMS, PITFALLS, AND REWARDS

Martin Reite, M.D. and J. Donald Pauley, Ph.D.

Department of Psychiatry, University of Colorado Medical Center
Denver, Colorado 80220

During the past several years our laboratory has been studying the physiological correlates of behavior in unrestrained, group-living monkey infants utilizing totally implantable multi-channel biotelemetry systems (1,2). During this time we have collected a rather extensive amount of data from over a dozen monkey infants, and we have acquired considerable experience in the operational use of implantable biotelemetry systems. We would like to discuss some of the highlights of our project from an operational standpoint in this paper, as well as to point out those situations in which implantable biotelemetry has unique value.

Our monkey subjects are naturally-born in the Department of Psychiatry Primate Laboratory and are raised by their mothers in social groups containing one adult male and six to eight adult females, several of whom usually have infants at any one time. When the infants are four to five months old and weigh approximately 1000 grams, they are surgically implanted with biotelemetry systems that transmit seven physiological data channels, including EOG, EMG, EKG, body temperature, and three channels of EEG. The telemetry system is constructed by hand from commercially available parts. It is a pulse amplitude modulated system that transmits in the FM band (3). The system consists of two packages connected by a 5 cm long cable which are implanted extraperitoneally in the anterior abdominal wall. The large package, which is about 3 x 2.6 x 0.7 cm, contains the amplifiers, multiplexer and RF oscillator. The smaller package, about 2 cm long by 1.5 cm in diameter, contains a 2.8 volt lithium cell.

All of the electrode lead wires (multistranded Teflon coated stainless steel) are assembled into a 1.5 mm OD silicone rubber tube to form a cable. The ECG electrode lead wires are initially led inside this main cable and exited where the cable will pass near the heart, where they are then bared and wrapped around the main cable to serve as the ECG electrodes. This is a useful technique as it requires no connection of electrodes to the wires at the time of implantation.

The electronic components are potted in epoxy, coated with a layer of a 50-50 mixture of molten beeswax and parafin, and then coated with medical grade silastic rubber. Proper sterilization is most important. Liquid sterilizing solutions such as Zephirin are not adequate as the germicidal solution cannot penetrate the silastic tubing containing electrode wires, and small bacteria pockets can persist. We have had excellent success with Ethylene Oxide sterilization, and routinely sterilize our implanted devices for 24 hours at room temperature, followed by 24 hours of out-gasing prior to implantation.

At the time of implantation, absolute sterility must be maintained. We recommend a dust-free operating room, with all participants involved wearing caps, masks and sterile gowns and gloves. Skin preparation includes a PhisoHex scrub followed by a Betadyne or Povidone-Iodine solution scrub. We routinely irrigate all implanted devices in situ with a sterile solution of ¼% Neomyocin before they are closed in place. A most important general rule is to never close the skin on top of an implanted device. Where ever possible, implanted devices should be placed in deep pockets that can be closed independently under the skin, and the skin incision should be placed at some distance from the subcutaneous device so that when the final skin closure is made, it does not rest over the device.

All incisions should be closed in layers. We recommend the use of polygycolic acid (Dexon) sutures (not catgut) for subcutaneous closure due to the bacteriostatic action of polyglycolic acid as it breaks down. Polypropylene (Prolene) or stainless steel can be used to close skin incisions. If a skin incision can be bandaged, then the sterile suture line after closure can be covered with collodion. Single interrupted sutures are used for subcutaneous tissue area closures whereas we usually close skin incisions with interrupted mattress sutures. Sutures should not be left in place longer than necessary for good healing of the wound as they may serve as a source of infection thereafter. If implanted devices are adequately sterilized and proper precautions are taken at the time of surgery, infections are rare.

Occasionally the skin may break down over implanted devices, with resulting local infections that can very quickly spread along the implant-tissue interface. Such skin breaks can be successfully managed providing prompt and vigorous treatment is instituted. We recommend the following treatment protocol:

(1) Sterilize the skin in the vicinity of the break, as for a normal surgical procedure.
(2) Use sterile instruments and sterile technique in all such treatments.
(3) Incise the infected area with a sterile scalpel and immediately swab the infected area to obtain a culture and sensitivity of the offending organism. Chances are high in most hospital-related environments that it will be a penicillin-resistant staphylococcus.
(4) Debride the wound (clean out all infected tissue).
(5) Irrigate the area copiously with a sterile solution

of Bacitracin (50,000 units in 30cc saline).

(6) Close with interrupted mattress sutures. Although it is against accepted surgical technique to close an infected wound, you should not leave a wound containing an implanted device open to drain.

(7) Apply a sterile bandage.

(8) Immediately begin intensive parenteral antibiotic treatment for a minimum of six to ten days. We normally begin treatment with cephaloridine (Loridine) or with a synthetic penicillin (e.g. Oxycillin) until the results of the culture and sensitivity are available.

Our treatment success rate when infections are caught early (e.g. an open wound through the skin to an implanted device but without evidence of major infection) is in excess of 50%. This is well-worth vigorous treatment considering the time and expense usually tied-up in such animal subjects.

If you have not clearly won the battle within five to seven days, the success rate drops dramatically as by that time there are usually deep abscess pockets that cannot be eradicated.

A frequent problem encountered in the operational use of implanted biotelemetry systems is that of RF signal reception. Where small, unreplaceable batteries are the sole power source, most of the power is usually consumed in the RF stage and radiated power is closely related to battery life. The lowering of RF stage current will result in a significant decrease in total consumption and considerable prolongation of battery life, but this of course entails a decrease in signal strength. We have found that effective RF radiated power can be reduced if several antennaes are utilized spaced at intervals around the group pen, and we have designed a device that automatically steps through six separate simple half wave length dipole antennaes, at a step rate of 100 msec, and which automatically selects the signal with the lowest noise (4). We would recommend that investigators using low power implantable telemetry systems consider an automatic antennae selector as one way of conserving power.

We formerly utilized certified cardiac pacemaker mercury cells as our power supply, but a recent change in their manufacturing process has resulted in their building-up an unacceptably high internal impedance when constant current drain approximates 600-700 micro amps. We have, however, had excellent results with a recently available 2.8 volt lithium cell (Matsushita LR-½A) which we now utilize routinely.

If battery life is to be extended with implantable devices, a method for remote switching must be incorporated. We have developed an RF actuated command receiver that is being described by Dr. Pauley elsewhere on this program. The advantages of such a command receiver are that it can be activated remotely, but the disadvantages are that it has a finite, albeit small, current drain even in the off condition. We have recently found a commercially available bistable magnetically activated latching reed switch that has proven to be most reliable in operational use (Hamlin MDLS-4). The switch is opened or closed by changing the magnetic field in its vicinity, and once this state is changed, it remains in that state until a new field is applied. It is not sensitive to G forces which have been a problem with older magnetically operated switches. The advantage of such a switch is that when it is off, the current drain is zero.

Implantable telemetry of course provides the benefits found with more conventional externally mounted systems. As most investigators working with telemetry are aware, the spectrum of physiological events recordable is often much wider than that obtainable with hard wire techniques, and we can often obtain data on states simply not available in other ways.

There are certain situations, especially in regard to certain cardiovascular variables, where for a variety of biotechnological reasons, implantable telemetry may be the method of choice. There is another major area however in which the benefits of implantable telemetry clearly outweigh the increased complexity, and this has to do with the experiments in which social behavior is an important part of the experimental data being evaluated. Clearly, when dealing with animal subjects with complex social behaviors, any artificial manipulation of the animal's appearance by methods such as the attachment of backpacks, head plugs or the like, is going to have some influence on the animal's individual behavior, and on the way it is perceived by other members of its group. In such situations, the resulting changes in social behavior, eventhough subtle, could significantly influence the experimental results. In such applications, totally implantable biotelemetry systems are uniquely valuable, and may in fact represent the only feasible way to collect artifact-free data.

REFERENCES

1. Reite M, Pauley JD, Walker S, et al: A systems approach to studying physiology and behavior in infant monkeys. J Appl Physiol 37: 417-423, 1974a.

2. Reite M, Kaufman IC, Pauley JD, et al: Depression in infant monkeys: physiological correlates. Psychosom Med 36: 363-367, 1974b.

3. Pauley JD, Reite M, Walker S: An implantable multi-channel biotelemetry system. Electroencephalog Clin Neurophysiol 37: 153-160, 1974.

4. Pauley JD, Reite M: Automatic antenna selector for use in biotelemetry applications. Submitted for publication.

(Supported by USPHS grant # MH 19514)

A TOTALLY IMPLANTABLE MULTI-CHANNEL TELEMETRY SYSTEM USING CUSTOM INTEGRATED CIRCUITS

James W. Knutti, Craig S. Sander, and James D. Meindl

Integrated Circuits Laboratory
Stanford, California 94305 U.S.A.

INTRODUCTION

A miniature totally implantable system has been developed for telemetering several channels of physiological information across the intact skin of free roaming animals. Custom integrated circuits permit realization of the system's complex signal processing circuits within the small volume of an implantable package.

Input signals are obtained simultaneously from four pressure transducers, a thermistor and EKG leads, with the system easily modified to accept alternate inputs. These physiological signals, multiplexed with internal reference and zero signals to compensate for drift, modulate a subcarrier which is F.M. telemetered to a remote data station for demodulation and storage or display.

The implanted package is implemented using two custom bipolar I.C. chips, two commercially available C-MOS chips, and a discrete F.M. transmitter-antenna assembly. In addition a radio controlled switch chip and battery are used to power the implanted electronics. Through micropower I.C. design, the system consumes less than 8 mW. of power while transmitting, and has a telemetry range greater than three meters.

NEED

Historically physiological research has been performed on anesthetized preparations to determine gross physiological features, and short term response. Further refinements in instrumentation enabled longer term measurement of deep body parameters in unanesthetized animals through use of implanted transducers and percutaneous leads, requiring either restraining the animal, or attaching a backpack. This disturbed the animal, precluding a natural response, and provided a constant source of irritation. In addition, percutaneous leads are susceptible to infection and damage from the animal.

Totally implantable telemetry packages have been developed to eliminate the need for percutaneous leads, by transmitting the physiological parameters across the intact skin (1). Implantation imposes serious constraints on the signal processing electronics. Many components must be tightly interconnected to realize the small package necessary for implantation.

In addition, most physiological experiments require measurement of several parameters from either several implanted packages or multi-channel systems, which allow telemetry of several parameters from a single telemetry package (2). These multi-channel systems require even larger numbers of discrete components individually assembled into the small implantable package.

Through use of custom integrated circuits, most of the devices necessary for a complex multi-channel system have been realized and interconnected on two silicon chips. Through bipolar integrated circuit technology large quantities of reliable I.C.'s can be produced, providing interconnection of the many devices at the chip level. Shallow devices and high value resistors allow micropower operation. Thus the use of custom I.C.'s allows a very small, low power realization of a totally implantable package with increased reliability, performance, and ease of assembly.

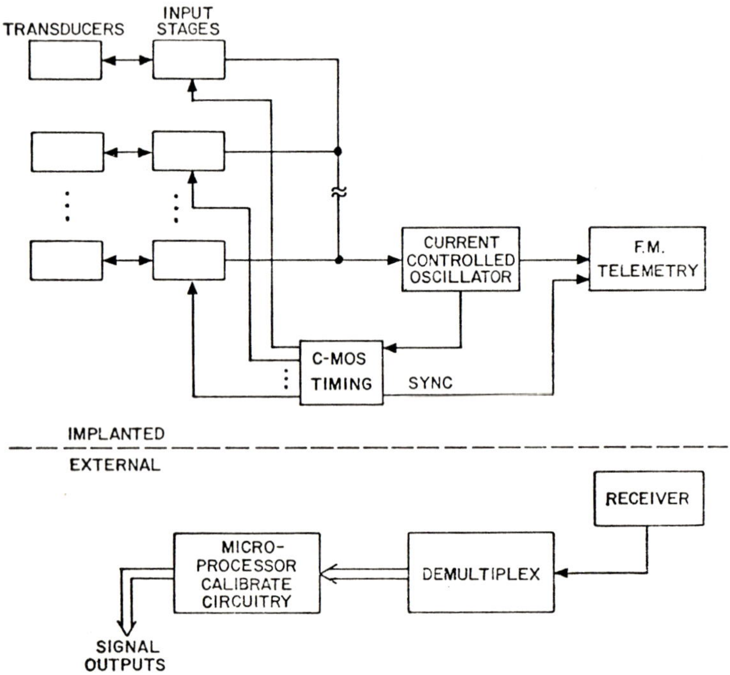

Figure 1. Block Diagram.

SYSTEM

Figure 1 shows a block diagram of the multi-channel system. A series of input stages sequentially convert the transducer differential voltage signals into a time division multiplexed current signal. This technique is very flexible, allowing any input signal to be converted to a current for sequential injection into the current controlled oscillator. Typical transducers may be pressure transducers, temperature sensors, or EKG electrodes. The time division multiplexed current signal is fed to a current controlled oscillator whose frequency is proportional to input current. This signal is transmitted to external electronics by an F.M. telemetry link. The current controlled oscillator output, deviating from 10 kHz to 13 kHz, is also used to drive the multiplexing circuitry, which utilizes two commercially available C-MOS chips. The multiplexing circuitry provides eight sequential outputs, activating one of the six input stages, or outputting a zero reference signal or a known calibrate signal. By dwelling on each signal for two cycles, the system allows one cycle whose period is dependent on input current, and one cycle to change inputs, as in figure 2. In addition, a sync pulse is added at the F.M. transmitter to indicate channel one for demultiplexing.

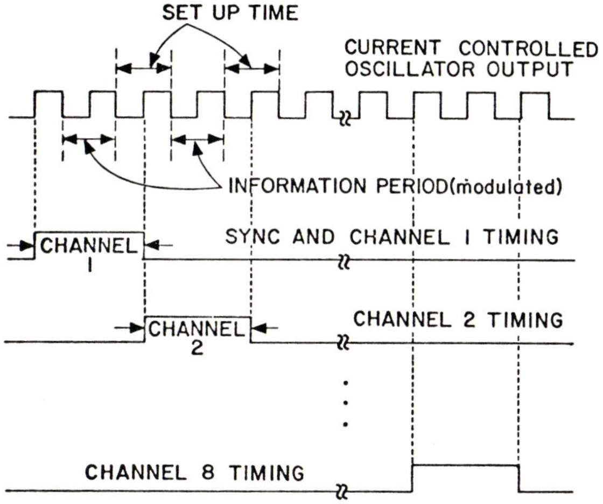

Figure 2. Timing Waveforms.

The internal circuitry consists of an input I.C. for driving transducers and receiving their outputs, and a current controlled oscillator I.C. Both of these circuits are currently realized on KITCHIPs, (arrays of bipolar devices optimized for low power operation) (3). In addition, 2 C-MOS chips for timing and a 2 transistor discrete F.M. transmitter are used. Overall power consumption is less than 3 mA at a 2.8 volt supply voltage, with an overall telemetry range greater than three meters.

An external F.M. receiver is used to recover the 10 kHz. subcarrier output, which is demultiplexed using a counter scheme similar to that of the internal multiplexer. The information interval for each time slot is converted to a digital signal, the zero reference subtracted, and the digital signal calibrated to the internally generated reference signal. Pressure signals are corrected for temperature drift using a telemetered temperature signal to look up correction factors in an inexpensive microprocessor system. Both digital and analog outputs are provided for further signal processing.

PERFORMANCE

The system currently is configured to transmit pressures between 0 and 200 mm Hg and temperatures between 35°C and 42°C. A silicon absolute pressure transducer with an anodically bonded glass cap is used to obtain the very low long term drift necessary for implantation (4). Overall channel bandwidth is set at 150 Hz. With a greater than 3 meter telemetry range, the system exhibits an on-time in excess of 200 hrs using a standard lithium cell. In its current configuration the system will transmit six channels of physiological information, as well as calibrate and zero signals. The number of channels may be easily expanded through an additional input chip, one more C-MOS chip, and a slightly higher subcarrier oscillator frequency.

CONCLUSION

Simultaneous remote transmission of six physiological parameters from this small totally implanted package permits valuable long term data collection without interacting with the experimental animal. The use of custom I.C.'s greatly reduces the number of components, achieving a small, low power implanted package.

ACKNOWLEDGMENTS

The authors wish to thank Stephen Combs, Douglas Karlson, Timothy Nunn and Owen Sharp for help in realizing this system.

This investigation was supported by PHS Research Grant No. 1 P01 GM17940-06 from the Department of Health, Education and Welfare.

REFERENCES

1. VanCitters L, Franklin DL: Telemetry of blood pressure in free ranging animals via an intravascular gauge. J Appl Physiology 21: 5, 1966
2. Fryer TB, Sandler H, Freund W, McCutcheon EP, Carlson EL: A multi-channel implantable telemetry system for flow, pressure and ECG measurements. J Appl Physiology 39: 2, 1975
3. Combs SR, Meindl JD: KITCHIP - A generalized integrated circuit for biomedical system design. Proceedings of 3rd International Symposium on Biotelemetry, Asilomar, Calif., 1976
4. Nunn TA: An IC absolute pressure transducer with built-in reference chamber. NIH workshop in "Indwelling Pressure Transducers and Systems," December 4 and 5, 1975

SINGLE FREQUENCY RF POWERING

Wen H. Ko, Jaroslav Hynecek, and Abraham H. Kou

Engineering Design Center
Case Western Reserve University
Cleveland, Ohio

INTRODUCTION

The need for smaller and smaller telemetry transmitters, which could be used to transmit physiological information from small animals used in various biomedical experiments, today imposes difficult restrictions on the construction of such devices. If the use of the implanted volume is analyzed, it can be found that the battery or equivalent energy storage generally occupies about 50% of the volume, the electronics require 25%, and the rest is usually taken by RF-tank circuit, antennas, connectors, and similar accessories. It is therefore apparent that miniaturization of the electronics itself will not be the complete answer to the problem. In the past few years we have investigated in our laboratory the possibility of using the RF field to power the implanted transmitter from the outside, especially when the short range telemetry is used for transmission of data from the animals strapped in chairs or moving in cages. The telemetry system using the external source of energy which is transmitted into the implant has one additional advantage in not being limited in operation by battery lifetime and can, therefore, operate for virtually infinite lengths of time. The description of the system based on this principle is given in the following sections.

SYSTEM DESCRIPTION

Since it is not desirable to increase volume of the implant further by adding the receiving tank circuit to the implant, it is necessary to use only one tank circuit for both functions, the RF energy absorption and the data transmission. This is possible to accomplish by time sharing multiplexing.

The first system built in our lab for the Marine Biomedical Institute in Galveston under NASA contract # NGR-36-027-053, worked well and had the following block diagram (Figure 1.)

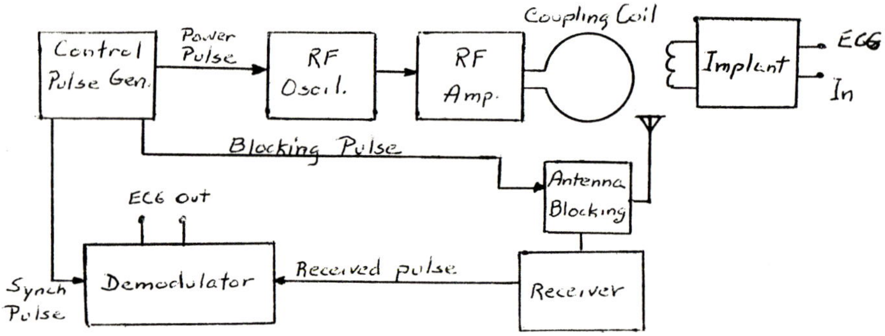

Figure 1. Telemetry System Block Diagram

The system was used to transmit the ECG signal from a Rhesus monkey strapped in a chair. The RF coupling coil was slipped on the chair so that the axis of the coil was parallel to the spine of the monkey. The operation of the system is apparent from the timing diagram given in Figure 2.

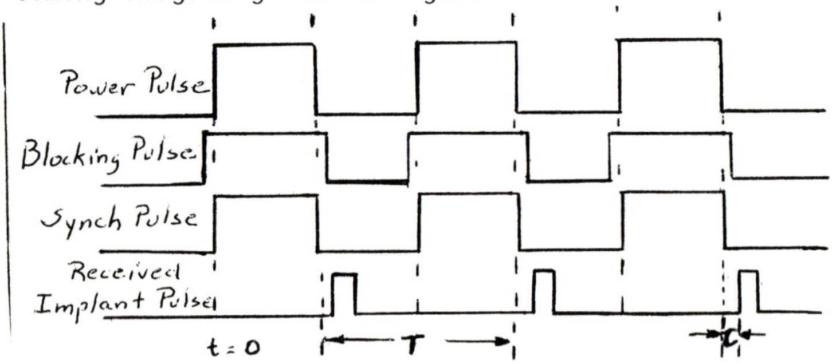

Figure 2. System Timing Diagram

The information about the ECG signal which is applied to the input of the implant is sampled once every period which has the length T and is converted into the delay τ of the transmitted pulse. Therefore, the position τ of the transmitted pulse is modulated (pulse position modulation) in correspondence to the input voltage. Since the synchronizing pulse for the implant is derived from the trailing edge of the RF power pulse, the accuracy of the transmitted information from the implant is not affected by any variations in the pulse widths. The described signal format can be easily extended into the multichannel operation simply by further time multiplexing. The three channel system is presently being investigated, where the following pulse format is employed. (Figure 2.)

Since all of the circuitry filling up the blocks in the diagram in Figure 1 is conventional, with the exception of the circuits for the implant, our description will be dealing in more detail only with the electronics of the implant.

DESCRIPTION OF THE IMPLANT

The most important section of the implantable transmitter is the oscillator. The complication here is the requirement that the same tank circuit has to be used for detection of the RF power and also for the transmission of the signal. In addition, the circuit has to generate a trigger pulse coinciding with the trailing edge of the RF pulse which is needed for synchronization. The complete circuit diagram for the oscillator is given in Figure 3.

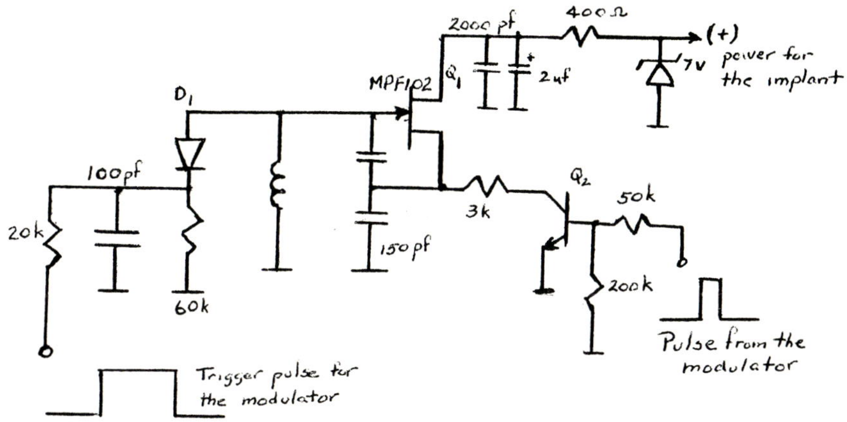

Figure 3. Oscillator Circuit Diagram

The FET MPF102 is used as the RF detector since the gate and drain serve as a diode when the transistor Q_2 is turned off. The charge is stored in the capacitor and the voltage is stabilized by the Zener diode. A separate RF detector is used to generate the trigger pulse for the modulator. When the transistor Q2 is turned on, the FET works as a conventional Colpitts oscillator. At this moment the RF voltage generated across the inductor L is not high enough to retrigger the modulator. The modulator circuit diagram is given in Figure 4.

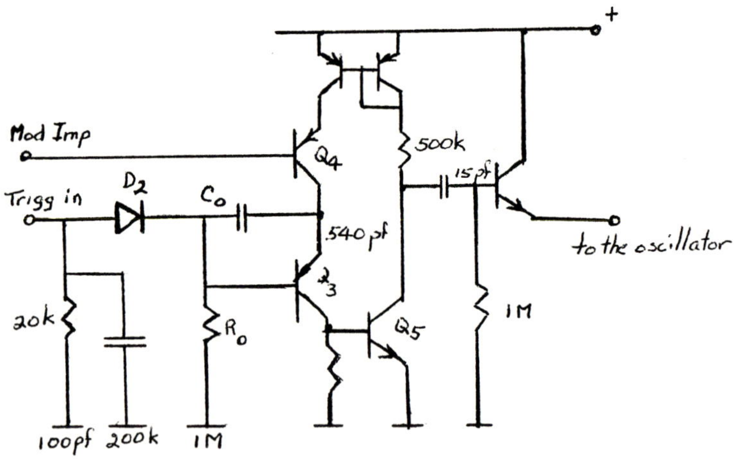

Figure 4. Modulator Circuit Diagram

Here the voltage to pulse position conversion is accomplished by means of a Miller integrator (Transistor Q_3). The voltage ramp generated at the collector of the Q_3 (after the trailing edge of the trigger pulse) is terminated as soon as the transistor Q4 is saturated. The time of integration is therefore proportional to the voltage at the base of Q_4, which is the input of the modulator. When the Q4 is saturated, the Q_3 and Q_5 are turned off and the pulse spike appears at the output of the modulator. This pulse then triggers the oscillator and RF burst is transmitted out from the implant. The signal which is used for the modulation can be obtained from a variety of sensors. In our case we used a standard ECG instrumentational amplifier to amplify the ECG signal up to the necessary amplitude which is required for the modulator. It should be stressed at this point that attention has to be given to the design of the input circuit of the high gain amplifier to avoid RF saturation during the powering. Usually a simple RC filtering network is sufficient. More complicated schemes using sampling and blocking techniques are under investigation. The implant was packaged in three flat packs with RF tank coil wound on a ferrite core.

ACKNOWLEDGMENT

Supported by NIH grant GM 14267-10

AN INDUCTIVELY POWERED IMPLANTABLE MULTICHANNEL TELEMETRY SYSTEM FOR CARDIOVASCULAR DATA

E. P. McCutcheon, R. Miranda, T. B. Fryer, and E. L. Carlson

Ames Research Center, NASA
Moffett Field, California 94035

The utility of multichannel implantable systems for telemetering cardiovascular variables has been demonstrated (1,2). Advances in miniaturized hybrid-circuit modules and use of inductively coupled power have made it possible to design a multichannel unit that can be implanted in monkeys or dogs as small as 5 to 10 kg. This paper describes the system design features and initial results with implantation.

SYSTEM DESCRIPTION

The basic operating principle is to convert analog data to a multiplexed, pulse-width-modulated (PWM) format that frequency modulates (FM) a radiofrequency (RF) transmitter (as described previously (1)). The system components placed internally include two pressure transducers, a temperature sensor, and an electrocardiographic (ECG) lead connected to a hermetically sealed mainframe package (6 cm long, 2.4 cm wide, 1.3 cm thick) containing the power converter, signal conditioner, and 88-MHz transmitter with integral transmitting antenna. An attached Silastic-covered coil receives power by inductive coupling from an external coil driven by a 250-kHz oscillator (fig. 1). The total weight of the

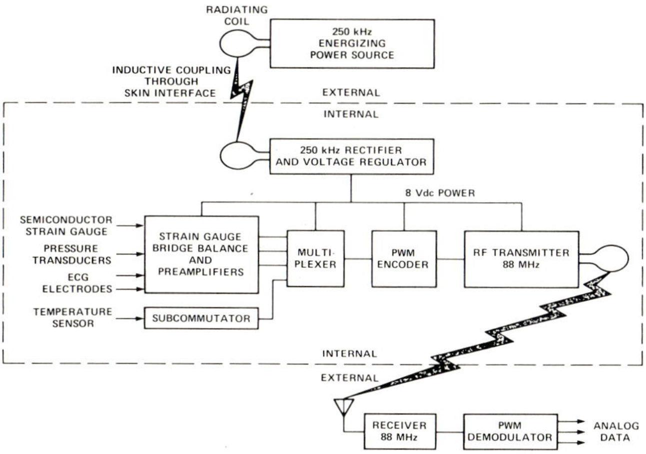

Figure 1. Inductively powered multichannel telemetry system functional elements.

implanted unit is approximately 40 g. The system is completed by a receiver and demodulator that convert the PWM signal to the original waveforms.

Transducers and Signal Conditioners

One of the pressure transducers is 3.5 mm in diameter, a size suitable for insertion in the descending thoracic aorta. Since long-term zero drift is accentuated with this small size, it was ac-coupled to provide a frequency response of 0.03 to 100 Hz. The second pressure transducer, 4.5 mm in diameter with a frequency response from dc to 100 Hz, is suitable for left-ventricular pressure measurement. The single ECG lead is a bipolar pair of stainless steel wires terminating in 3-mm uninsulated loops. Frequency response is 0.1 to 100 Hz. Two additional dc to 100-Hz channels are available if desired.

Signal conditioning includes bridge completion and balancing for the pressure cells, and amplification of pressure and ECG signals prior to multiplexing.

The sixth channel is subcommutated to provide a total of eight additional subchannels, four of which are active. One is used for temperature and the other three for system monitoring. The thermistor temperature sensor is positioned to project approximately 1 cm from the main package; the stainless steel case required for sealing results in a 30-sec time constant. The engineering variables monitored are the unregulated voltage from the internal receiving coil and the ±3.7-V regulated output from the power supply. After subcommutation, these channels are amplified and multiplexed.

Pressure transducer signal conditioning requires approximately 11 mA at 3.7 V, and the other signals require about 2 mA.

Multiplexer, Transmitter, Receiver, and Power

The basic clock rate for system timing is 10 kHz. Time frames containing eight PWM pulses or "words" are generated at 1250 Hz. One is a blanked pulse denoting the initiation of the sampling sequence; a second is a system zero reference. The other six words are used for the physiological and system monitoring data. The five high-frequency channels available are each sampled at 1250 Hz. The sixth word samples each of the subcommutated channels at 156 Hz.

The train of PWM information then frequency modulates (FM) the 88-108-MHz RF carrier of the transmitter, which broadcasts from the internal loop through the tissue to an external antenna and telemetry receiver. The receiver bandwidth is 500 kHz, to provide an acceptable rise time for the PWM signal. Despite the legal constraints on field strength, the transmitter achieves a 5-m range even in a noisy RF environment.

No active power source is implanted. All energy to operate the system is inductively coupled from an external energizing coil (7-cm o.d.) to an internal receiving coil (5-cm o.d.). The power source for the external coil supplies 250 kHz and operates at 20 V, 150 mA. The internal coil and power rectifier-regulator combine to provide an 8-V dc, 15-mA supply to operate the internal system components. Total power consumption is approximately 150 mW.

Demodulator

The received signal is coupled to a demodulator that reverses the encoding process (3). The receiver output is clipped and limited in the demodulator to remove effects of amplitude variations. The individual decoded analog waveforms are then available for display or processing in a standard manner.

System Accuracy

At 80% modulation, 5-m range, in a relatively noisy environment noise at the output is less than 0.1% of full scale, nonlinearity 0.25%, and zero and gain instability less than 1%. The nominal accuracy of the total system is approximately ±2%, excluding transducer drift. Assessing transducer stability requires periodic calibration and comparison with a reference standard.

Implantation

A completed unit ready for implant is shown in figure 2. The entire unit is placed within the thoracic cavity, using surgical techniques similar to those reported previously (1,2). In our laboratory, the pressure transducers are coated with TDMAC-heparin complex (Polysciences, Inc.) prior to implantation, to minimize the possibility of thrombus formation. The main unit is stabilized on the rib cage in the intrapleural space deep in the posterior thoracic gutter. The internal coil is positioned just cephalad to the main unit, where the chest musculature is minimal. The tissue changes are those expected with thoracic surgery and placement of an implanted device, and the unit is satisfactorily tolerated.

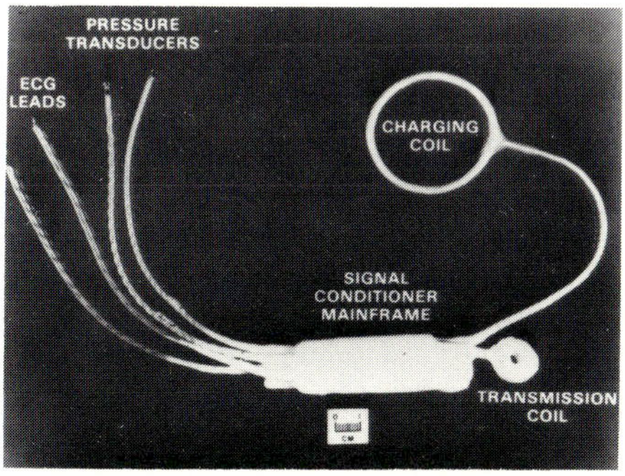

Figure 2. Multichannel telemetry unit prepared for implantation.

RESULTS AND DISCUSSION

To date, units have been implanted successfully in 8 monkeys (9 to 11 kg) and 3 dogs (10 kg), and have been in place for periods ranging from 3 to 14 months. The postoperative course has been smooth. Data have been obtained during studies of dogs on the treadmill and monkeys in chair restraint (fig. 3). The studies in monkeys have been both ground-based and in an airborne simulation of Spacelab test conditions. The distance between the coplanar external and internal coils can vary from about 2.5 to

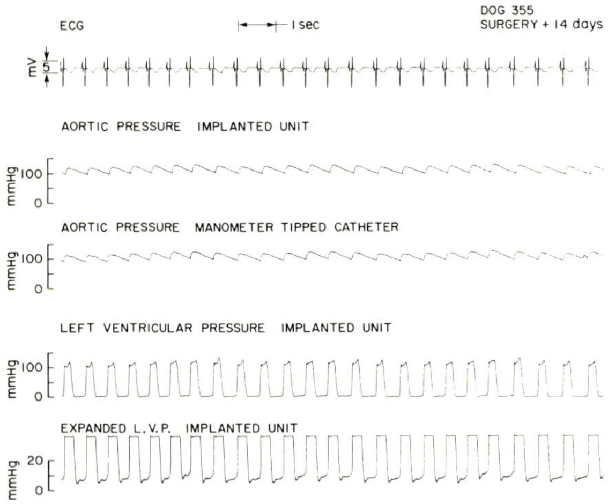

Figure 3. Representative data from a 10-kg dog, 2 weeks after implantation, with comparison aortic pressure waveform from a manometer-tipped catheter.

4 cm, and small amounts of relative motion can be tolerated without loss of signal.

This system expands the capability for the study of awake animals in a wide variety of conditions, particularly onboard future spaceflights and in similar stressful environments. Other applications include study of the effects of drugs, analysis of psychophysiological interactions, and similar investigations requiring long-term observation with minimal interference. The unit is available commercially (Konigsberg Instruments, Pasadena, CA; Model T-21).

The major improvements for future development are more stable pressure transducers and circuitry with even less power consumption. Lowered power requirements would decrease the amount of energy to be transferred and would permit greater latitude in coil placement.

REFERENCES

1. Fryer TB, Sandler H, Freund W, McCutcheon EP, Carlson EL: A multichannel implantable telemetry system for flow, pressure and ECG measurements. J Appl Physiol 39:318-326, 1975.

2. Sandler H, Stone HL, Fryer TB, Westbrook RM: Use of implantable telemetry systems for study of cardiovascular phenomena. Circulation Res 39:II:85-100, 1972.

3. Fryer TB, Sandler H, Datnow B: A multichannel telemetry system. Med Res Eng 8:9-15, 1969.

TISSUE CHANGES ASSOCIATED WITH LONG-TERM THORACIC INSTRUMENTATION

H. L. Stone, Ph.D., M. N. Golarz de Bourne, Ph.D.*, H. M. McClure*, G. H. Bourne, Ph.D.*, & H. Sandler, M.D.**

Marine Biomedical Inst., Univ. Texas Med. Br., Galveston, Texas
*Yerkes Primate Center, Emory Univ., Atlanta, Georgia
**NASA Ames Research Center, Moffett Field, California

The implantation of cardiovascular sensors into the thoracic cavity of experimental animals has been accomplished for many years and became popular with the advent of flow probes and pressure transducers in the 1950's and early 1960's (1,2,3). The advent of implantable telemetry units seemed to favor their use in cardiovascular instrumentation for long-term chronic studies (4). The size of the sensors were small enough for dogs and some primates; but, when coupled with the necessary electronics and power packages, the overall telemetry implant size became a serious problem for small animals. Reduction in size of the implantable unit has been achieved, but there are no studies that define the anatomical changes associated with the implantation of the unit in the chest cavity. In this study, the macroscopic and microscopic tissue changes associated with implantable telemetry systems have been examined in a small primate for up to 14 months.

METHODS

Six Rhesus monkeys (Macaca mulatta), weighing 3-4.5 kg, were used in this study. The animals were anesthetized with Ketamine hydrochloride and were intubated. The level of anesthesia was maintained with a mixture of halothane, oxygen, and nitrous oxide. The chest was entered in the left 5th intercostal space. A section of descending aorta approximately 4 cm in length was dissected and freed from surrounding tissue. The pericardium was incised, exposing the apex of the left ventricle. A solid-state pressure cell (6 mm diameter) was implanted into the chamber through a stab wound in the apical region. An arteriotomy was performed in the middle of the section of aorta freed from surrounding tissue. A pressure cell (3.5 mm diameter) was placed in the aorta, and the arteriotomy was closed with 6-0 cardiovascular suture. A small piece of dacron graft material was sutured over the arteriotomy. Electrocardiograph leads were attached to the pericardium. The electronic package was positioned in the lower left thoracic cavity and anchored to adjacent ribs. The antennae from the telemetry unit were anchored to the ribs adjacent to the incision. The chest was closed and evacuated. Two of the animals were sacrificed 7 months after implantation, and four animals were sacrificed 14 months after implantation.

Macroscopic and microscopic examinations were performed on all animals. The microscopic examination included detailed light microscopic examination of cardiac muscle tissue surrounding the implant and adjacent areas. The thoracic aorta was removed, and the tissue around the pressure cell implant was examined microscopically.

RESULTS & DISCUSSION

The autopsy revealed no significant changes in the head or abdominal regions. The adrenal glands that might indicate continuing undue stress appeared normal and were within the normal weight range for these animals (0.3 gm). In four of the animals, the kidneys were normal. In the fifth animal, the kidneys appeared to have an increased connective tissue content while in the sixth animal they were slightly congested but of normal appearance. Both conditions may or may not be the result of the sensor implantation into the thoracic aorta. If thrombi were to be shed by this device with any consistency, the kidneys would be one area where marked damage would have been expected. The gastrointestinal tract showed no signs of residual infarcts from any source. The livers appeared normal, with one animal having scattered pinpoints of whitish spots in the parenchyma. Abdominal lymph nodes and spleen also were grossly normal.

No abnormal amounts of fluid were found in the right thorax. In one animal, an apparent old tuberculosis adhesion was found. A completely different picture was observed in the left thorax. At the point of surgical entry, some lung adhesions were found. The lung had become adherent to the chest wall. The telemetry unit with associated antennae was completely overgrown with fibrous-like connective tissue. The wires leaving the unit were encased with tissue to form a sinus to the end of the wire. Figure 1 shows the telemetry unit (white object behind marker) removed from the tough fibrous pocket. In each animal this pocket was found to contain a small amount of fluid. The pressure transducer placed in the thoracic aorta also can be seen in the lower part of Figure 1. In all animals, there was no streaming clot associated with this foreign body in the blood stream. In most cases, the pressure transducer was overgrown with an endothelial cell layer that effectively partitioned it away from the blood stream. The endothelial cell layer was found to be contiguous with the intima of the aorta. The lumen of the vessel was not noticeably constricted at the point of the arteriotomy.

The pericardium was found to be adhered to the heart over the area exposed during the surgery.

The average heart weight in these animals was 20.3 gm. The left ventricular wall appeared to be slightly thicker than normal but would not be considered hypertrophic. The left ventricular

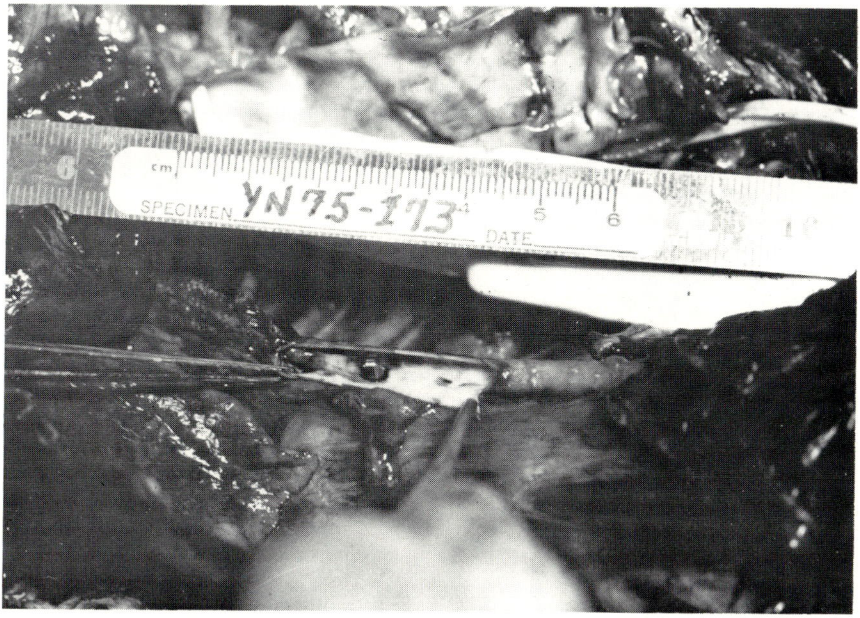

Figure 1

transducer was fibrosed into the left ventricular wall and overgrown with endothelial tissue.

Microscopical study of the left ventricle indicated an increase in fibrous tissue together with a thickened myocardium. In the areas of increased fibrous tissue, cardiac muscle fibers showed evidence of thinning and fragmentation. There was a good deal of infiltration by macrophages. There was replacement of myocardium by dense collagenous tissue with adherence of a thickened fibrous pericardium. These areas contained scattered accumulation of mononuclear inflammatory cells with some areas showing numerous polymorphonuclear cells. Some of the small coronary vessels near the heart apex showed intimal and medial proliferation with virtual or complete occlusion of the lumen in some cases. In some areas in one animal the myocardial fibers were packed with droplets of fat. All cardiac valves were normal, macroscopically and microscopically.

In conclusion, the careful placement of an implantable telemetry unit into the left thoracic cavity of the monkey causes minimal changes. Endothelialization over the surface of the intravascular transducer seems to protect the animal from embolic episodes that would tend to obscure the usefulness of the implant. The long-term use of this type of chronic animal preparation is highly desirable for many types of biomedical studies, but the results obtained must be interpreted with knowledge of microscopic changes.

ACKNOWLEDGMENT

This work was supported, in part, by NASA grant NGR 44 088 002.

REFERENCES

1. Franklin DL, Schlegal WA and Rushmer RF: Blood flow measured by Doppler frequency shift of backscattered ultrasound. Science, NY 134:564, 1961

2. Kolin A: Blood flow determination of the zero setting of an electromagnetic flowmeter without interruption of flow. Rev Sc Inst 24:178, 1952

3. Van Citters RL and Franklin DL: Radio telemetry techniques for study of cardiovascular dynamics in ambulatory primates. Ann NY Acad Sci 62:137, 1969

4. Stone HL, Sandler HH and Fryer TB: Implantable telemetry in the chimpanzee. Proc Inter Telemetering Conf 8:464, 1972

BIOTELEMETRICAL DETECTION OF RADIONUCLIDE TRACERS IN THE STUDY OF BIOLOGICAL RHYTHMS IN RENAL FUNCTION

Steen Vadstrup and Jørgen Bojsen
Dept. of Clinical Physiology and The Finsen Laboratory
The Finsen Institute, Copenhagen, Denmark

INTRODUCTION

Physiological experiments on renal function in intact animals have always been hampered by experimental stress, which may be largely eliminated by biotelemetrical methods.

Direct measurements of renal functions have been performed by Rader et al. (1), who monitored the renal blood flow radio-telemetrically in unrestrained dogs.

Parameters indicating renal functions may, however, be determined indirectly by means of telemetry. We have for this purpose used an implantable radio-telemetrical GM detector unit (2). This unit detects, when implanted subcutaneously (sc.) in rabbits, the whole-body disappearance of i.v. injected radionuclide tracers. If the tracer is excreted exclusively via the kidneys, it is possible to calculate the renal clearance.

METHODS

Albino rabbits (4-6 kg) were housed in a radiofrequency shielded room, with controlled light cyclus. Radio-telemetrical GM detector units were implanted sc. on the upper part of the back (Fig.1). The tracer studies were initiated within one week from the time of implantation.

The tracer was injected i.v. during a period of less than one minute, while the rabbit was restrained in a box. When the rabbit was replaced in the cage and left alone, the continuous GM detector output was recorded in a neighbouring room.

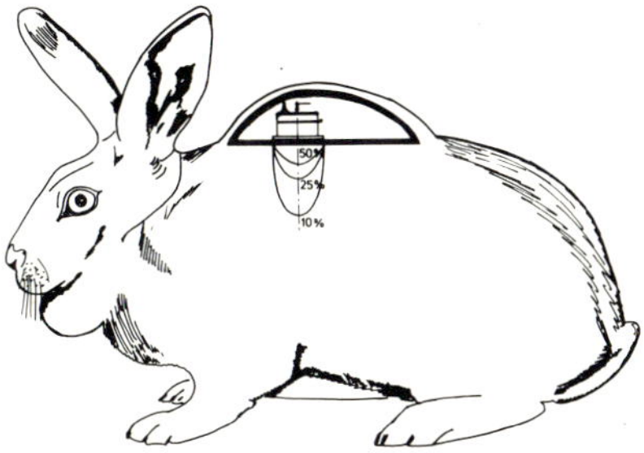

Fig.1. The GM detector unit with 125-I-isocount curves.

The following tracers were used: 125-I-iodide (iodide clearance), 125-I-iothalamate (GFR) and 125-I-hippuran (renal plasma flow (RPF)). Carrier iodide was administered simultaneously i.v. in order to prevent thyroid uptake of free 125-I.

The experiments included:A. Studies of the diurnal variation in renal function. B. Studies of the effect of different hormones on the measured parameters.

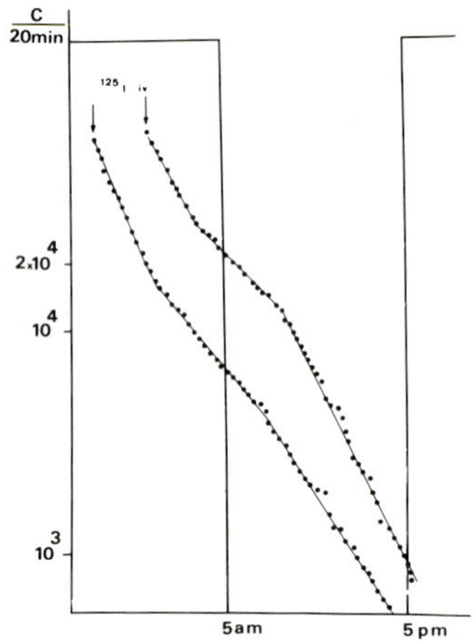

Fig.2. 125-I-iodide disappearance curves from 2 rabbits.

When the diurnal variation in renal function was studied, 125-I-iodide was chiefly used, due to its relative low renal clearance, which permitted continuous measurements over 24 h. (Fig.2). The continuous measuring period following administration of a similar tracer dose of iothalamate and hippuran (100-400 µCi) was only 1-3 h.

The studies concerning the effect of different hormones on the renal function in unrestrained rabbits, were performed with all 3 tracers. The hormones were administered i.v. or sc. in graded doses during a very short period of restraint and the effect on the tracer disappearance compared to control values either before or after the hormone effect (Figs. 3 and 4).

A few experiments have been performed with a 2 channel GM detector unit. This unit may be used with an end-window absorber in front of one of the GM tubes, which permits detection of two different labelled tracers simultaneously or it may be used without an absorber, which permits detection from two different tissue volumes at the same time.

RESULTS

A diurnal variation in the renal iodide clearance is present in rabbits as in humans. Detailed studies in 5 rabbits (3) have shown that rabbits in our experimental set up exhibit a 50 % lower iodide clearance during the last part of the dark period (Fig. 2).

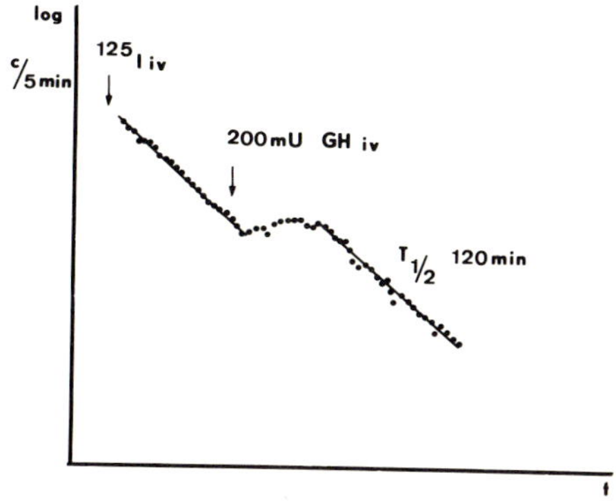

Fig.3. The effect of GH on the iodide excretion.

The lower iodide clearance at night could be caused by hormones, with a similar cyclic pattern of secretion and in order to test this hypothesis, we have studied the effect of different pituitary hormones on the iodide clearance, GFR and RPF. Some results concerning the renal effect of TSH have been published (4), but the further experiments have shown that GH lowers the iodide clearance temporarily just like TSH (Fig.3).

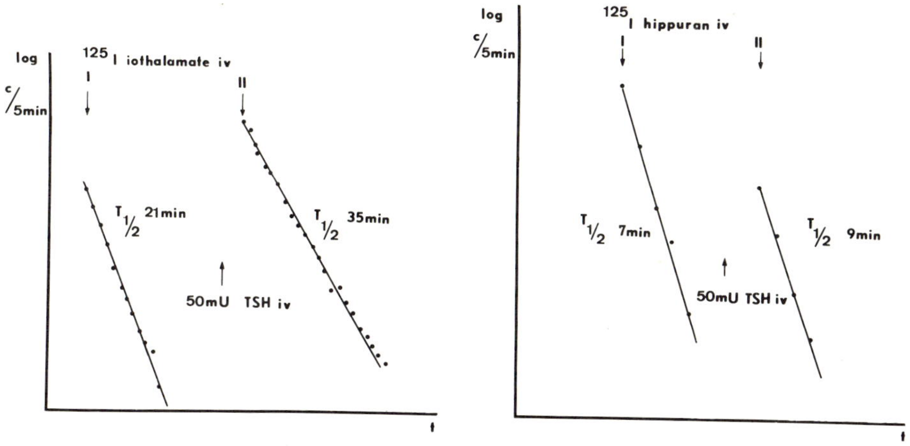

Fig.4. The effect of TSH on GFR and RPF.

The studies with hippuran and iothalamate show that TSH and GH lower both GFR and RPF, when administered i.v. in small doses (Fig.4).

DISCUSSION AND CONCLUSION

The most suitable radionuclide for this kind of study is 125-I. It is detected with a relatively high efficiency. The volume of tissue seen by the detector is appropiate and consists primarily of the dorsal muscle layer and to a lesser extent lungs and liver, whereas activity from the kidneys and the bladder is not detected.

The volume of tissue seen by the detector is a constant part of the whole animal. It is evident from the analysis of iodide as well as iothalamate and hippuran disappearance curves, that they are monoexponential, shortly after the tracer injection. In spite of many theoretical objections, this experimental evidence permits us to apply the disappearance curves as fractional whole-body disappearance curves. The disappearance rate constants are proportional to the renal clearances and if the distribution volume is known, the renal clearance may be determined in absolute values.

It is our impression that biotelemetrical monitoring of renal function in experimental animals by the method outlined in this paper may be an alternative to more unphysiological experimental conditions especially where experimental stress should be avoided, such as in studies of hormonal influence on renal function.

REFERENCES

1. Rader RD, Stevens CM, Meehan JP, Henry JP: Telemetered renal responses to avoidance and aggression in dogs. Biotelemetry 1:1, 1974.

2. Bojsen J, Wallevik K: A radiotelemetrical measuring device implantable on animals, for long term measurements of radionuclide-tracers. Int J Appl Radiat Isotopes 23:505, 1972.

3. Vadstrup S, Bojsen J: The diurnal variation in renal iodide excretion rate in rabbits. Acta Endocr. (Kbh.) 81:716, 1976.

4. Vadstrup S, Bojsen J: The immediate effect of TSH on renal iodide excretion rate in rabbits. Acta Endocr. (Kbh.) 81:723, 1976.

ACKNOWLEDGEMENT

The authors wish to thank Mr. F. Haslev for excellent technical assistance.

A COMMAND RECEIVER FOR USE WITH LOW POWER IMPLANTABLE BIOTELEMETRY SYSTEMS

J. Donald Pauley, Ph.D. and Martin Reite, M.D.

Department of Psychiatry, University of Colorado Medical Center, Denver, Colorado 80220

Implantable biotelemetry systems have unique advantages in many experimental situations, but in systems where battery recharging or replacement is not possible, the non-renewable power supply must be conserved whenever possible. Magnetically activated latching reed switches may be used for this purpose, but a strong field must be applied near the experimental subject to change their state. For those situations in which the implanted telemetry unit must be turned on and off at a distance, an RF activated command receiver is preferred. Previously described command receivers (1) are designed for implanted biotelemetry systems utilizing a moderate current flow, and have not worked well with very low power (e.g. 500 μA) devices.

We have designed a command receiver circuit for use with very low power implanted telemetry systems. Although not yet completely problem-free, we believe the basic system design may well be of value to researchers working in the area, and so are describing our system here.

DESIGN AND CONSTRUCTION

The command receiver system consists of a remote control box, a slightly modified armature transmitter, and an implanted radio controlled switching circuit. This is illustrated in block diagram form in Figure 1. The switching circuit is latched on by addressing it with a 3.5 to 4.0 MHz carrier amplitude modulated by a 1000 Hz sine wave. When addressed using 100 Hz modulation, the implanted switch is turned off. The control box containing the modulating sine wave oscillator and a transmit key is handheld and connected via a cable to the MIKE and KEY inputs to the transmitter. Panel controls are provided for adjusting the oscillator frequency and keying the transmitter. The control box circuit diagram is given in Figure 2.

The transmitter is a Heath model DX-60B armature transmitter with three minor modifications. The coupling capacitors in the speech amplifier have been increased to permit easy passage of the 100 Hz modulation signal. A simple 10 volt zener regulated power supply has been added to operate the modulating oscillator, and the two-terminal KEY phone jack has been changed to the three-terminal type to provide an output for the 10 V power supply. The

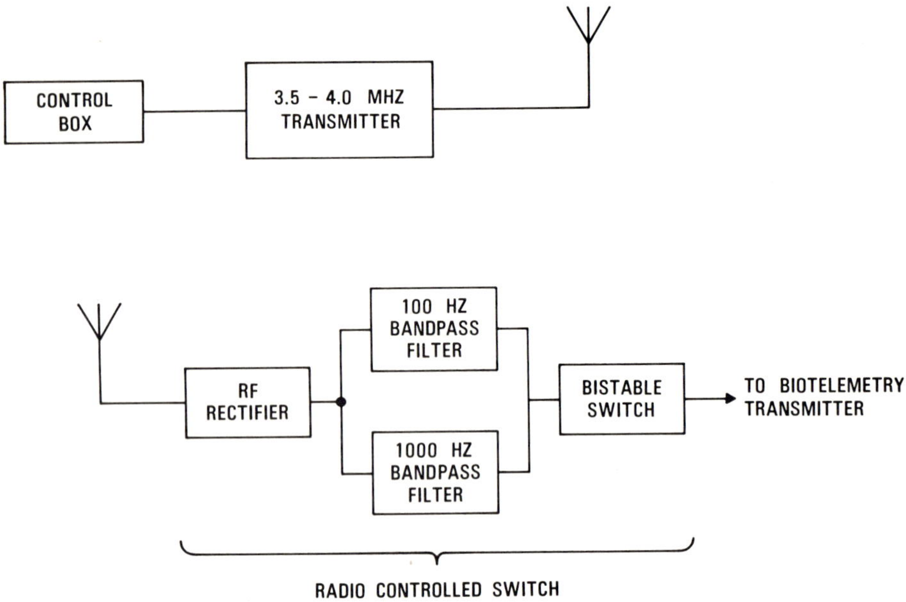

Block diagram of the command receiver system. Figure 1.

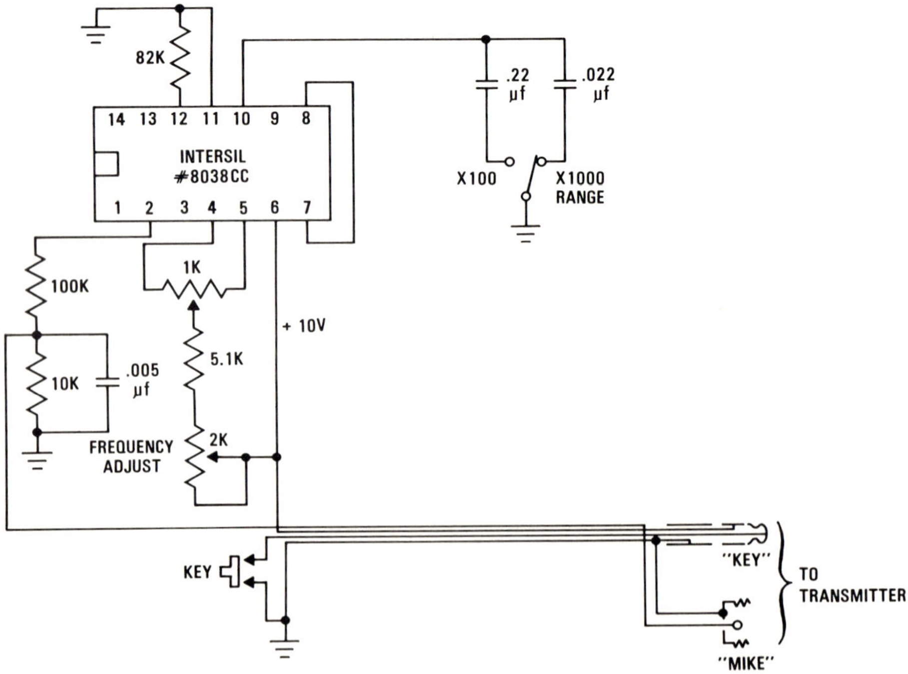

Schematic diagram of the control box. Figure 2.

transmitter is operated in the AM mode at full rated power and 100% modulation into an open-ended wire antenna imbedded in the masonry walls of our 3 M x 5 M enclosure.

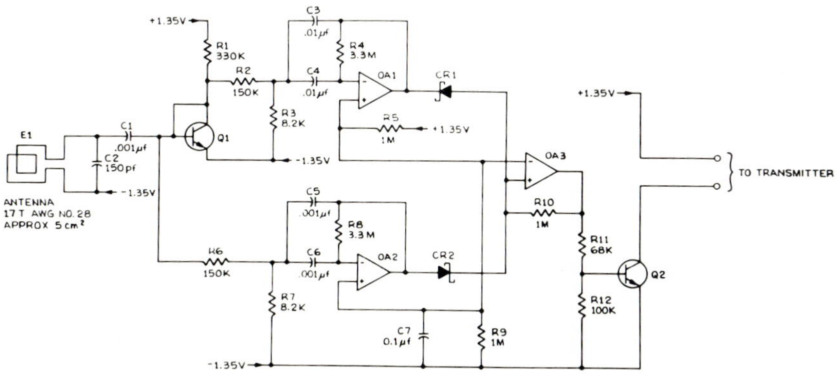

NOTES -
1 ALL RESISTOR VALUES IN OHMS.
2 OA1,OA2,OA3 ARE 1/3 SILITRONIX L144C WITH 680K BIAS RESISTOR.
3 Q1,Q2 ARE G.E. 26C3.
4 CR1,CR2 ARE MOT. MB1-101.

Schematic diagram of the radio controlled switch. Figure 3.

The circuit diagram of the implanted radio controlled switch is shown in Figure 3. The incoming RF signal is rectified by the base-emitter junction of Q1. The modulation signal is fed to two bandpass filters (OA1 and OA2). One filter has a bandpass at about 1000 Hz and the other at 100 Hz. If the modulating signal is in the bandpass of the 1000 Hz filter, current passes through CR2 to OA3 which operates in a bistable mode due to R10. Current from CR2 causes the output of OA3 to latch at a positive voltage, thus turning the implanted transmitter on via Q2. If the modulating signal is in the bandpass of the 100 Hz filter, current passes through CR1 to OA3 causing its output to latch at a negative voltage which biases Q2 off. The amount of current that can be switched is determined by the characteristics of Q3 and the amount of base current provided. Switching capability of several milliamperes is easily attained.

The switch has a current drain of 50 μA in the off state and 60 μA in the on state. It can be operated from two mercury cells or a single lithium cell. As fabricated by us, the circuitry is epoxy encapsulated within a dome 0.5 cm high by 1.6 cm in diameter. This mounts conveniently on the end of a ½ AA size lithium cell or an RM640 mercury cell stack. The antenna is wrapped around the longitudinal axis of the package forming a loop of several turns. When fabrication is complete, the unit is tested to determine the frequency of maximum RF sensitivity and the center frequencies of the bandpass filters. These frequencies are recorded so that when in use, the switch can be optimally addressed.

OPERATIONAL RESULTS

This system has been used in a number of experiments with mixed success to date. When operating properly, the switches change state reliably and only when addressed, and allow optimum utilization of battery life. In several cases, however, potted command receiver units have developed internal short circuits, draining the batteries. We believe this has been due to mechanical design (construction) problems, but we have not yet been able to isolate the problem with certainty, and we would therefore recommend thorough bench-testing of units to be implanted. This would not be a serious problem in external systems, where other construction techniques could be utilized.

It should be remembered that any command receiver system such as this, if not used in enclosed shielded facilities which contain the radiated RF carrier, will require an FCC station license for the controlling transmitter, and certain design changes, such as the RF carrier frequency, may be necessary.

ACKNOWLEDGMENT

This work was supported by USPHS grant # MH 19514.

REFERENCES

1. Hudson PH, Meindl JD: A monolithic micropower command receiver. IEEE J Solid State Circuits SC7: 125, 1972

INTEGRATED CIRCUITS
FOR AN IMPLANTABLE CW BIDIRECTIONAL DOPPLER SYSTEM

R. D. Davies and J. D. Meindl

Integrated Circuits Laboratory
Stanford University
Stanford, California

A family of four monolithic integrated circuits (ICs) has been developed for realization of a chronically implantable CW bidirectional ultrasonic Doppler blood velocity meter. The block diagram of Figure 1 illustrates this system, which is the result of a group effort within the Stanford Integrated Circuits Laboratory. Our goal was to draw upon the previous implantable blood flowmeter experience to implement a workable and reliable CW system that is easy to produce. Success under these criteria is evidenced by the fact that the ICs described here have each been reproduced several dozen times, with excellent yield and excellent uniformity. In vivo data from fully implanted units has yielded a signal-to-noise ratio greater than 30 db. An important aspect allowing development in a few months time was the KITCHIP, described in these proceedings by Combs and Meindl (1). The KITCHIP is a generalized array of uncommitted integrated active and passive components, optimized for micropower circuits, that the designer is free to configure as he wishes via the metalization photolithography mask. This permits convenient design iteration because only one mask is involved per circuit; the underlying five masks defining the KITCHIP are never altered.

The configuration of the system follows a general scheme for implantable units described previously by Gill and Meindl (2), and Frescura and Meindl (3). Partitioning is done into implanted and external subsystems (Figure 1) to minimize size and power drain of the implanted package, and to preserve flexibility in choice of (external) demodulation technique. The significant points here are:

(1) Use of synchronous (baseband) quadrature detection to avoid direction ambiguity while retaining circuit simplicity and clutter rejection inherent in baseband detection. A double balanced detector circuit (4,5) is employed to best reject both electrical interference coupled from the exciter, and non-Doppler shifted acoustic-path signals. The detectors are DC biased by the preceding RF amplifier stage after Gill (6), with RF amplifier cascade output to prevent oscillations due to input transformer-RF tuning coil interaction (7) (Figure 2).

(2) Use of a crystal controlled exciter oscillator to achieve a very narrow excitation spectrum (significant sidebands are

Figure 1.

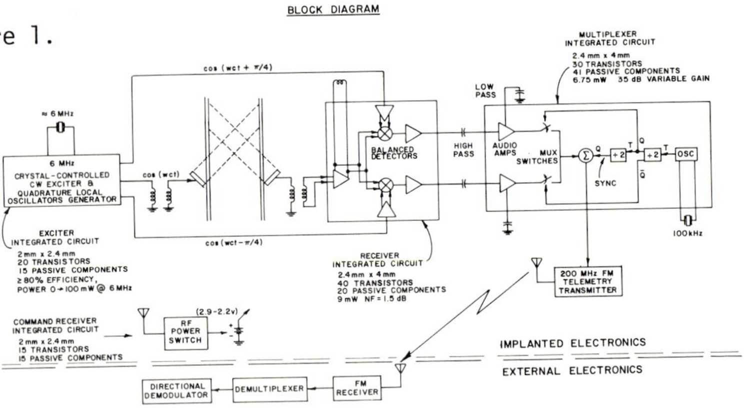

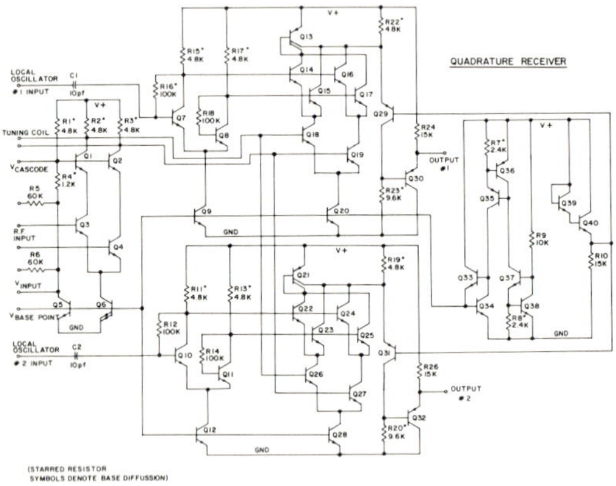

Figure 2. Quadrature Receiver.

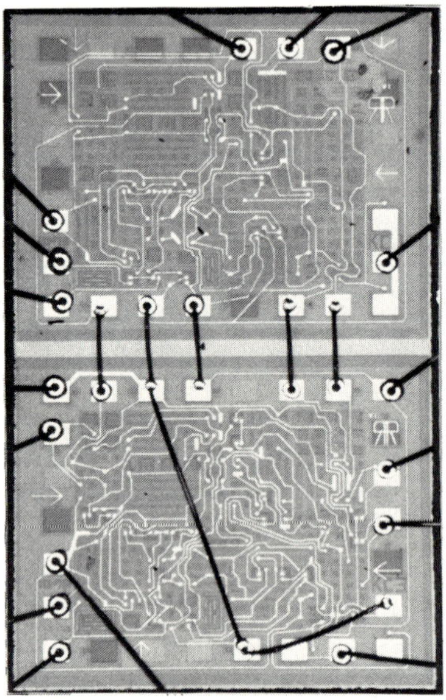

Figure 4. Exciter.

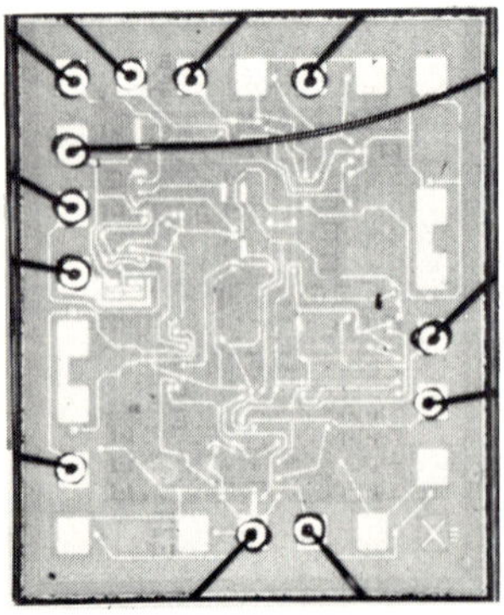

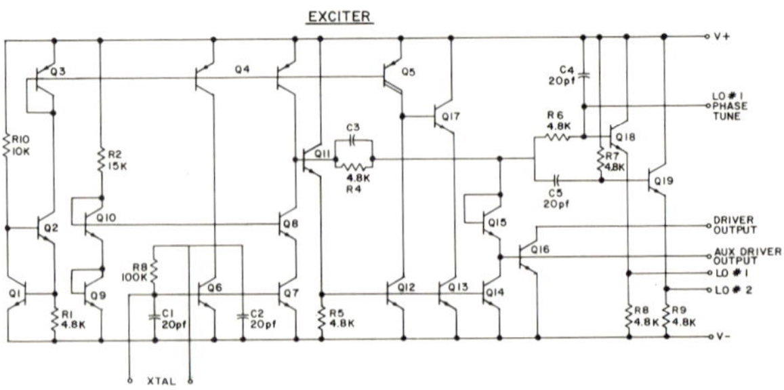

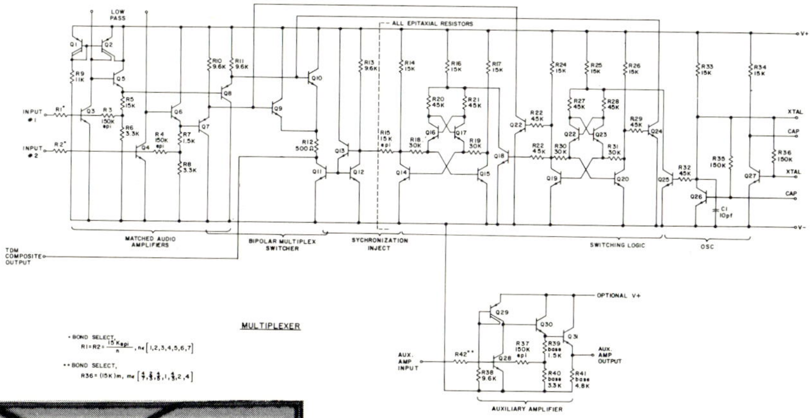

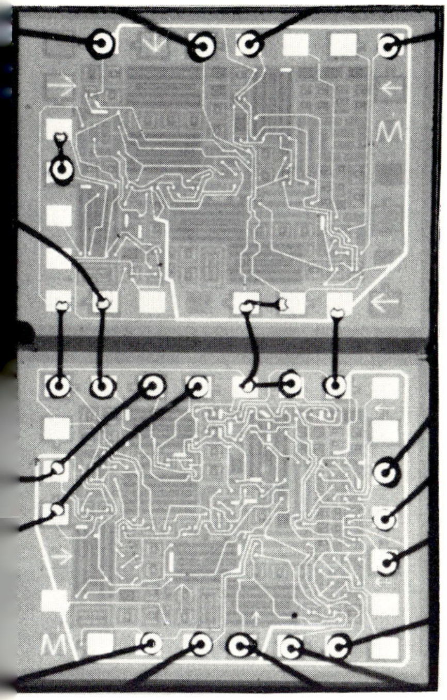

Figure 3. Multiplexer.

Figure 5. RF Switch.

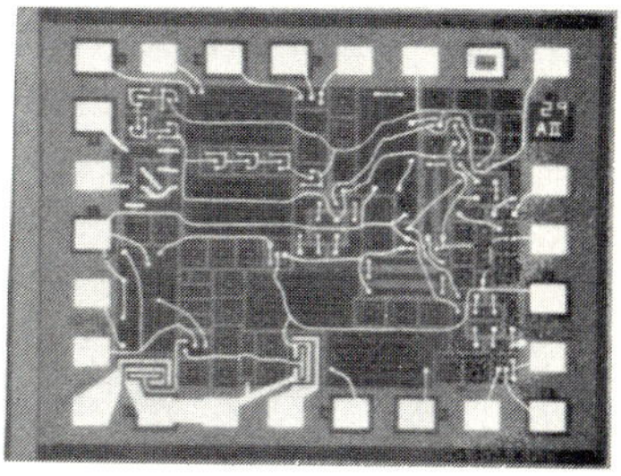

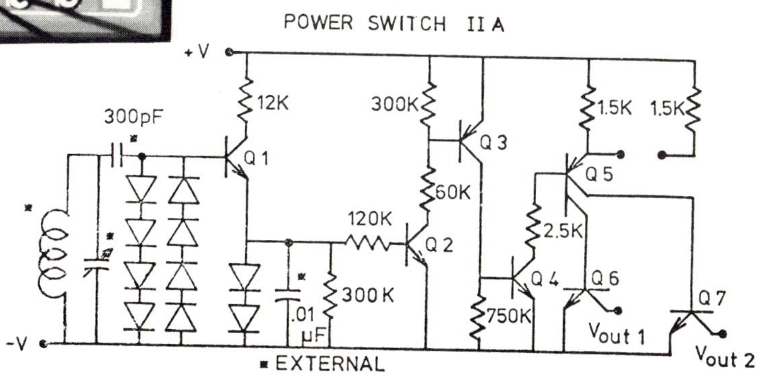

greater than 60 db down). This makes signal-to-noise ratio a direct function of driver power level and receiver input noise only.

(3) Flexibility in setting (post detector) amplifier gain to accomodate a variety of driver power levels and tranducer configurations, Figure 3. The matched pre-multiplex switch audio amplifiers may be set over a 20 db gain range at assembly by selective wire bonding. In addition, an optional post-multplex amplifier with 15 db gain range is provided; if not used the amplifier power lead is not connected during assembly. The bipolar multiplex switches are a parallel combination of cascaded PNP-NPN emitter followers giving 50 db isolation.

(4) The monolithic RF power switch (Figure 5) does not use RF amplification; this allows zero quiescent power drain but results in short range (10 cm) capability. A pulse-powered RF switch capable of greatly increased range is described in these proceedings by Pettengill and Meindl (8).

All the circuits operate over a 2.2v to 5v or higher supply range and therefore allow maximum operating life from a nominal 3v battery. Biasing of critical stages is done with tracking diode-current sources set by a resistor paralleling a base-emitter junction and buffered by an emitter follower (Q_{36}, R_7, Q_{35} in the receiver, Figure 2, and Q_2, R_1, Q_1 in the exciter, Figure 4). Supply current is then nearly constant with supply voltage except for a small effect of the resistive connection to the current setting network; circuit biasing is essentially independent of supply voltage.

ACKNOWLEDGMENTS

The authors wish to thank W. Foletta and S.R. Combs for many helpful suggestions regarding design and fabrication of the ICs. H.V. Allen provided the exciter circuit design, and R. Pettengill designed and fabricated the RF power switch. The help of the entire Stanford IC Lab flowmeter group is appreciated.

REFERENCES

1. Combs SR, Meindl JD: KITCHIP - A generalized integrated circuit for biomedical system design. Proceedings of 3rd International Symposium on Biotelemetry, Asilomar, Calif., 1976
2. Gill RW, Meindl JD: Micropower integrated circuits for an implantable pulsed-Doppler ultrasonic blood flowmeter. ISSCC Digest, 1975
3. Frescura BL, Meindl JD: Micropower integrated circuits for an implantable bidirectional blood flowmeter. ISSCC Digest, 1976
4. Jones HE: Dual output synchronous detector utilizing transistorized differential amplifiers. US patent 3241078, June 1963
5. Bilotti A: Applications of a monolithic analog multiplier. JSSCC, December 1968
6. Gill RW, Haase WC, Meindl JD: An implantable pulsed Doppler ultrasonic flowmeter. Proceedings of 24th ACEMB, 1971
7. Allen HV: Private communication
8. Pettengill RC, Meindl JD: A pulsed powered micropower monolithic command receiver. Proceedings of 3rd International Symposium on Biotelemetry, Asilomar, Calif. 1976

KITCHIP - A GENERALIZED INTEGRATED CIRCUIT FOR BIOMEDICAL SYSTEM DESIGN

S.R. Combs, M.S. and J.D. Meindl, Ph.D.

Stanford Integrated Circuits Laboratory
Stanford, CA 94305

INTRODUCTION

The advantages of integrated circuits (ICs) in terms of performance, power consumption and size are important to a number of medical electronics systems. A quick cost-effective alternative to the design and fabrication of custom ICs is presented. The Stanford Integrated Circuits Lab KITCHIP is a monolithic array of unconnected bipolar transistors, resistors, and capacitors. Devices on the chip are not committed to a particular circuit design but are positioned to serve a wide range of applications. Except for definition of a metal interconnect pattern, all IC fabrication steps are complete. Thus, only a single photolithography and metal etch are required by the user to yield a finished integrated circuit.

The concept behind a KITCHIP IC is to cover a range of potential applications with proper process design and device layout. The circuit described in this paper has been tailored for implantable medical telemetry systems. These systems require ICs with both good high frequency performance and minimal power consumption. Commercial ICs to fill these needs are not available.

ADVANTAGES

Circuit fabrication using the KITCHIP offers a number of significant advantages over both discrete component and custom integrated circuit designs. A KITCHIP design allows a substantial reduction in size and power consumption over a discrete component circuit while achieving higher levels of performance and device matching. Since the KITCHIP begins as a collection of unconnected parts, it provides the ease of interconnection and simplicity of modification seen in discrete designs. It also separates the tasks of device fabrication and circuit design as in the discrete case. The KITCHIP is processed thousands of die at a time by an IC fabrication specialist, then stored for later use by circuit designers. It allows the most advanced processing techniques to be employed without requiring this expertise of potential users.

Since the KITCHIP is fully fabricated at the time of storage, samples of the stock may be tested and characterized. Circuit designs may be developed with precise knowledge of device performance. Alternatively, stock can be selected to meet specific circuit needs. Finally, extracted Gummel-Poon (1) transistor model parameters are available for all the active devices on the chip. These may be used with known resistance and capacitance values to form accurate computer simulations of circuit designs. Predictions of the circuit performance using program SPICE (2) have been reliable enough that some designs are integrated without first building a discrete component breadboard.

The clearest advantage of a KITCHIP over a custom IC is the time required to achieve working circuits. Mask generation and IC fabrication for a custom circuit take at least an order of magnitude more time than specifying a KITCHIP interconnect pattern. A second time advantage stems from the fact that KITCHIP devices are known to work and their performance is well-characterized. These advantages are compounded whenever a design requires modification. For a custom circuit, new masks must be generated and the entire fabrication procedure must be repeated. Since many custom circuits require several iterations before settling on a design, the net time advantage of the KITCHIP is at least 10^n where n is the number of required iterations.

Another feature of IC design with the KITCHIP is flexibility of the final circuit configuration up to the point of packaging. Since most circuit designs require only a fraction of the total available devices, remaining devices may be used for circuit options or modifications. In this way, IC performance can be tailored to compensate for variations in other inflexible system parameters. In the case of an implantable blood flowmeter built from KITCHIP stock, receiver gain can be modified to match specific properties of a piezoelectric transducer used with each unit.

With the KITCHIP, several versions of a simple circuit design can be fabricated on the same die. It allows effects of design modification to be accurately determined since all devices are very well matched and in the same thermal environment.

In cases where the complexity of a circuit design is high, two or more KITCHIP die can be used in conjunction. For the Implant I flowmeter (3), two circuit blocks were constructed with multiple chips in a single package. To provide good electrical matching, adjacent die were etched using different metal interconnect masks. Expandability of the KITCHIP allows integration of nearly every complex circuit design.

DESIGN AND FABRICATION

A process and devices were needed to yield high performance ICs at low current levels. For this reason, a high frequency bipolar process was chosen. Starting with a p<100> substrate, a 40 $\Omega/\square$, arsenic doped buried layer is diffused to a depth of 5μ. An arsenic 5μ, 5 Ω-cm epitaxial layer is then grown. This layer acts as the npn transistor collector and as a 10 k$\Omega/\square$ resistor. A boron doped, 1.8μ deep p-type diffusion is made next. This serves as the npn base and a 200 $\Omega/\square$ resistor. It is followed by a 10 $\Omega/\square$, 1.2μ deep phosphorous diffusion forming the npn emitter and collector contact. After holes are opened, the wafer is covered with a 1.5μ thick aluminum layer and stored.

Performance typical of this shallow bipolar process is flat npn beta over six decades of current. The value of beta may be set from one to four hundred with a minor process variation. Due to their small size, the frequency response of the npn transistors is excellent at low currents. For higher currents, the transistors show a maximum f_T of 1.2 GHz at 3 mA. Lateral pnp transistors have a typical beta of 100 which falls off above 100 μA due to large base resistance. Substrate pnp transistors with an average beta of 300 show good performance at all current levels.

Constraints for device design come from photolithographic tolerances and need for flexible structures in the KITCHIP layout. NPN transistors are built with two collector contacts allowing unused transistors to act as low resistance cross-unders. By using a horseshoe shaped collector diffusion, the transistor area is increased by only 5% while the cross-under resistance is reduced by a factor of 3.

Often, in low power circuits, pnp transistors are used as active loads and current mirrors. To facilitate this application, lateral pnp transistors have been designed with a split collector contact. This allows a single device to act as a differential active load or current repeater.

Resistors in the KITCHIP appear in groups, scaled in value, to provide a large number of series and parallel combinations. Standard resistors are built in a 200 Ω/□ diffusion while high value resistors, necessary for micropower applications, take advantage of the 10 kΩ/□ epitaxial layer.

Junction and MOS capacitors are both available on the KITCHIP. In order to conserve area, MOS capacitors are formed by placing a diffusion under bonding pads. The two sides of the capacitor formed by the diffusion and pad aluminum are separated by 1000Å of oxide. Each pad produces 7 pf of capacitance.

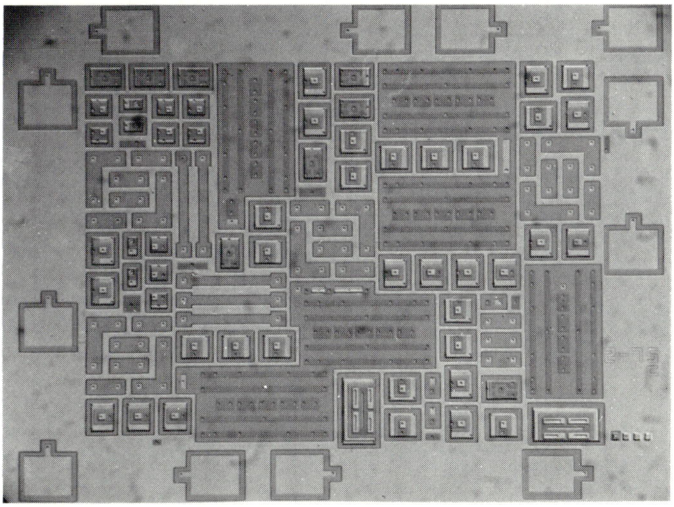

FIGURE 1. KITCHIP IC.

The KITCHIP topology exhibits a definite block structure. It is intended to provide a high degree of flexibility in the general case while allowing easy metalization for some specific applications. As seen in Figure 1, thirty-four standard geometry npn transistors are distributed uniformly across the die. Eight lateral pnp transistors occur primarily on top and bottom edges. Nine minimum geometry npn transistors are located in the upper left corner. This area best serves as input or high frequency sections of a circuit. Two high current substrate pnp's and two large npn's occupy the lower right corner. This corner has been designed for output sections. The orientation of groups of transistors within the die allows for easy interconnection into current mirrors and multiplier cells.

APPLICATIONS

Over the past year, the KITCHIP has been used in more than a dozen applications. In every case, the circuits operated on the first attempt. It has proven useful in complex systems. Integrated circuits for an implantable flowmeter described by Davies and Meindl (3) were developed in 30 days. This system employs six die with over 200 active and passive devices.

For most applications, utilizations over 50% of the active devices are readily obtainable. However, the ease of expansion to multiple die obviates the need for higher percentage use. Because of close device matching on adjacent die, performance of circuit blocks requiring more than one die is equal to a single die realization.

CONCLUSIONS

The Stanford Integrated Circuits Lab KITCHIP represents a new way to achieve advantages of custom ICs. It offers higher performance with less power consumption and size than discrete circuit designs. The KITCHIP is often preferable to a custom IC approach since it allows quick integrated and rapid modification of a wide range of discrete circuit designs. Its precharacterized devices allow easy and accurate circuit simulation.

The KITCHIP described has been tailored for low power implantable telemetry systems. Devices and process have been designed for good frequency performance at low bias levels. Over three megohms of resistance available on chip allows micropower circuit operation.

Several significant medical systems have been fabricated with KITCHIP ICs. Their rapid development has been aided by its reliability and ease of use. Future systems will rely heavily on this approach.

REFERENCES

1. Gummel HK, Poon HC: Charge Control Model of Bipolar Transistors. Bell System Technical Journal May: 827, 1970.
2. Nagel LW, Pederson DO: Simulation Program with Integrated Circuit Emphasis. Symposium on Circuit Theory, Ontario. April 1973.
3. Davies RD, Meindl JD: Integrated Circuits for an Implantable CW Bidirectional Doppler System. This Journal.

A PULSE POWERED MICROPOWER MONOLITHIC COMMAND RECEIVER

R.C. Pettengill, J.D. Meindl

Integrated Circuits Laboratory, Stanford University
Stanford, California

Chronically implantable biomedical instrumentation is growing in its contribution to medical research (1,2). The long life and small size required of chronic implants can best be met by the use of custom micropower monolithic integrated circuits. However, the stringent performance requirements of some types of instrumentation, where power drain is relatively high, preclude continuous operation. In such cases, the operating life of battery powered implanted electronics must be greatly extended by the use of a radio controlled power switch, or command receiver, to disconnect the battery when data are not needed. Such a command receiver must combine ultra-low power operation with reasonably long range. Micropower monolithic circuits for this function have been developed (3), but their use has been limited by short range and high cost problems. Continuous micropower operation limits these command receiver circuits to a carrier frequency of less than 1 MHz which results in a range of about 3 m when a minimum volume receiving antenna and hand-held transmitter are used. By using a novel pulse powered (4) command receiver, the carrier frequency can be increased to 30 MHz. This results in an improved immunity to false triggering and a range in excess of 15 m with no additional receiver power drain or antenna volume and no increase in command transmitter size. Cost reduction is also achieved through the use of a KITCHIP requiring only a custom metal mask to implement the receiver.

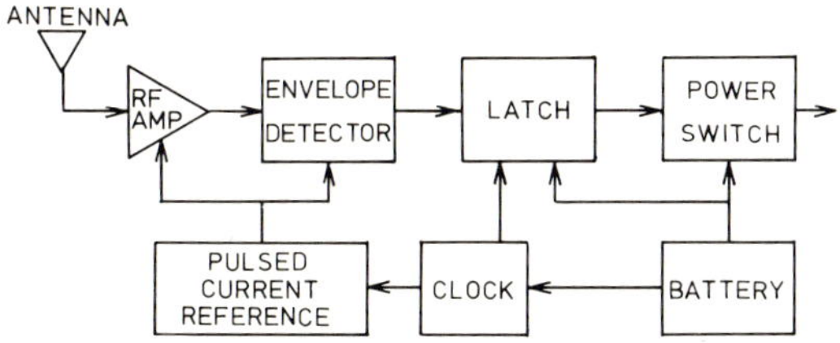

Fig. 1. Pulse Powered Command Receiver Block Diagram.

CIRCUIT DESCRIPTION

A block diagram of the new pulse powered receiver is shown in Figure 1. The RF amplifier and detector (Figure 2) are pulse powered reducing power consumption from 15 mW to 20 μW. The proba-

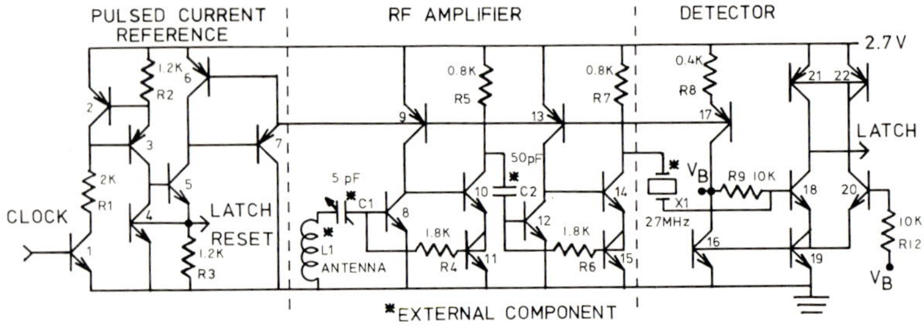

Fig. 2. Schematic of the Pulse Powered Circuitry. (Resistance in kΩ, Capacitance in pF).

bility of false triggering is similarly reduced. The clock, latch, and power switch are continuously operating micropower circuits (Figure 3). The specifications of the complete receiver are given in Table 1.

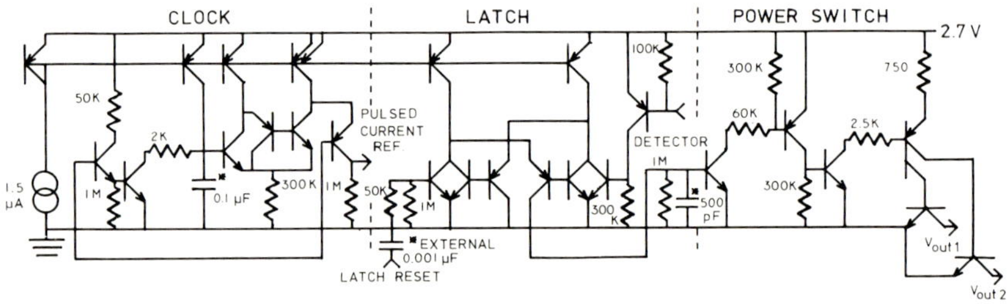

Fig. 3. Simplified Schematic of the Continuously Powered Receiver Electronics.

Table 1. Specifications for the Radio Controlled Power Switch

Range - 15 m (with 500 mW transmitter)
Volume - 4 cm^3 (with 1 cm^3 antenna)
Supply Voltage - 2.5 to 3.0V
Current Drain - 20 µA (1 year life from 1 cm^3 Li battery).
Transmitter - 27 MHz, CW modulation, hand-held
Output - Switch to ground (1-6V, 0-20 mA)
Sensitivity - 50 µV
Selectivity - 5 kHz channel set by crystal filter
Bandwidth - 10 Hz signal bandwidth

While monolithic circuits can be built with useful gain at 30 MHz, the inductor tuning or feedback methods typically required to reduce power dissipation and bandwidth in a receiver are not practical in a monolithic realization (5). However, operation of the receiver in a sampling or pulse powered mode allows the average power dissipation to be reduced along with bandwidth very easily. Figure 4 shows how this occurs. Previously this technique has been applied to digital circuits and to receiver circuits in a standby mode (4,6). In a pulse powered receiver bandwidth can be traded for lower power dissipation simply by increasing t_C (reducing the clock frequency). The probability of false triggering from random noise is reduced by β, however the selectivity is not affected.

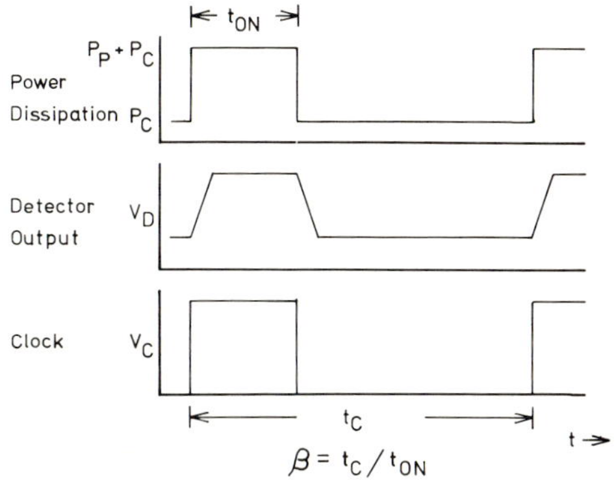

Fig. 4. Pulse Powered Operation.

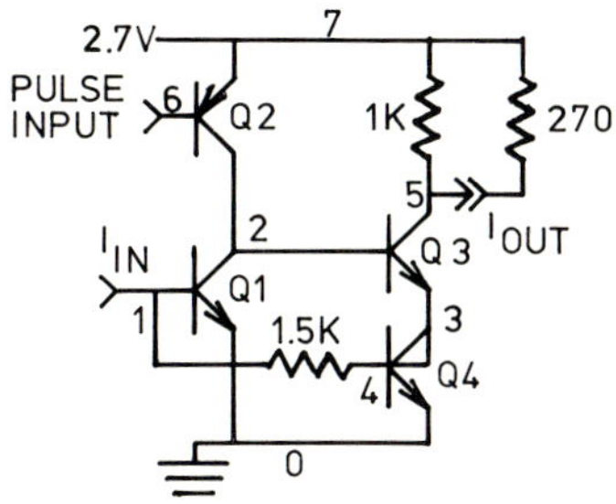

Fig. 5. Pulse Powered Shunt-Series Gain Cell.

An investigation of possible receiver circuit configurations and power switching techniques determined that a current-switched shunt-series pair configuration offers many advantages in this application. Figure 5 shows a schematic of the modified shunt-series gain cell. Both RF amplification and detection are achieved with this gain cell whose salient advantages are: 1) All of the circuit nodes, 1 through 6, exhibit low impedance. No high value biasing resistors or capacitors are required. This minimizes switching transients, which is essential for pulse powered operation. 2) Current flow in the entire gain cell is controlled by the low power current switching of the base of a single transistor, Q_2. 3) Since the shunt-series pair is a current amplifier, the input and load impedances are very low. This minimizes the effects of parasitic capacitances in the non-optimal KITCHIP layout. The collector current of Q_1 is about 0.5 mA and that of Q_3 1 mA. The current gain-bandwidth product is about 1 GHz and the total power dissipation only 4 mW. The detector, a common emitter envelope detector, is transistor 18 in Figure 2. The optimum collector current of Q_{18}, for maximum small signal conversion gain, was found to be 400 μA by computer simulation. Q_{18} is biased as the second transistor in a shunt-series pair with unity feedback. R_{12} and Q_{20} allow adjustment of the detection threshold. The total continuous power required by the complete receiver is 40 μW. The power dissipation of the receiver was reduced 375 times by pulse powered operation.

The carrier frequency can be selected for maximum signal-to-noise ratio from a plot of the ratio of receiving antenna gain to RF noise. Figure 6 shows the optimum carrier frequency to be between 10 and 30 MHz when a miniature ferrite loop antenna is used (7,8). A constant RF field strength is assumed.

The high cost of existing custom integrated circuits for implantable instrumentation is directly tied to the economics of scale in integrated circuit production. A promising way of reducing this cost is offered by the use of KITCHIP circuits. These integrated circuits are arrays of monolithic components on which particular circuits are delineated by custom metalization alone (9). The Stanford KITCHIP is a modular array of monolithic components specifically designed for micropower circuits (10). A photograph of a KITCHIP containing the RF section of the receiver is shown in Figure 7.

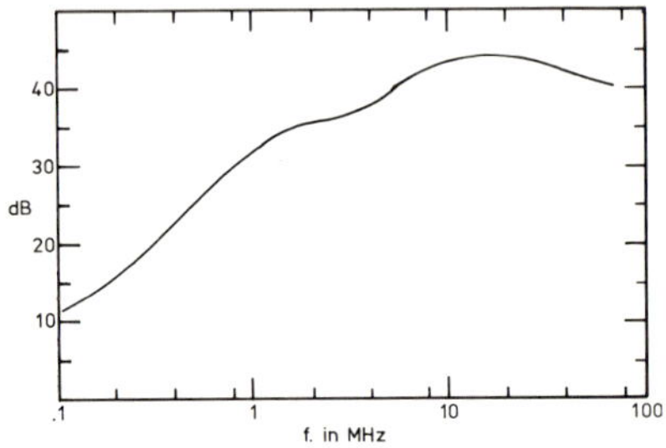

Fig. 6. Antenna Gain to Noise Ratio.

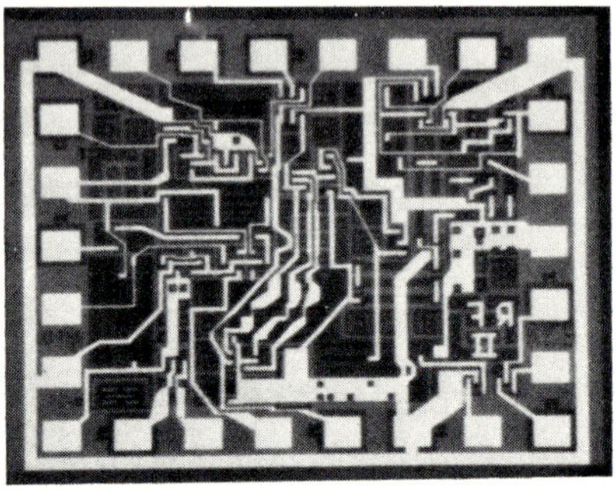

Fig. 7. RF KITCHIP.

REFERENCES

1. Meindl JD: Integrated Electronics in Medicine. ISSCC Digest of Technical Papers: 160-161, 1974
2. Gill RW, Meindl JD: Micropower Integrated Circuits for an Implantable Pulsed Doppler Ultrasonic Blood Flowmeter. ISSCC Digest of Technical Papers: 182-183, 1975.
3. Hudson PH, Meindl JD: A Monolithic Micropower Command Receiver. ISSCC Digest of Technical Papers: 108-109, 1971.
4. Baker RH, et al: Pulse Powered Circuits, IEEE Trans. Electron Computer 15: 321-323, 1966.
5. Meindl JD: Micropower Circuits. New York, Wiley, 1969, Chap 5.
6. Wycoff K: Battery-Saving Techniques for Radio Paging System. ICC Conference Records: 37.17-37.20, 1975.
7. Foletta WS: Development of a Micropower Radio Command Link. Engineer's Thesis Stanford University: 5, 1974.
8. Pettengill RC, Garland HT, Meindl JD: Receiving Antenna Design for Miniature Receivers. Integrated Circuits Laboratory, Stanford University, 1975 (to be published).
9. Interdesign, Inc., Monochip Designer's Handbook, Sunnyvale, CA.
10. Combs SR: KITCHIP - A Generalized Integrated Circuit for Bio-Medical System Design. Proceedings, 3rd International Symposium on Biotelemetry, Asilomar, Calif., 1976.

SESSION III

PATIENT MONITORING,

CLINICAL TELEMETRY

Chairman

E.L. NAGEL, M.D.

EDITOR'S OVERVIEW

The true significance of the miniaturization and micro-miniaturization discussed in the preceding sessions on Human Research and Implantable Applications comes to light in the next two sessions on clinical applications of biotelemetry. These papers present discussions of a wide range of uses to which the equipment now being developed has been applied. Applications which are discussed include: monitoring of patients in operating rooms, baths, physical therapy wards, general care hospital rooms, and in-transit to the hospital via emergency vehicles. The measurements made include the usual vital signs (heart rate, respiration, body temperature) and specialized items such as intra-ocular pressure, EEG from ambulatory epileptics, fetal heart beat, muscular activity and gait from cerebral palsy patients, and pressure distributions from the ball of a hip-joint prosthesis. The last measurements, not yet made in vivo, represent a truly unique effort in micro-miniaturization in that 14 pressure transducers, a transmitter, and an inductive power unit were enclosed in a metallic ball to measure the pressure distribution on the joint.

In nearly every paper, such benefits of miniaturization permit measurements to be made in otherwise inaccessible locations and monitor patients without having them carry heavy, cumbersome equipment and trailing cables. These latter advances will permit much more realistic data to be obtained by minimizing or eliminating the physiological and psychological interferences associated with the use of previously available equipment.

The final group of four papers in this session, including an editorial contribution, discuss some of the recent concerns with the operation and regulation of Emergency Medical Systems in which voice and patient data are transmitted from remote locations to a central health care facility.

It is noteworthy that a large number of papers have been presented here that describe work in progress. Researchers are reporting their progress and sharing ideas that are not fully harvested. They are sharing ideas that may be used and exploited by other researchers in their own work. Most workers in this area tackle jobs that are tremendously complex; they reach into areas of engineering, electronics, biology, and medicine. This session is to be commended for involving so many workers who are willing to expose themselves early or mid-course in their research work.

Two pieces of work reported in this session tie together. The work on fetal monitoring may result in fewer cerebral palsy children needing gait analysis for therapy.

The real measure of the importance of biotelemetry is not in the number of dollars involved but in its performing an indispensable function to a group of patients.

COMBINED RADIOTELEMETRY OF ECG, ARTERIAL BLOOD PRESSURE, PULMONARY ARTERY PRESSURE AND CARDIAC OUTPUT

Reinhard Zerzawy, Hermann Fleischer and Kurt Bachmann

Med. Poliklinik der Universitaet
D-8520 Erlangen, Fed. Rep. Germany

Radiotelemetric measurement of direct arterial and pulmonary artery blood pressure has proven to be a valuable method in analyzing cardiovascular response to daily life and sports activities in both normal persons and patients with cardiovascular diseases. Still, the problem of estimating total peripheral resistance in man outside the laboratory could not be solved satisfactorily. This necessity of a method for estimating total peripheral resistance in unrestricted persons lead to the development of a telemetric device, which allows on-line registration of cardiac output, derived from the calibrated impedance cardiogram, direct arterial blood pressure and total peripheral resistance. In the first part of this paper, a description of this method will be given. The second part presents some results of combined radiotelemetry of impedance cardiogram, direct arterial pressure and pulmonary artery pressure in normal persons, hypertensive patients and coronary patients.

METHOD

According to Kubicek's formula for the calculation of cardiac output from the impedance cardiogram,

$$CO = p\frac{L^2}{Z_o^2}\ T\ (dZ/dt)_{max}\ \frac{HR}{1000}, \qquad (1)$$

quantitative information about L, Z_o, T, $(dZ/dt)_{max}$ and HR must be provided and computed electronically (CO = cardiac output in l/min, p = 135 Ohms x cm = specific resistivity of blood, L = distance between inner electrodes in cm, Z = mean thoracic impedance in Ohms, T = ejection time in seconds, $(dZ/dt)_{max}$ = peak value of the first derivative of the impedance cardiogram in Ohms/sec, HR = heart rate in beats/min).

In the patient-borne unit (fig. 1) a 51.2 KHz current source drives a constant 4 mA rms sinusoidal current through the thorax. To reduce power consumption, this current source is gated by a 200 Hz pulse frequency with a duty cycle of 25%. The differential signal from the inner electrodes is amplified and demodulated. The demodulator delivers a smoothed DC-signal proportional to Z_o and an AC-signal $dZ_{(t)}$, which represents fast resistance changes during the cardiac cycle. An ECG-amplifier and trigger unit controls a program generator, which performs the following operations: a) gating of current source and demodulator, b) zero-setting of

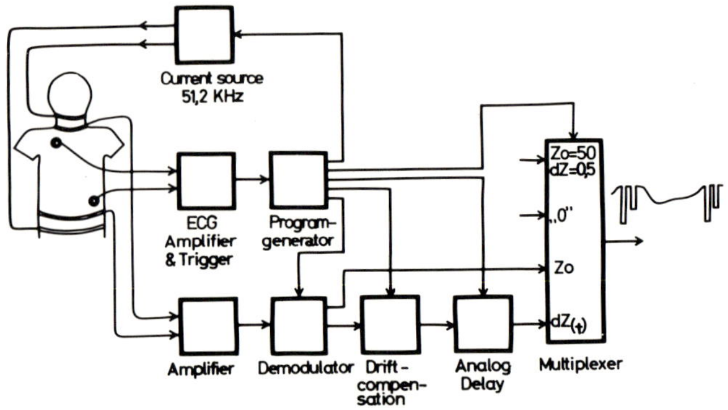

fig. 1 Block diagram of patient borne unit for telemetric transmission of calibrated impedance cardiogram by standard telemetry systems.

the $dZ_{(t)}$-signal synchronous with the R-wave, c) 80 msec analog delay of the drift-compensated $dZ_{(t)}$-signal, d) controlling of the output-multiplexer. The time division multiplexed output signal contains the following informations: a) a 20 msec calibration pulse (Z_o = 50 Ohms, dZ = 0.5 Ohm), starting at the trigger point on the R-wave, b) a 20 msec zero-interval for reference, c) a 20 msec pulse representing the value of Z_o, d) the delayed $dZ_{(t)}$-signal.

In the receiver unit (fig. 2), a trigger detects the beginning of the cardiac cycle by the leading edge of the calibrating pulse. This trigger controls a beat-to-beat heart rate integrator and a program generator. After demultiplexing of the input signal, the four signals enter a sample and hold circuit, which generates calibrated Z_o- and $dZ_{(t)}$-signals. A differentiator with peak-detector then delivers $(dZ/dt)_{max}$. The ejection time is found by one of two ways, depending on the signal conditions. If a characteristic point indicating the end of ejection time is detected in the second derivative of $dZ_{(t)}$, this point is used to measure T_s. If this determination of T_s does not fit the condition

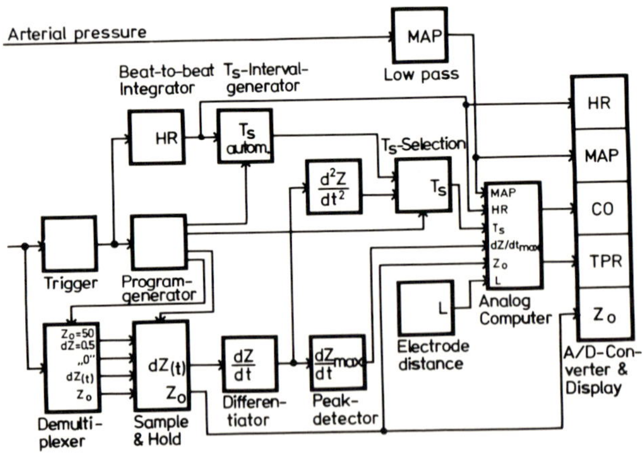

fig. 2 Block diagram of receiver unit for computing cardiac output and total peripheral resistance from telemetrically transmitted impedance cardiogram and direct arterial pressure.

$$0.05 + \frac{12.8}{HR} \quad T_s \quad 0.09 + \frac{21.3}{HR} \qquad (2)$$

the T_S-selection unit switches to a preprogrammed empirically found function

$$T_{s\ autom.} = 0.05 + \frac{18.6}{HR} \qquad (3)$$

The computation of cardiac output and total peripheral resistance is performed by an analog computer. The input signals to this analog computer are shown in fig. 2. The distance between inner electrodes L is entered manually by coded switches. Mean arterial pressure (MAP) is the filtered signal from the pressure channel of the telemetry system. Heart rate, mean arterial pressure, cardiac output, total peripheral resistance and mean thoracic impedance are displayed by numeric indicators.

RESULTS

Fig. 3 a shows results of radiotelemetry of ECG, direct arterial pressure and impedance cardiogram in 15 healthy subjects during rhythmic isometric exercise with a steel torsion bar, 4 kg-weights-holding with each hand, static knee-bends and bicycle ergometry. All three gymnastic exercises provoke hypertensive mean values from 178-185 mm Hg systolic and 111-115 mm Hg diastolic, at moderate heart rates from 128-150 beats/min. At equal heart rates,

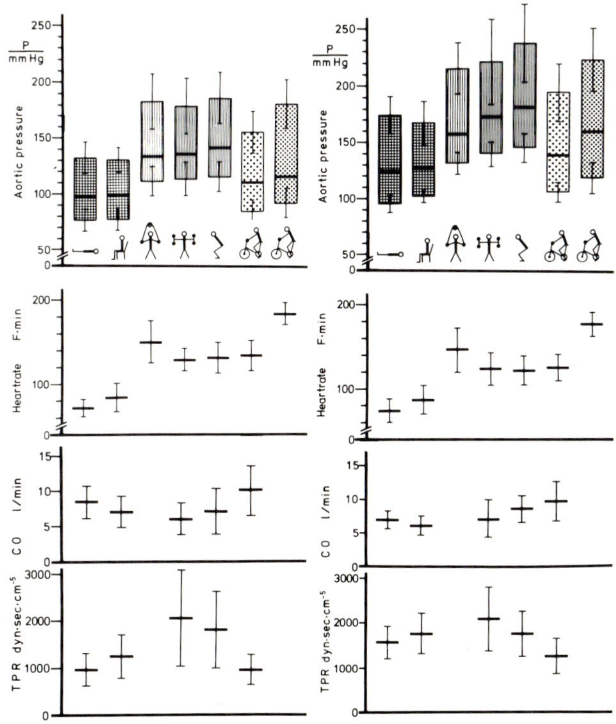

fig. 3a (left)
fig. 3b (right)
Mean values of combined radiotelemetry of direct arterial blood pressure and impedance cardiogram with computation of cardiac output and total peripheral resistance in normal persons (fig. 3a) and hypertensive patients (fig. 3b). From left to right: recumbent, sitting, torsion bar exercise, weights-holding, static knee-bend, ergometry at equal heart rate as during knee-bends, maximum effort cycling.

systolic, mean and diastolic pressures remain significantly lower during ergometry (6th column) than during isometric exercise. Mean TPR is significantly higher (p 0.001) during isometric exercise (2090 resp. 1840 dyn sec cm^{-5}) than during ergometry (985 dyn sec cm^{-5}) at equal heart rate. The increase of TPR during isometric exercise compared to the recumbent position is also significant (p 0.001). Whereas systolic pressure during maximum effort cycling shows no difference to isometric exercise, mean and diastolic pressures were distinctly lower (p 0.001). Isometric and rhythmic isometric exercise must be regarded as unsuitable methods for cardiovascular endurance training.

In 12 fully compensated hypertensive patients (fig. 3b) blood pressure reactions during isometric and ergometric exercise show a similar pattern at higher absolute values compared to normal persons. The mean values of 237/144 mm Hg during knee-bends indicate the extreme pressure load of heart and arteries during static exercise in hypertensives. TPR shows less variability in hypertensives than in normals.

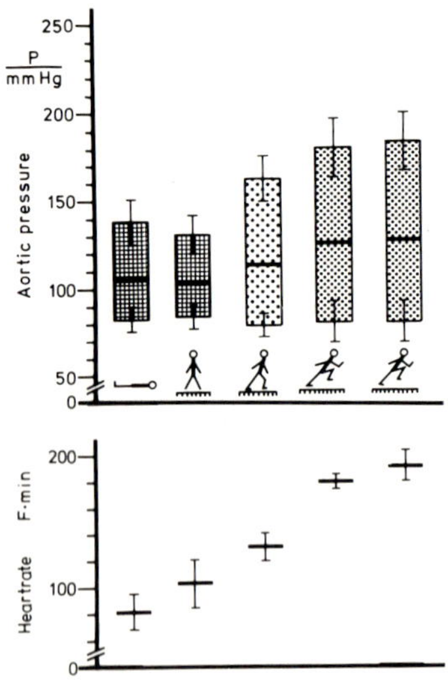

In fig. 4 (left), mean values of direct arterial blood pressure telemetry in 11 normal subjects show nearly unchanged diastolic pressures during rest, standing, walking at a speed of 6 km/h, after the first minute and in the last minute of 800 m-running (from left to right). Therefore, mean arterial pressure is only slightly increased during these dynamic exercises despite of the distinct increase in blood pressure amplitude.

In 20 coronary patients, pulmonary artery pressure, arterial pressure and heart rate were compared during ergometric exercise and cross-country walking (fig. 5). Walking uphill, maximum physical workload was 23% higher, systolic pressure-rate product was identical and diastolic pulmonary artery pressurewas significantly lower. Good correlation is found between coronary score and hemodynamic parameters.

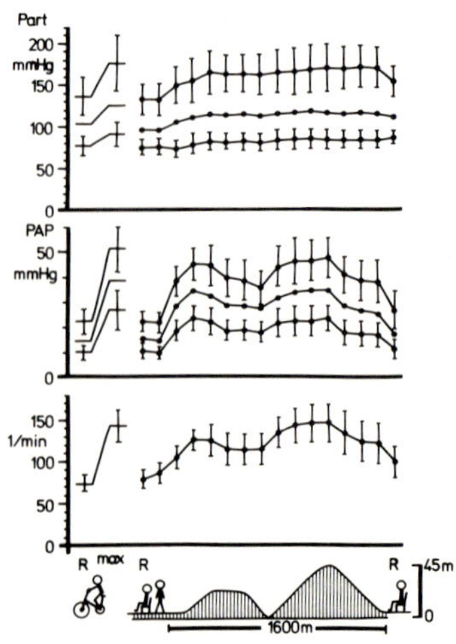

Practical Operating Room Monitoring by Multichannel Radio Telemetry

Fritz Klein and David A. Davis
Department of Anesthesiology
Duke University Medical Center
Durham, N.C.

In the world of clinical monitoring, radio telemetry usually invokes thoughts of added complexity, difficult situations, implants, microminiturization, expense, and limitation of monitoring to a single function. It is fair to state that nearly all clinicians and most researchers are unaware of the potential advantages and applications of radio telemetry. It is even more obvious that the manufacturers and salespersons of clinical monitoring equipment have, for many reasons, done little to provide customers with monitoring systems which furnish a true picture of the capabilities and merits of radio telemetry. It is hoped that presentations of experiences like ours will stimulate interest in others, for it is our sincere opinion that the future of radio telemetry in physiological monitoring lies in the application of its principles to common rather than uncommon situations.

Some seventeen years ago one of us (D.A.D.) began to use radio telemetry of the electrocardiogram in the operating rooms of the North Carolina Memorial Hospital. At first this was a novelty. Soon it became apparent that radio offered distinct, though subtle, advantages so it became and still remains the technique of choice for electrocardiographic monitoring in these operating rooms. In 1972 a similar experiment in radio transmission of multiple physiological signals was launched at Duke University Medical Center. Once more the enthusiastic reception of radio telemetry by relatively unsophisticated personnel dictated the utilization of this technique in all the operating, delivery, and recovery rooms of this institution. Some 26 systems are in daily use; new systems are in the design stage. These two experiences form the basis for further argument.

THE ADVANTAGES OF RADIO TELEMETRY

Those unfamiliar with patterns of traffic and activity in busy operating, delivery and recovery rooms usually fail to appreciate the tremendous advantages of radio in spanning the few feet between the patient and recording and display equipment. Convenience is a principal asset of radio systems. Many readily recognize how radio can minimize the contribution of monitoring equipment to patient electrocution. Others see how a physiological signal, encoded on an audio subcarrier at a significant voltage, lends itself to storage and processing by conventional audio techniques as tape recording and transmission by telephone or radio relay. All who are acquainted with radio telemetry will agree that its principles are well established and require no major technological advances.

DESIGN CONSIDERATIONS - DISPLAY AND RECORDING EQUIPMENT

Most "medical" monitoring systems employ oscilloscopes as display devices. These are usually equipped with special amplifiers tailored to a specific physiological signal. Manufacturers tend to encourage their positioning in carefully selected fixed locations and the incorporation of permanent wires in hospital constructions. Instead, we selected maximum mobility of a rugged general-purpose industrial oscilloscope as a prime consideration. Experience has vastly strengthened this concept in all our applications. Additionally, we have become well aware of the advantages of occasional utilization of other types of recording and display devices and thus have equipped our Department with mobile cart-mounted radio receivers which feed signals into telephone lines, tape recorders or general-purpose chart recorders. All these are readily available and their location is dictated by convenience rather than length of wires. We have lost all faith in the virtues of fixed locations for routine monitoring systems.

DESIGN CONSIDERATIONS - TRANSMITTERS AND RECEIVERS

The initial consideration in multichannel transmission was that of functions to be monitored. Of the many signals compatible with radio telemetry we chose four. These were the electroencephalogram, the electrocardiogram, the peripheral pulse and arterial blood pressure. The last parameter required telemetry of a very stable DC signal and utilization of the transmitter's power supply to excite the pressure transducer, problems which yielded to careful design. Other and perhaps unconventional design considerations included concentration on minimal radiated power with batteries of maximum storage capacity. This factor, combined with selection of micropower components and a pressure transducer of unusually high bridge impedance has resulted in battery power expectancy of four to five months of ordinary use.

Initial considerations of bandwidth and FM deviation led to selection of a stable coil oscillator instead of a crystal. In addition the ready availability of entertainment-type radio receivers led to concentration of prototype efforts to the 88-108 mHz FM band. The minimal power radiation and limited institutional applications of these systems made them readily tolerable by the Federal Communications Commission. Though initially and presently serving well, it is anticipated that future systems might benefit by crystal control of transmitters and choice of other tranmission frequencies.

While conceding advantages to fixed transmission frequencies, we are much more positive in our attitude that variable tuning of receivers is highly desirable. Preservation of this variable function permits the rapid substitution of equipment in case of failure of either transmitter or receiver. We have already mentioned the advantages of bringing occasionally-used recording equipment into or near operating rooms; variable tuning in these systems is definitely advantageous.

Size and construction of transmitters has received much attention. We have had no cause to reconsider our early concepts of concentration on ruggedness, shock-mounting, easy access for repair and the firm (solder) fixation of batteries of maximal amperage. Ours has been a relatively simple situation.

SYSTEMS EXPENSE

This is a subject of frequent questions but few answers, for there is little basis for comparison or analysis. Those familiar with principles and practices of radio telemetry will recognize that component costs are relatively small. The technology involved could be much less demanding than that incorporated in amateur radios and pocket calculators. Were the demand for radio telemetry to increase, it is quite reasonable to believe that cost benefits would almost immediately make radio telemetry the method of choice in operating room monitoring. Certainly our years of experience support this statement.

COMMENTS, RECOMMENDATIONS AND SUMMARY

Our principal objective in this presentation has been that of offering our own experiences as evidence that the monitoring of physiological functions by radio telemetry is eminently practical in operating rooms. We make no claims that systems which we have designed and used so effectively represent a major technological advance. On the contrary, we recognize their inadequacies and strongly urge further system design in the development of more appropriate hardware.

A FULLY INTEGRATED MULTICHANNEL TELEMETRY SYSTEM FOR PATIENT MONITORING

Wolfgang Steinhagen

Institut für Theoretische Elektrotechnik, TH Aachen
Aachen, W. Germany

Clinical surveillance for instance in an intensive care unit requires the continuous monitoring of physiological parameters in patients like ECG, pulse, blood pressure etc. Important features are the reliability of such a monitoring system and the safety under fault conditions. It is desirable to strain the patient as little as possible. The phychological effect of being tied with cables must not be neglected either. These problems can be reduced by a small, light and battery-powered telemetry transmitter system, which can be worn by the patient. To perform this task an integrated complementary MOS multichannel telemetry transmitter has been developed, which is able to transmit a set of physiological parameters. The C-MOS technique was choosen to get a low power consumption and to have offset-free analog switches. The telemetry transmitter chip has been constructed as a universal building block so that seven available channels can be connected with different signal sources. With an incorporated power output stage the transmitter is able either to drive an inductive transmitting coil or to serve as suboscillator in a high frequency system. An inductive transmission avoids the difficulties in respect to federal regulations of telemetry frequencies.

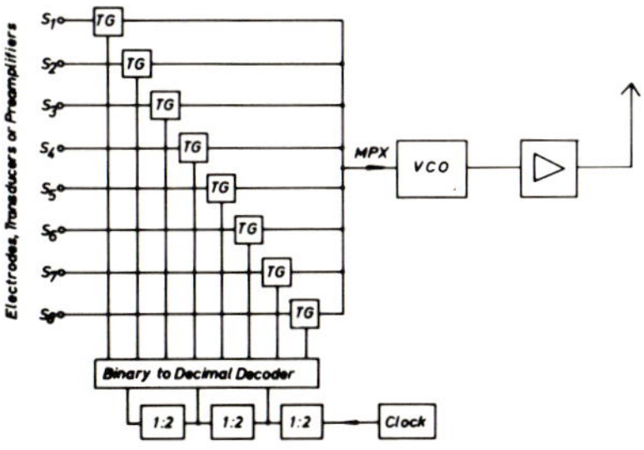

Fig. 1. Transmitter Block Diagram

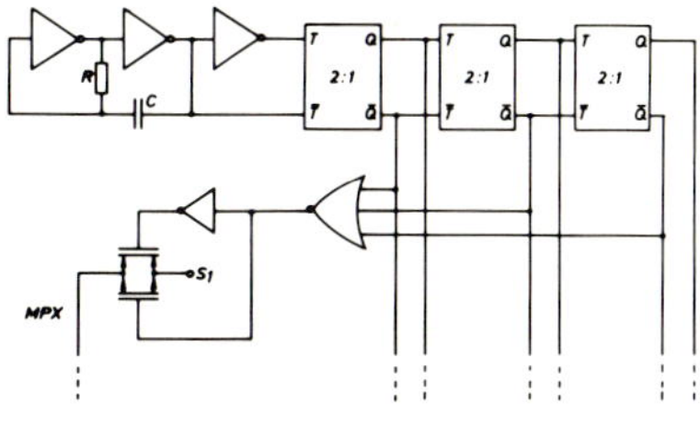

Fig. 2. Multiplexer

Fig. 1 shows the block diagram of the telemetry transmitter. The whole system consists of a multiplexer, which switches the seven channels successively to the input of a voltage controlled oscillator. The output of the VCO is connected to a power stage, which drives an inductive transmitting coil. By this transmitting coil the frequency modulated carrier is inductively coupled to the receiver coil, which for instance could be fixed under the patient's bed.

Fig. 2 shows the schematic diagram of the multiplexer. Here only the signal path for one channel is drawn. The multiplexer consists of a clock generator realized by three inverters. The output signal is divided by a 3-stage asynchonous frequency divider and decoded by a binary-to-decimal decoder. The so formed pulses drive the analog switches realized by transmission gates. For this digital part of the circuit consisting of 136 transistors the major advantage of the C-MOS technique i.e. the low power consumption is obvious. The on-resistance of the transmission gates is about 5 kΩ so that the resulting time constant with the input capacitance of the VCO is negligible.

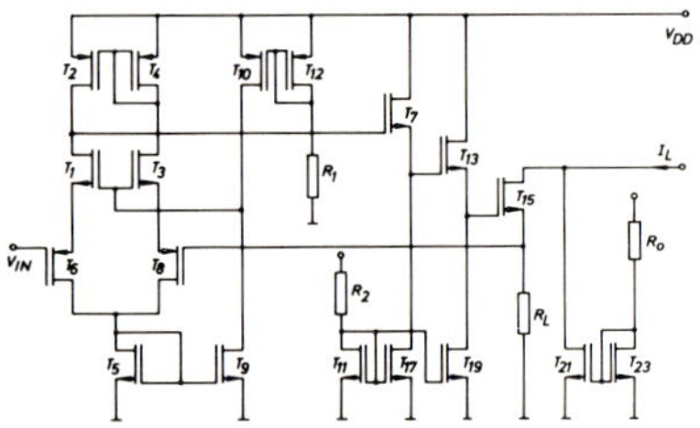

Fig. 3. Voltage to Current Converter

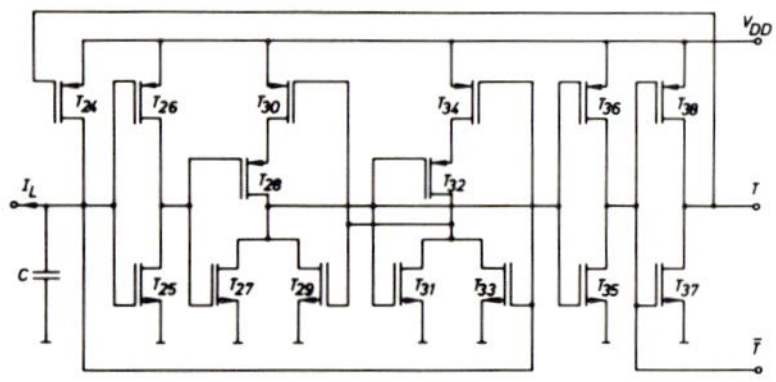

Fig. 4. Current Controlled Oscillator

The most essential part of this work was concentrated on the development of a suitable voltage controlled oscillator. The problem was to find new circuit configurations which could be realized in integrated C-MOS technique. These circuit configurations had to fulfill the requirements concerning linearity, stability and dynamic behavior. The VCO was composed of two parts. First a voltage to current converter produces a current, which is proportional to the input voltage. Fig. 3 shows its schematic diagram. The conversion is performed by a high gain differential amplifier in a voltage follower configuration so that the voltage drop across the resistor R_L is equal to the input voltage. Since there exists no gate current, the output current at the drain is identical to the current through R_L. The differential amplifier is designed to handle signals down to zero with a single supply voltage. The so produced current with an additional offset current source form the charging current for the multivibrator (Fig. 4). The multivibrator consists of a Schmitt-Trigger and a charging capacitor. The capacitor is discharged by the output current of the voltage to current converter, whereas the charging is done by the fast switching transistor T_{24} whose gate is driven by the output of the Schmitt-Trigger. In order to reach a good temperature stability of the frequency it is necessary to have a stable hysteresis. The transistor geometries of T_{25} and T_{26} were chosen according to a symmetrical transfer curve. So a well temperature compensated switching point is obtained. The charging is done fast enough so that the voltage of the capacitor reaches always V_{DD}. The measured temperature coefficient of the hysteresis is about 60 ppm/^{o}C. The same temperature dependence results for the frequency.

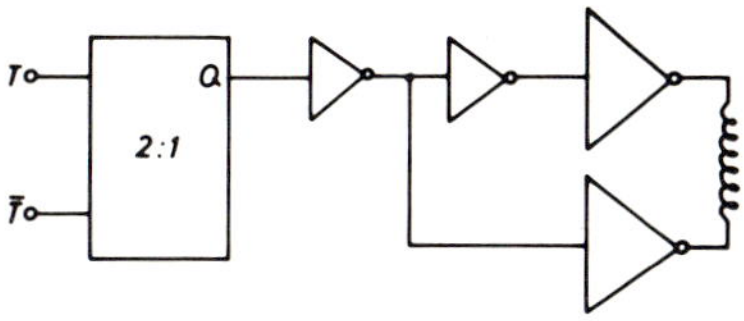

Fig. 5. Frequency Divider and Output Stage

In order to obtain 50% duty cycle the output is followed by a frequency divider. This is shown in Fig. 5. Two further inverters drive the power stage where the transmitting coil is connected between two power inverters so that only alternating current flows through the coil.

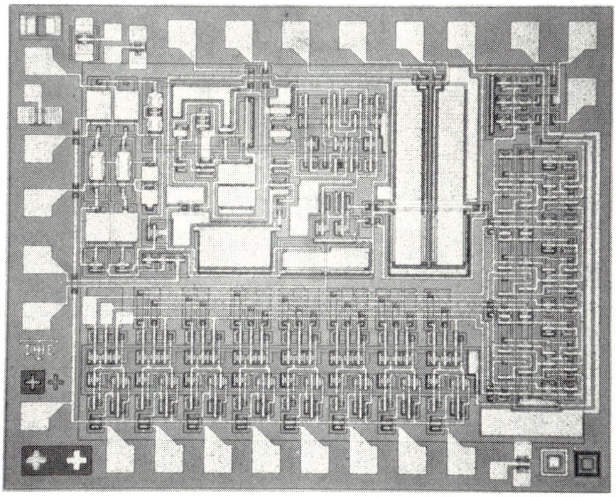

Fig. 6. Photomicrograph of the Transmitter IC-Chip

RESULTS

The photomicrograph of the complete integrated multiplex telemetry transmitter is shown in Fig. 6. The chip size is 1.8 by 2.0 mm. Since this chip is a prototype where a maximum of parameter variations was desired to investigate the circuit behavior the two required capacitors are not yet incorporated. With a value of about 50 pF each they need an additional space of 1/3 mm^2 to be integrated. Eight channels (7 measuring signals and one synchronization signal) are scanned with a maximum frequency of 8 kHZ. The measuring bandwidth of each channel is 500 HZ with an input voltage range from ground to nearly half the supply voltage. The nonlinearity at a carrier frequency of 100 kHZ within a sweep of $\pm$ 5 kHZ is less than 0,2 %. The temperature drift related to the frequency sweep is about 700 ppm/^{o}C so that the stability is sufficient for a limited temperature range.

The power consumption of the whole system is some hundred µAmps so that the battery life is practically determined by the desired transmission range. The output stage is able to deliver an output current up to 30 mA at 8 V supply voltage. With a current of 15 mA for example a transmission range of about 6 meters has been realized.

The described telemetry transmitter may be employed as a versatile building block for different telemetry applications either for direct inductive coupling or to serve as a suboscillator in a high frequency system.

ACKNOWLEDGMENT

This research was supported by the Bundesministerium für Forschung und Technologie of the Federal Republic of Germany.

CLINICAL AND RESEARCH APPLICATIONS OF A MULTICHANNEL EEG RADIOTELEMETRY SYSTEM IN PATIENTS WITH PARTIAL EPILEPSY

J.P. Lieb, A.M. Dymond, G.O. Walsh, and P.H. Crandall

Reed Neurological Research Center
UCLA School of Medicine
Los Angeles, California

The primary objective of electrographic procedures in the diagnosis of partial epilepsy is the localization of a delimited region (or regions) of the brain from which spontaneous clinical seizures arise. The attainment of this objective is vital in those patients for whom surgical treatment is considered. The non-invasive technique of scalp EEG recording is often useful in obtaining this information but it is important to realize the limitations of this procedure. For instance, scalp recordings are often obscured with artifact during the course of a seizure; additionally, the localization of focal onset of seizure activity may occur in the depths of the brain and not be clearly decidable from scalp recordings. These and other difficulties led to the use of stereotactically implanted depth electrodes at UCLA hospital in 1961 (2). Localization of the seizure focus in these patients is done by interpreting the EEG record that accompanies spontaneously occurring clinical seizures. However, since spontaneously occurring seizures are often unpredictable with respect to their time of occurrence, it is necessary to record continuously in order to assure acquisition of a sufficient number of such records during the patient's stay in the hospital.

Radiotelemetry monitoring of EEG data is a practical solution to the above problem; such monitoring permits essentially continuous recordings to be taken from unrestrained patients. In addition, such recordings are more immune to movement artifact than conventional "hardwired" recordings. Multichannel biotelemetry has been a routine clinical tool at UCLA hospital since 1970 (3). The initial equipment was an 8 channel IRIG proportional bandwidth FM subcarrier system (7) utilizing AM modulation at about 90 MHz. The FM subcarrier composite signal from the receiver was recorded on a direct recording audio tape machine. The recordings were made continuously and selected epochs which contained electrographic seizure activity were played back from the tape through FM discriminators onto a multichannel chartwriter for clinical evaluation. This system was later modified to record 12 channels onto an FM tape recorder.

This system was not primarily designed for clinical applications and the equipment was built based on mid 1960's technology. Clinical interpretation of seizure records was, in some cases, impaired by unequal frequency response across the 12 channels, due to the proportional bandwidth subcarrier system. Further processing of this data, using automated and computer-oriented techniques, was also hampered by the unequal frequency response. Additional problems with this system included limited transmission range, excessive weight, and the expense and difficulty of acquiring special mercury batteries to power the system. Finally, as the equipment aged, its reliability declined to unacceptable levels.

A new multichannel biotelemetry system has recently been developed which is intended for clinical and research application and which reflects our 5 years of experience with this type of telemetric monitoring (4). The system can easily be implemented by ward staff, is small and light enough to be continually worn on the head, has good transmission range, is reliable, and can be powered for 24 hours by a standard 9 volt battery. The system contains 14 channels of high input-impedance differential amplifiers which are multiplexed and pulse-width encoded for transmission. Transmission can be either FM or phase modulation and is normally done in the TV band. After reception, the signal is demultiplexed, decoded, and each channel is then stored individually on an FM tape recorder. System parameters such as frequency response, gain, and cross-talk are appropriate for clinical and research electrophysiological applications. One special feature of the system is the inclusion of a 'systems status' channel which continuously reports on system performance. This allows easy identification of system faults, such as might occur if the patient had walked out of range of the receiver or if the receiver had been detuned. This has proven to be an important feature for the selection of epochs containing true physiological signals.

Figure 1 shows a block diagram of the complete monitoring system as it is presently used. The patient is on the ward and the telemetry pack is mounted on his head. The transmitted signal is received by an antenna mounted in the ceiling directly above the patient's bed. The receiver and other equipment is located in a separate room. The output of the receiver goes through a PWM demodulator and onto individual tracks of the tape recorder. Up to 12 channels of the demodulator are also connected to an on-line seizure detection circuit (1). The output of the seizure detector enables a printer which lists the day, hour, minute, and second of each detected seizure. This time is generated by a digital clock which is also recorded on FM tape. Videotaping of the patient is sometimes employed when the patient is in his bed on the ward. This is implemented by a camera which is mounted on the wall opposite and above the patient's bed. The output of the camera goes to a monitor and videotape which are located remotely with the other equipment. The videotape and the FM tape are synchronized by means of the digital clock. This permits correlation of the patient's behavior during seizure episodes with the electrographic record. Output from the tape recorder can go either online or off-line to a multichannel polygraph. Eventually split-screen recording of the patient's behavior and the electrographic

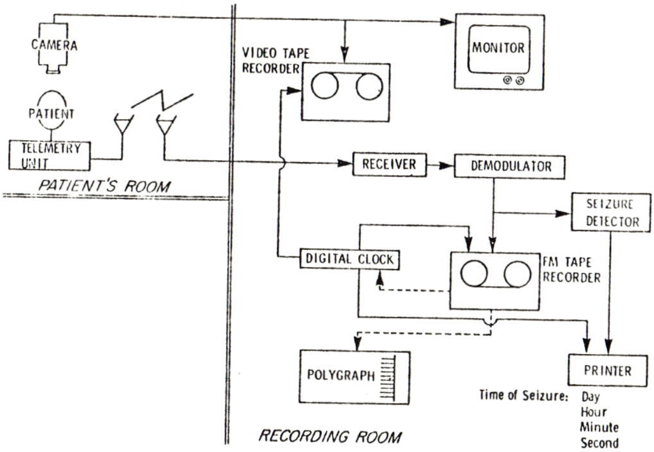

Figure 1. Block diagram of monitoring system

seizure record will be done (6). Selected epochs containing seizures can be located for review by using the information from the printer and the clock channel on both the FM recorder and the videotapes. Figure 2 shows the start of a spontaneous clinical seizure that was recorded with this apparatus.

This system has proven to be highly cost-effective since it permits recording of almost all of the spontaneous clinical seizures which might occur in a 24 hour period, thus shortening the number of days the patient must stay in the hospital in order to acquire the necessary data for clinical evaluation. Besides its

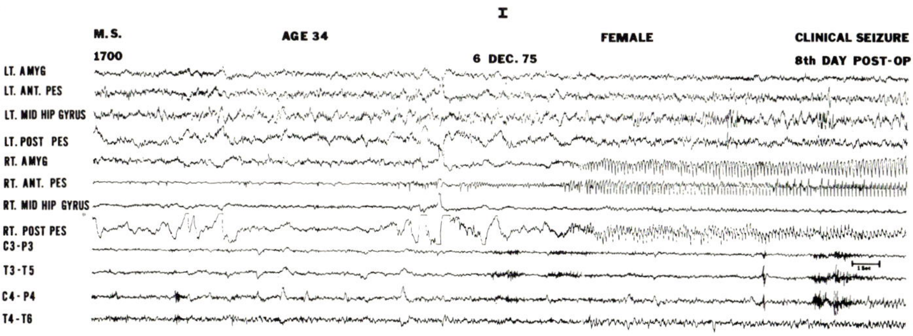

Figure 2. Start of a spontaneous seizure record.

use on the ward, we have found telemetry to be a very effective technique for obtaining recordings in the operating room (5).

REFERENCES

1. Babb TL, Mariani E, and Crandall PH: An electronic circuit for detection of EEG seizures recorded with implanted electrodes. Electroenceph clin Neurophysiol 37: 305, 1974.

2. Crandall PH, Walter RD, and Rand RW: Clinical applications of studies on stereotactically implanted electrodes in temporal lobe epilepsy. J Neurosurg 21: 827, 1963.

3. Dymond AM, Zweizig JR, Crandall PH et al: Clinical applications of an EEG radiotelemetry system. Biomed Instrum 8: 16, 1971.

4. Dymond AM, Fryer TR, Coger RW, et al: A multichannel radiotelemetry system for clinical and research use. Med Instrum 9: 41, 1975.

5. Ferrer-Allado T, Brechner VL, Dymond AM et al: Ketamine induced electroconvulsive phenomena in the human limbic and thalamic regions. Anesthesiology 38: 333, 1973.

6. Walsh GO: Correlation of the aura and behavioral sequence with the electrical activity recorded by depth electrodes from the hippocampus of patients with seizure disorders. Am EEG Soc 28th Ann Meeting, 1974.

7. Zweizig JR, Kado RT, Hanley J, et al: The design and use of an FM/AM radiotelemetry system for multichannel recording of biological data. IEEE Transact on Biomed Engr, BME 14: 230, 1967.

Supported by USPHS Grant NS 02808 and Veterans Administration.

FETAL MONITORING BY RADIO TELEMETRY

Michael R. Neuman, Jacques F. Roux*, Jaroslav Hynecek, and Scott F. Boettcher

Engineering Design Center and Department of Reproductive Biology
Case Western Reserve University, Cleveland, Ohio USA
*Department of Obstetrics and Gynecology, Ste. Justine and Hotel Dieu Hospitals, Montreal, Quebec, Canada

Continuous recordings of the fetal heart rate (FHR), maternal intrauterine pressure (IUP), and in some cases the fetal electrocardiogram (FEKG) during labor is a clinically accepted procedure in the management of the high risk obstetrical patient. In many instances fetal monitoring can lead to the early detection of fetal distress which can in turn improve perinatal mortality and morbidity statistics. Fetal monitoring is currently carried out using direct wire connections between a monitoring device and an electrode placed on the fetal presenting part along with an open-ended fluid filled catheter coupling the amniotic cavity with an external strain gauge pressure transducer which is also connected to the monitoring apparatus by means of wires.

Two-channel radio telemetry systems for fetal monitoring have been developed and applied clinically. The basic system is illustrated in Figures 1 and 2. The intrauterine pressure signal is obtained from a specially packaged commercially available silicon piezoresistive strain gauge bridge pressure transducer. The bridge is excited by a pulse having a nominal 5% duty cycle, thereby, reducing its power consumption. Its output is sampled and held and modulates a voltage controlled oscillator providing a pulsed subcarrier signal at a rate ranging from 1.3 to 4 KHz with a 20 μs pulse width. This frequency modulated subcarrier in turn supplies the exitation signal for the pressure transducer and controls the sample and hold circuit. The current drain for the circuit ranges from .3 to .5 mA.

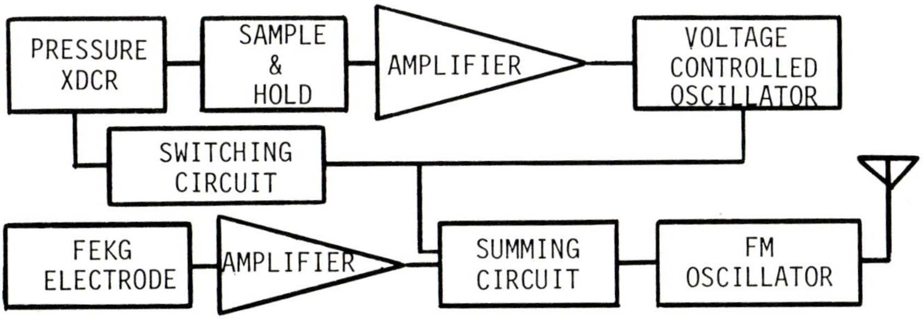

Fig. 1. Block diagram of transmitter

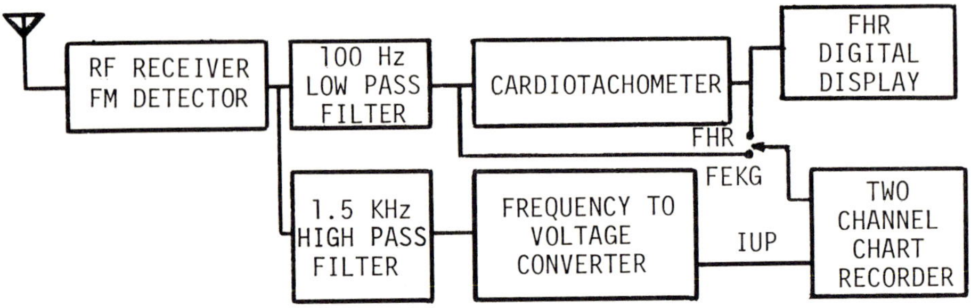

Fig. 2. Block diagram of receiver system.

The FEKG is obtained from a conventional clip or helical electrode placed on the fetal presenting part. The 100 to 700 μV signal is amplified and filtered to cover the range of from .5 to 100 Hz. This signal is added to the subcarrier and the resulting signal frequency modulates a Hartley oscillator operating in the 110-130 MHz band. The overall transmitter circuit is powered by 2 mercury cells and has a nominal power consumption of 5 mw. Larger models of the transmitter have an RF amplifier stage following the oscillator to improve transmission range.

The receiving system consists of a radio frequency receiver having a 1 MHz bandwidth and an FM detector. The IUP subcarrier and FEKG signals are separated by a 1.5 KHz high pass filter and a 100 Hz low pass filter respectively. The FHR is determined from the FEKG by a beat-to-beat cardiotachometer, and the output is displayed on a digital readout as well as on one channel of a two-channel chart recorder operating at the standard speed of .5 mm/sec. The FEKG can be displayed on the recorder by depressing a front panel switch which increases the recorder speed to the standard 25 mm/sec and replaces the FHR signal with the FEKG.

The pressure subcarrier is demodulated using a frequency to voltage converter and is displayed on the second channel of the chart recorder. The receiver system is compactly packaged in a small cabinet with a minimum of front panel controls as shown in Figure 3.

The telemetry transmitters have been packaged in three different forms. The smallest form illustrated in Figure 4 consists of three hybrid integrated circuits, batteries, and a magnetic switch in a cast epoxy package. The package has a mass of 5 g and occupies 2.7 cm^3 and has a range of 5 to 10 m. After the pressure sensor has been placed within the amniotic cavity and the FEKG electrode attached to the fetus, the transmitter can be placed within the vagina so that there are no external connections to the patient, and following application she is no longer aware that she is being monitored.

A somewhat larger transmitter consisting of replaceable batteries, hybrid integrated circuits and some miniature discrete components as well as calibration controls is illustrated in

Fig. 3. Receiver Unit

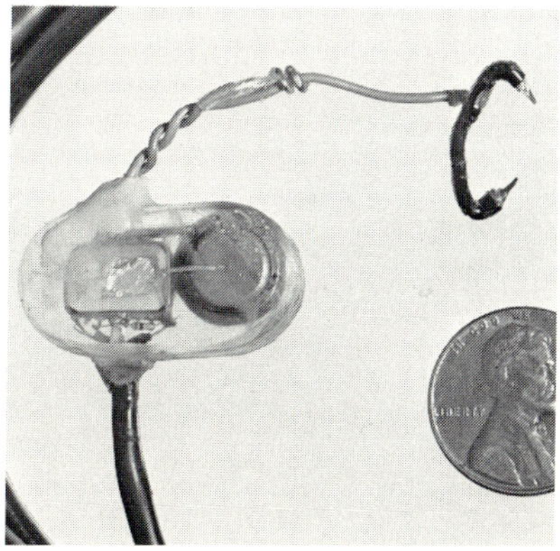

Fig. 4. Intravaginal Transmitter

Figure 5. The circuit packaged in cast epoxy measures 5 X 2.3 X 1.5 cm and weighs 17 g. In use it is placed on the patient's upper thigh with surgical tape, and lead wires from the FEKG and IUP sensors run from the vaginal introitus to the transmitter.

The third transmitter, illustrated in Figure 6, is fabricated from discrete and commercially available integrated circuits using conventional miniature electronic techniques. In addition, it has the radio frequency amplifier to increase its range to from 20 to 30 m. This unit packaged in a machined plexiglass case measures 6.7 X 7.6 X 2.5 cm and weighs 165 g. It can be attached to the upper thigh or lower abdomen and connected to the lead wires for the sensors.

The operating characteristics of the telemetry systems are

Fig. 5. Intermediate Xmtr.

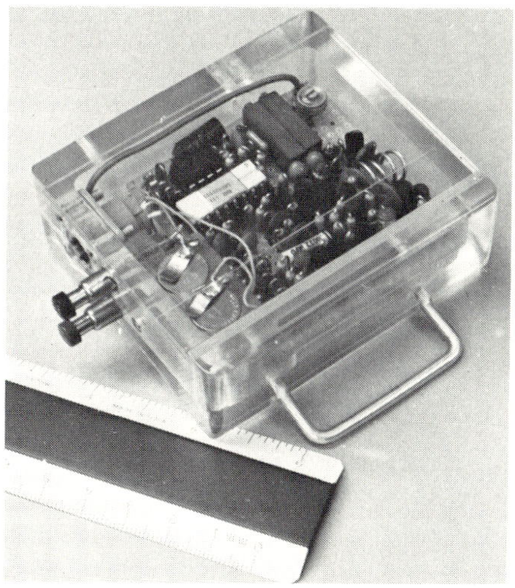

Fig. 6. Discrete component Xmtr.

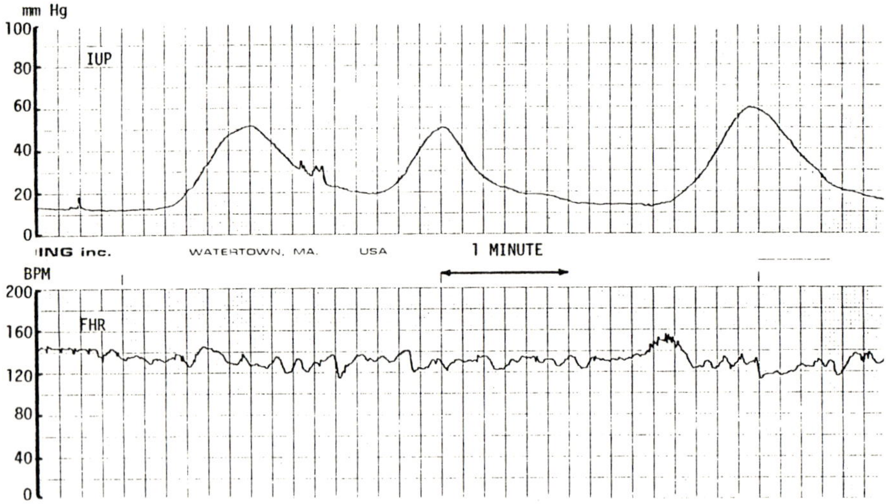

Fig. 7. Typical telemetered data from a patient in labor.

summarized in Table I. A typical recording from a patient being monitored during labor is shown in Figure 7. In applying this telemetry system to a series of 10 consecutive patients and comparing them with 26 similar patients monitored by hard wire techniques, it was found that the radio telemetry system provided readable FHR data 95% of the time compared to 75% for the hard wire system. Readable IUP data from the former was seen 92% of the time compared to 97% for the latter.

Table I. SPECIFICATIONS FOR FETAL TELEMETRY SYSTEM

FHR & FEKG:	Noise (Input Shorted)	5 μV p. to p.
	Frequency Response	1-600 Hz
	Input Impedance	100 KΩ
PRESSURE:	Range	0-100 mm Hg
	Linearity	± 5%
	Drift	<2 mm/5 hr.
	Frequency Response	0-10 Hz

Thus, radio telemetry used for fetal monitoring can have some advantages over conventional hard wire techniques as far as the FHR is concerned. In addition, the patient can be moved quickly without having to disconnect anything. The system completely isolates the patient from the power lines, thereby, eliminating the microshock hazard and makes it easier for the clinical staff to work with the patient due to the absence of wires and catheters. The system is currently being applied in three hospitals and is being found to be acceptable both to patients and to staff.

Supported by NIGMS Grant # GM 14267 and the CWRU General Res. Fund.

TELEMETRY MONITORING OF INTRAOCULAR PRESSURE

L. A. Couvillon, Jr.
C. D. Baker
T. P. Grover
*E. Konigsberg**

Utah Biomedical Test Laboratory, Salt Lake City, Utah; and
*Konigsberg Instruments Inc., Pasadena, California

INTRODUCTION

Our objective has been to develop a new instrument for the measurement of intraocular pressure which might provide a more thorough understanding of the physiologic changes in the eye in glaucomatous patients. A particular purpose of this system is to provide a continuous monitoring in order to evaluate more precisely the diurnal variations in the pressure of those glaucomatous patients who have progressive field loss despite apparent adequate control of the intraocular pressure. In this manner, more accurate and more adequate diagnosis might be made and more timely pharmacologic therapy administered.

Our system is based on well-known principles of applanation tonometry which have previously been exploited on the cornea in the McKay-Marg Electronic Tonometer (1). In our application, the tonometric measurement is made on the sclera rather than the cornea, in order that the monitoring not interfere with vision.

METHODS

The applanation sensor consists of a miniature pressure transducer mounted in a ring-shaped hydrogel holder which fits in the conjunctival cul-de-sac. The output of the transducer is recorded via a radio telemetry system consisting of a transmitter held adhesively on the temporal region of the head, radiating to a nearby receiver console.

The pressure transducer, developed especially for this application, is mounted in a ring-shaped hydrogel holder (Figure 1). Three 0.25 mm diameter teflon-insulated stranded stainless steel wires are used to connect the transducer to the telemetry package. These wires cross the eyelid and may exit the eye near either the nasal or lateral canthus. The telemetry package is approximately 4 cm diameter and 1 cm thick, and contains batteries for four weeks operation. CMOS timing circuitry in the transmitter excites it periodically (presently one minute out of each ten minutes). The radiated signal is received and recorded at a nearby receiver console which also records the time of day and a reference pressure.

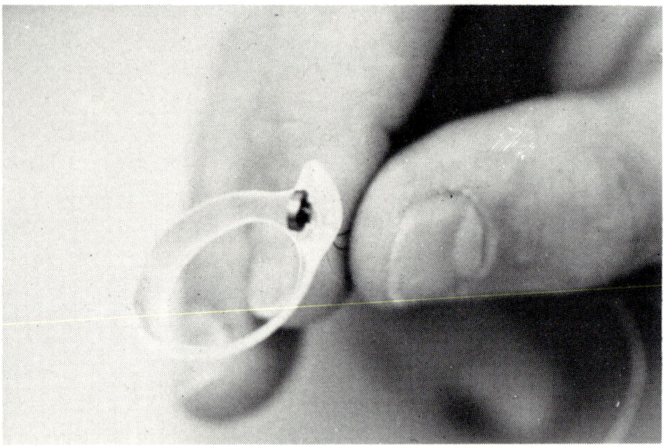

Figure 1. The applanation pressure transducer is held under the eyelid in a hydrogel fixture. Output wires in this version cross the eyelid to connect to an external telemetry transmitter.

Transducers for this project were based on the design of units which had been developed for use in the cardiovascular system (2). The transducer components are machined from titanium and consist of an outer ring (5 mm dia) supporting a thin small diameter diaphragm (0.06 mm thick x 3 mm dia). Two approaches to transducer case sealing were used. In one group of transducers, the metal components (whose radial clearance was 0.01 to 0.03 mm) were bonded with an epoxy compound. In a second group, the major interfaces were welded with an electron beam. Both techniques seemed to work well and resulted in transducers with zero drift of less than 0.5 mm Hg per day.

The best results in terms of eye tolerance and reliability of reading have been obtained using a holder made from the hydrophilic polymer material conventionally used for soft contact lenses. This material is obtained in a dry state and is machined using special tools on a milling machine and a hand grinder. The holder is polished and then hydrated in sterile saline. Its dimensions change 20 to 30% during hydration.

The telemetry transmitter consists of a voltage-to-frequency converter, a modulated oscillator and doubler, and timing and control circuitry. The present version, packaged on miniature printed circuit boards, is shown in Figure 2.

The voltage-to-frequency converter forms a subcarrier oscillator for the telemetry system. The transducer charges a capacitor in the feedback path of an operational amplifier. The capacitor discharge is controlled by a voltage sensing circuit (3). The radio frequency circuits consist of a crystal controlled oscillator operating at 51 MHz, and a tuned frequency doubler stage. The antenna is a printed loop around the circumference of the circuit board.

Figure 2. Electronics package, including signal processing, timing circuits, and radio transmitter.

RESULTS

This system has been evaluated in a series of comparison experiments with dogs in which a manometric measure of intraocular pressure was obtained by cannulating the animal's eye. Figure 3 shows the raw result of an animal evaluation experiment.

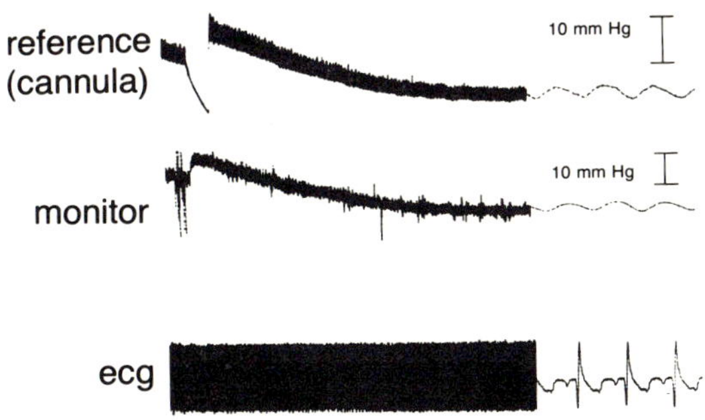

Figure 3. Invasive pressure (top) and noninvasive pressure (center) data. Cardiac pressure pulses are seen, correlated with ECG pulses (bottom).

Data reduction is accomplished by reading pairs of values, one each from the invasive measurement and the monitor system. These pairs of numbers are then analyzed using standard statistical regression techniques.

We have recently begun a trial of the tolerance of this system in human eyes (Figure 4). The two purposes of this trial are: to assess the tolerance of the holder by humans for periods of time up to 24 hours; and to ascertain the effects of wires which cross the eyelid. If the wires crossing the eyelid are not well tolerated, then it will be necessary to redesign the system so that the telemetry unit is self-contained in the holder under the eyelids.

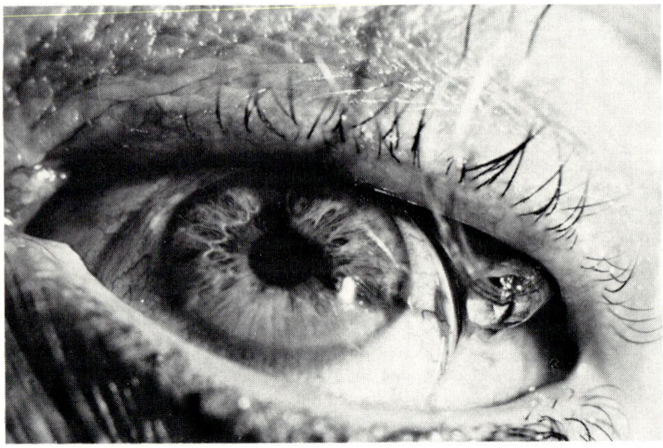

Figure 4. System (with connecting wires) in place in a human eye.

ACKNOWLEDGMENT

This project was sponsored by the National Eye Institute under Contract N01-EY-3-2115. The authors gratefully acknowledge the participation of H. Smith, M.D. and R. Head, O.D.

REFERENCES

1. Mackay RS, et al: Automatic tonometer with exact theory: Various biological applications. Science: 131, June 1960

2. Konigsberg E: Program to develop improved implantable pressure transducers. Report #NIH-69-2241-1, U.S. Govt. Printing Office, Nov. 1969

3. Grover TP: Precise voltage to frequency converter for telemetry applications of strain gage pressure transducers. IEEE Transactions on Biomedical Engineering: 441, 1975

DESIGN FEATURES OF A PRESSURE TELEMETERING HIP PROSTHESIS

Charles E. Carlson
Massachusetts Institute of Technology
Cambridge, Mass.
Robert W. Mann
Massachusetts Institute of Technology
Cambridge, Mass.
William H. Harris
Harvard Medical School
Boston, Mass.

The object of an ongoing project at MIT is to determine the spatial and temporal variations in the pressures generated between the load-carrying cartilage surfaces in the human hip joint. The purpose of this project is to investigate the stresses to which cartilage is subjected during normal, daily activities and the mechanical response of cartilage to these stresses. Information gained from these surface pressure studies will provide a better understanding of the behavior of normal cartilage and may suggest mechanisms that cause cartilage wear and degeneration.

PRESSURE MEASURING TECHNIQUE

The cartilage surface pressures will be measured experimentally by replacing the ball half of the hip joint with an instrumented prosthesis. One type of standard prosthesis, shown in Fig. 1, replaces the head and neck of the femur and functions in contact with a normal, cartilage-lined acetabulum. This particular prosthesis has a hollow ball, which is an ideal location for the necessary pressure-measuring instrumentation.

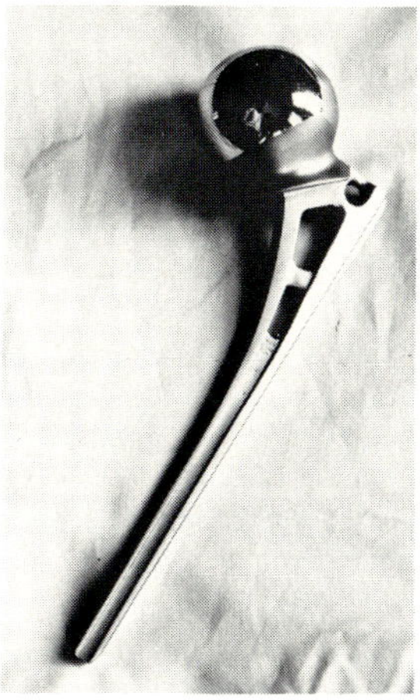

Fig. 1. Standard hip prosthesis.*

The pressure distribution generated on the ball of the prosthesis is measured by an array of fourteen pressure transducers distributed over the upper half of the ball. The signals from the transducers are transmitted to external receiving equipment by a radio telemetry device, also located inside the ball of the prosthesis. The pressure distribution can then be reconstructed by interpolating between the fourteen measured values.

PRESSURE TRANSDUCER DESIGN

A cross section of one of the fourteen identical transducers is presented in Fig. 2. The pressure sensing element of the transducer is a thin, spherical diaphragm that is machined directly into the wall of the hemi-

* See acknowledgements.

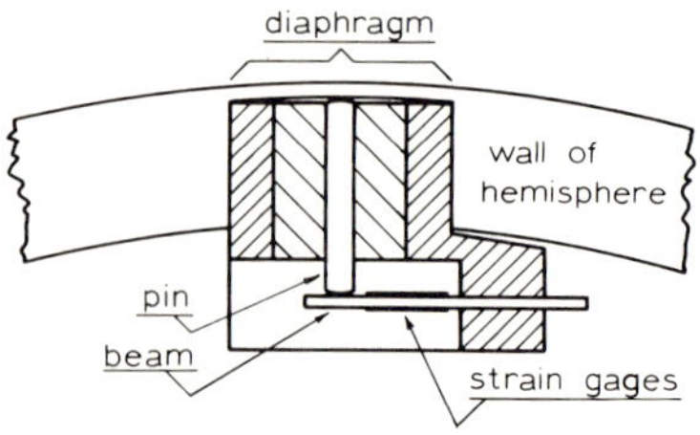

Fig. 2. Cross section of pressure transducer.*

sphere. The diaphragm bends inward in response to pressure on its external surface, and a sliding pin transmits the motion of the center of the diaphragm to the end of a silicon cantilever beam. This beam has four piezoresistive strain gages diffused into its surface and connected in a bridge configuration, thus producing an output voltage proportional to the applied pressure.

TELEMETRY DEVICE DESIGN CONSIDERATIONS

The telemetry device has to accommodate at least fourteen signal channels, fit inside the ball of the prosthesis and function reliably under intermittent usage for at least one year and preferably longer.

The circuit design process was greatly simplified by two initial decisions. First, in view of the relatively large number of signal channels, time division multiplexing was chosen to minimize circuit complexity and the number of components. The second decision was to power the telemetry device from an external source by means of a transformer power induction link. Powering the telemetry device from an external source eliminates the need for internal batteries with their limitations, such as low energy storage density, limited lifetimes and low voltages. The disadvantage is that some form of external primary transformer winding is then needed to power the device. For the instrumented hip prosthesis, however, this disadvantage is outweighed by the advantage of having available relatively large amounts of power at almost any desired voltage level.

PAM TIME MULTIPLEXING

The telemetry unit employs a pulse amplitude modulation form of time multiplexing in which the pressure information is carried in the pulse amplitudes. One frame of an idealized PAM waveform is shown in Fig. 3. Each frame has 16 channels. Fourteen are used for the pressure transducers and two are used for synchronization and calibration purposes. The beginning of each frame is identified by a synchronization pulse, a fixed amplitude pulse that is larger in amplitude than the 15 succeed- pulses. Immediately

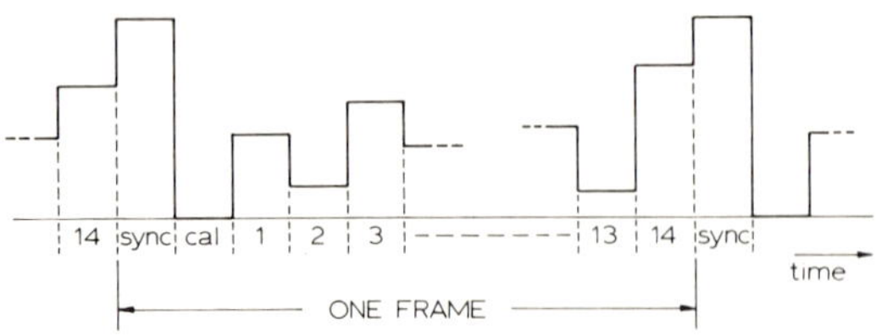

Fig. 3. Time multiplexed PAM signal.*

* See acknowledgements.

following the synchronization pulse is the calibration or zero level reference pulse from which all pulse amplitudes are measured. The remaining fourteen channels are assigned to the pressure transducers.

The height of the synchronization pulse as measured from the calibration pulse level corresponds to a known pressure. The pressure applied to a particular transducer is determined by scaling off that transducer's pulse amplitude against the known pressure level as represented by the synchronization pulse. By this technique zero baseline and reference signals are transmitted in each frame.

TELEMETRY UNIT BLOCK DIAGRAM

Fig. 4 shows the functional elements of the telemetry unit. The heart of the unit is the multiplexer, an IC device that sequentially connects one of 16 lines to a common line. The multiplexer is controlled by a pulse generator which sets the basic sampling rate. Each input pulse causes the multiplexer to advance one channel. The multiplexing is performed ahead of the pressure transducers, switching the inputs to the strain gage bridges rather than the outputs. This technique allows one 16-channel multiplexer to multiplex 16 bridges, whereas two 16-channel multiplexers would be required to handle the 16 differential output pairs from the bridges.

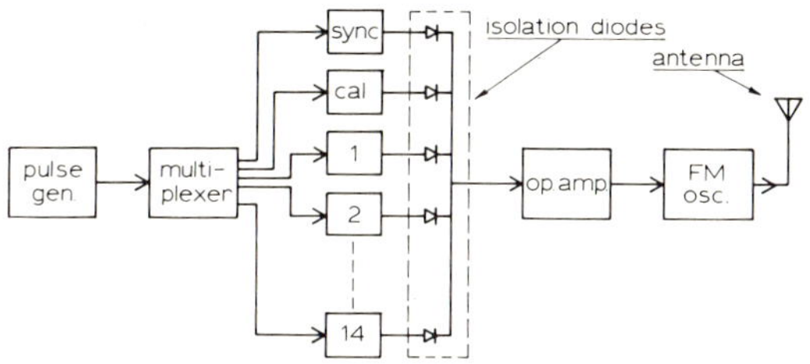

Fig. 4. Block diagram of telemetry unit.*

In order to isolate the output of the one bridge that is 'on' from all the other bridge outputs, a diode is put in series with each of the bridge output leads. The bridge that is 'on' forward biases its two diodes, while the diodes associated with the other transducers are reverse biased, effectively isolating the 'on' channel from the 'off' channels.

The output of the diode isolation stage is the PAM signal. This PAM signal is amplified by an op amp and then frequency modulates a 100 MHz Colpitts oscillator by varying the bias on the base of the oscillator transistor. This modulation technique is simple and gives adequate linearity. The oscillator stage does not have to be highly stabilized against drift, since the zero pressure reference is transmitted in each frame. The modulation level is adjusted so that the transmitted signal substantially fills the 200 kHz IF bandwidth of a standard hi-fi tuner, and the PAM signal is recovered directly from the ratio detector stage of the FM tuner. A more detailed description of the telemetry unit

* See acknowledgements.

Fig. 5. Assembled 16-channel telemetry unit.*

is presented in reference 1.

The 16-channel telemetry unit consists of five double-sided printed circuit boards. Eight IC devices and one discrete transistor are used, and chip resistors and capacitors are employed wherever possible. The assembled unit is shown in Fig. 5.

TRANSFORMER POWER INDUCTION

The telemetry unit is powered by means of a simple transformer arrangement that consists of an oscillator and amplifier operating at 100 kHz, a primary winding approximately 8 inches in diameter that can be slipped around the person's thigh, and a secondary winding that is mounted on the tip of the prosthesis stem inside a protective sleeve. The secondary winding also serves as the transmitting antenna for the 100 MHz signal. The instrumented prosthesis and the primary coil are shown in Fig. 6. Both primary and secondary coils are connected in parallel resonant circuits tuned to the driving frequency. The secondary winding is wound on a ferrite rod core to increase coupling efficiency.

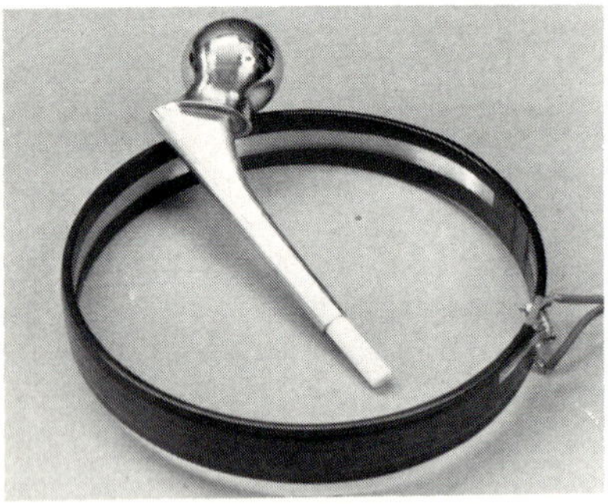

Fig. 6. Instrumented prosthesis and primary winding.

CONCLUSIONS

It should be emphasized that the telemetry unit used in the pressure measuring hip prosthesis was designed for a specific application, and no claim is made for its usefulness in other applications. Some of its design features, however, may be applicable to other telemetry devices, specifically a simple time multiplexing scheme to accommodate several signal channels, and transformer power coupling, which greatly eases the low power, low voltage constraints on circuit design.

ACKNOWLEDGEMENTS

This project has been supported in part by the National Institute of Arthritis, Metabolism, and Digestive Diseases of the National Institutes of Health, grant no. AM16116.

* Figures 1, 2, 3, 4 and 5 are reprinted by permission and were originally published in reference 1.

REFERENCES

1. Carlson CE, Mann RW, Harris WH: A Radio Telemetry Device for Monitoring Cartilage Surface Pressures in the Human Hip. IEEE Trans on Biomed Eng BME-21 no. 4: 257, July 1974

CLINICAL PEDIATRIC GAIT-BIOTELEMETRY

Fran Ford[1], Sal Rositano[2], and Gene Schmidt[3]

1. Children's Hospital at Stanford, Palo Alto, CA., 2. Medical Engineering, NASA-Ames Research Center, Moffett Field, CA., 3. Stanford Biomedical Applications Team, Palo Alto, CA.

INTRODUCTION

Electromyography is the recording of the electrical activity produced during muscular contraction. One use of electromyography has been to define periods of muscle activity during ambulation. The purpose of this paper is to discuss the application of gait biotelemetry to clinical pediatrics. A relatively inexpensive 4-channel FM-FM telemetry system is described to illustrate the utility and the limitations of biotelemetry in this setting. A review of the clinical experience with this equipment served as the basis for setting the performance characteristics for a more advanced, miniaturized multi-channel system.

BACKGROUND

Analysis of a particular walking abnormality by visual and hands-on inspection alone is frequently difficult even for the experienced medical clinician. This is particularly true in the case of children with cerebral palsy. The difficulties presented are found in determining which muscle movement pattern is contributing most to the child's walking problem.

During the past 2½ years the muscle activity pattern of 26 normal children and over 250 muscle groups in children with cerebral palsy were studied. The general approach using telemetry has proven to be extremely useful in evaluating walking patterns for pre-operative information, evaluating the potential of various types of braces, and in a pharmacological evaluation of a drug to decrease spasticity. Admittedly, there are many established approaches for conducting these tests. Our approach was aimed at immediate application in a clinical setting.

The main advantage of telemetry vs. hard-wire recording of EMG lies in the freedom of movement afforded the monitored subject. This is particularly important in children with neuromuscular impairment. The child with cerebral palsy not only has uncoordinated muscle movements which affect the walking pattern, but he also lacks a sense of balance and the muscle control used to protect himself when he falls. Telemetry, therefore, affords the opportunity for the unencumbered recording of the child's true gait pattern. Even the best designed hard-wire suspension system will cause alternations in the child's natural giat.

More importantly, that approach has a psychological impact, in that the child is constantly aware that he is being observed and tends to cover up his handicap and walk in a way unnatural to him. The results of this "laboratory walk" could lead the physician to recommend tendon lengthening surgery which would not produce optimal results.

MATERIALS (See fig. 1)

A 4-channel 15 to 350 Hz, FM to FM telemetry system with a range of 30 meters has been used. Three channels were used for the EMG transmission and the fourth for foot contact data. The foot contact data is the temporal reference for the muscle activity patterns recorded. The EMG potentials were sensed by miniature bio-potential skin electrodes and immediately fed into an isolation preamplifier to reduce artifacts. The pre-amp had a minimum input impedence of 200,000 ohms and an output impedence of 1,000 ohms. The amplifier was connected via shielded cable to the transmitter package which was worn on a waist belt. Within the package the potentials were amplified by a telemetry amplifier and multiplied onto an FM transmitter with a voltage control oscillator.

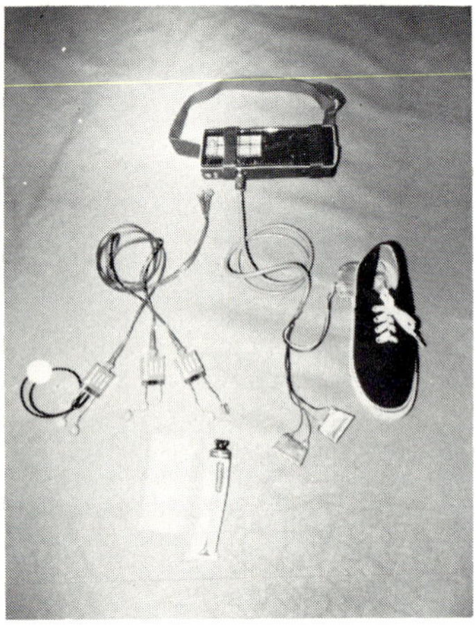

Figure 1. Four-channel waist belt transmitter with electrodes and foot switches.

Foot contact information was obtained by three small insulated brass or copper switches that closed when a minimum pressure of 5 lbs. was placed upon them. Closure of the switches modulated a 350 Hz oscillator to produce predetermined outputs for any combination of switch closure. This stairstep amplitude pattern was amplified and transmitted in the same manner as were the EMG potentials.

The signals transmitted from the child's waist pack were received and demodulated with sub-carrier discriminators. The demodulated signals were printed on a photographic strip chart recorder at a paper speed of 10 cm/second.

APPROACH

Children's Hospital at Stanford serves as a regional referral facility for pediatric patients referred by physicians throughout California. In particular, children with cerebral palsy are referred to the motion analysis laboratory for evaluation of gait abnormalities. The child is instrumented with the EMG telemetry and walks about. During ambulation a strip chart recording is made (see figure 2).

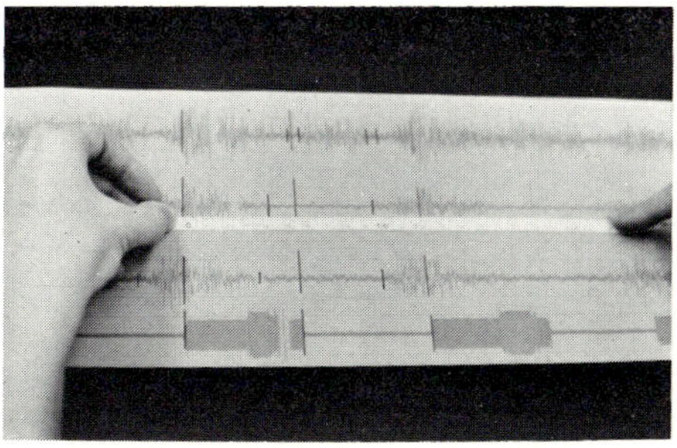

Figure 2. EMG and foot switch wave forms recorded via telemetry.

The upper three tracings show the phasic activity of the EMG wave form of the muscles under study. The bottom most tracing shows the foot switch signal. Zero amplitude indicates the foot was in the swing phase of gait and the various levels of amplitude indicate the part of the foot contacting the floor. Children with cerebral palsy often show excessive muscle activity or spasticity and this is shown by the prolonged EMG signal seen in the top line in the figure.

The data has been analyzed by hand. The muscle activity and inactivity periods and the foot contact sequence were identified for each gait cycle. The lengths of the muscle activity periods were then related to the length of the gait cycle as identified by the foot contact channel.

EQUIPMENT PROBLEMS

The size and weight of the transmitter package was a problem in our application. Remember the child with cerebral palsy not only has uncoordinated muscle activity which places his feet in awkward positions for walking, but he also often has a decreased sense of balance. The 1.6 kg weight of the waist pack transmitter was too much for some children. It's 17.5 x 8.8 x 3.8 cm size was too large to fit comfortably on an average 3 or 4 year old. During walking, the packet was jarred by the jerkey, uncoordinated movements of the child, causing signal artifacts and frequently breaking connectors and cables. The cables running from the leg sensors to the transmitter presented another problem. Wires long enough to fit a teenager had to be coiled up for use on a younger child; thereby the smallest child was carrying the greater load.

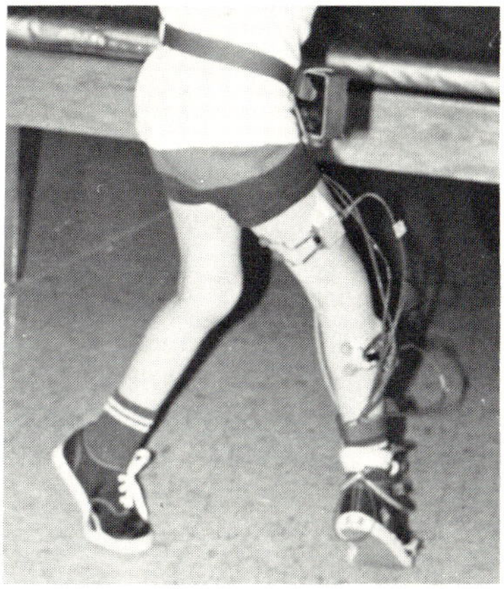

Figure 3. Transmitter in use showing cabling problem.

Another drawback of the system was the limit of 4 channels. This allowed the retrieval of gait data from only one leg at a time. This increased the recording time, increased the data analysis time, and decreased the total amount of information that could be retrieved.

Probably the most consistent problem involved the wires and connectors which were needed with the central transmitter. Because of the stresses imposed on the equipment by the characteristic--jarring and jerking of the child's walking pattern--there were repeated failures of both connectors and wires. This problem emphasizes the need to test electronic bio-monitoring equipment on the type of person to be evaluated, not on the work bench and not even on a normal subject.

SUMMARY

Analysis of the gait of children with cerebral palsy by the use of EMG is fraught with difficulties. Even the best designed hard wired suspension system will cause alternations in the child's natural gait patterns. More importantly, that approach has a psychological impact, in that the child is constantly aware that he is being observed and tends to cover up his handicap and walk in an unnatural way. The results of the "laboratory walk" could lead the physician to recommend a tendon lengthening surgery which would not produce the optimal results.

Experience over the past 2½ years with a simple telemetry system has shown that biotelemetry is indeed particularly suitable to monitoring the gait of children with cerebral palsy. The needs of the clinician can best be met by a telemetry system which: 1) can be rapidly applied and removed from the patient, 2) needs only to be turned on and off, 3) is small and lightweight enough to allow complete freedom of movement, and 4) can withstand repeated rough use.

Although the 4-channel waistbelt system described here has not met all of these criteria, it has proven the feasibility of applying telemetry to pediatric gait analysis and has helped to better define the specifications for an improved system utilizing individual miniaturized transmitters.

FUTURE DIRECTION

Based on our experience at Children's Hospital at Stanford, NASA-Ames Research Center engineers have designed a new wide-band, crystal controlled biotelemetry system for EMG analysis which employs individual miniature transmitters at the muscle group being monitored. The paper, A Wide Band EMG Telemetry System by Westbrook, Fryer, and Rositano, in Session VIII of this proceedings describes the engineering aspects of this new approach.

ACKNOWLEDGEMENTS -- The authors gratefully acknowledge the contributions of the United Cerebral Palsy Foundation, Edward P. Luzzi of L & M Electronics Co., and NASA-Ames Research Center. Manuscript preparation by Carla Dunham.

PATIENT MONITORING DURING BATH THERAPY

Leonid I.Tatarenko, Vladimir V.Poltoranov, Vadim I.Lebedev, Alexander A.Vorobyov

Research Biotelemetry Laboratory, Kislovodsk, USSR

A well-developed network of health resorts in the USSR provides large facilities for millions of people who need rest and sanatorium treatment. A typical treatment given to patients in health resorts is the treatment by baths and waters of mineral springs.

Balneotherapy (bath therapy) has obvious strain effects on cardiovascular system which explains the necessity of permanent control of heart activity in patients with both prevalent and concomitant heart diseases. That implies that monitoring devices which enable control of patients during balneotherapy are certain to command a ready market.

Monitoring systems that have been worked out take into account, along with arterial pressure, also heart contraction frequency, the indicator we consider more informative of patient's condition than peripheral pulse rate. Following requirements must be met when monitoring devices of control of patients with heart diseases are elaborated:

1. Heart contraction frequencies must be checked on the basis of bioelectrical manifestations of heart activity (for instance, by ECG);
2. Methods of bioinformation taking should not cause discomfort for the patient and must be sufficiently simple;
3. Electric security of the patient must be ensured;

4. signal registration methods must be reliable.

We have made an experimental model of a 4-channel monitor; a simplified block diagram of a monitor channel is given in Fig.1.

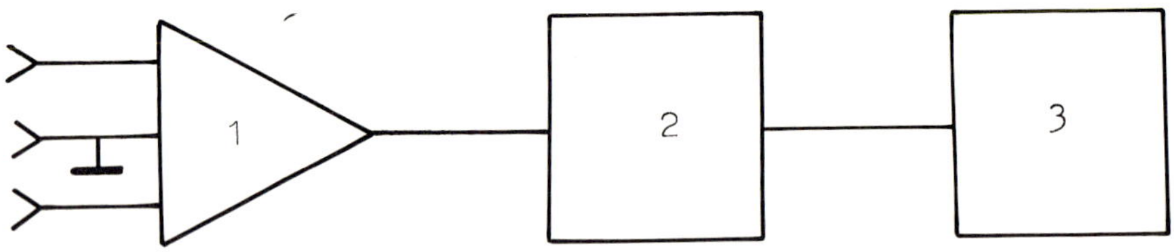

Fig.1. A simplified functional diagram of a monitor channel. 1 - Selective Amplifier and Signal Former; 2 - Patient Electric Protection Block; 3 - Information Analysis and Signalling Block.

The first and the second requirements are met by registration of bioelectrical manifestation of heart activity with the help of electrodes placed directly into the liquid medium of the bath where a patient is lying. By application of a selective band amplifier (with the signal frequency range between 14 and 21 Hz), a rejector filter for the industrial current frequency and an impulse signal former we obtained a reliable filtration and a reliable bioelectric signal selection with the amplitude of about 20.10^{-6}V.

To ensure electric security of the patient, a Patient Electric Protection Block was used (see Fig.1). The block includes high-frequency transformers that supply power to the amplification and preliminary formation of the signal and also transfer data on heart activity to the Information Analysis and Signalling

Block. The transformers have been put to an additional test by a voltage of 4 kV.

Other measures that contribute to the reliable registration of bioelectric signals include

1. careful choice of material for electrodes (Ag/AgCl electrodes have been used),

2. coating of active electrode surface with a thick conductive jelly and

3. careful placing of electrodes in the bath (Fig.2).

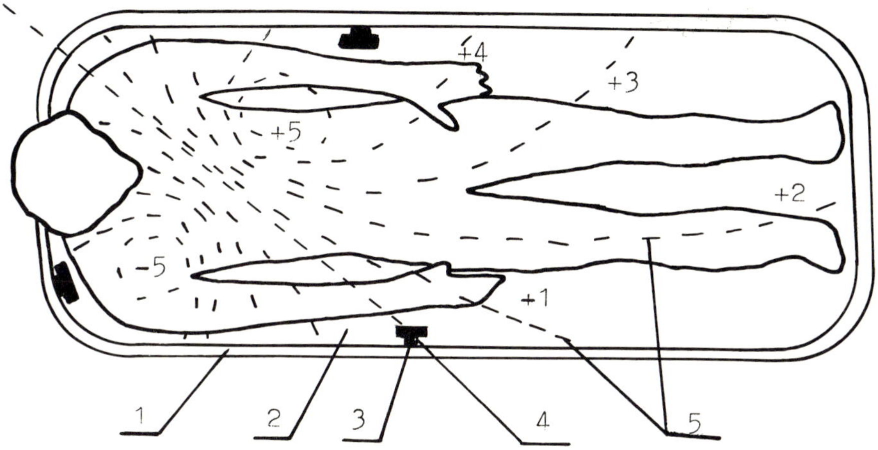

Fig.2. Diagram of electrode arrangement in the procedure bath. 1 - bath, 2 - procedure medium, 3 - electrodes, 4 - electric axis of the heart, 5 - equipotential lines of the heart generator.

ACKNOWLEDGEMENT

The authors thank Prof. Vladimir E. Manoylov whose consultations in problems of electric protection proved most helpful.

EMERGENCY MEDICAL SERVICES

Eugene L. Nagel

Department of Anesthesiology, Harbor General Hospital
Torrance, California

The need for EMS was summarized in a National Research Council report in 1966. There were several attempts to provide EMS both abroad and in this country. These initial attempts were, in essence, experimental probes in which essential elements of the system were identified and their interrelationships studied.

For purposes of the national design, it is estimated that some 350-400 regional EMS systems will be needed to entirely cover the country. The average size of the regions will be approximately 10,000 square miles and the average population about 500,000 persons. At the extremes, there is Los Angeles County with approximately 7 million persons living in about 2,000 square miles and, by contrast, the State of Idaho with about 700,000 people in an area nearly eight times larger than the idealized region.

An estimated 65 million persons request services annually; 80% of these require some type of primary care, the remaining 20% are true emergencies which require time dependent treatment. One-fourth of this latter group are critically ill or injured and would not be salvagable a few years ago. Available data suggest that an additional significant portion of these may be saved if definitive emergency care is provided early in the course of their illness or post-traumatic period. Specific high risk categories demanding specialized care include: care of accident victims; specific care of the burn victim; acute coronary care; care of high risk infants; patient management for poisonings, drug overdoses, acute alcohol intoxication, and psychiatric emergencies.

As of July 1975, some 89 regions entered EMS planning phase, 93 regions began operating their systems, and 27 regions are now in the expansion and improvement phase (1). On a cumulative basis, 120 systems or 30% of the national goals of 400 systems that serve a population estimated at 60 million are in a development phase. Since there are systems that receive no federal support, the final figures may be 10-15% more than these.

Monitoring systems in basic life support are designed to provide essential diagnostic data without resorting to invasive methodology. The modalities include observation, limited history and physical diagnosis, use of the stethoscope and sphygmomanometer for palpatory, oscillometric and auscultatory blood pressure. There are a number of therapeutic modalities which could be considered invasive but which are increasingly employed to basic life support systems. These include oxygen-powered ventilation equipment (2), oropharyngeal suctioning, esophageal obturator airway, military anti-shock trouser (MAST) (3), and pneumatically-operated cardiac assist device (4).

Advanced life support is considered to be emergency medical treatment, including invasive techniques, that are necessary to support life. The care is delivered by an advanced level emergency medical technician termed EMT-paramedic. There are approximately 20 states that have enacted enabling legislation so that advanced level care can be offered to accident victims. There are concurrent programs for continuing education curricula for emergency department nurses. Finally, there are over 30 residency programs to train physicians to deliver emergency medical care.

Table 1. Common Modalities of Treatment

Airway	naso & oropharyngeal airways oxygen-powered ventilators esophageal obturator airway orotracheal intubation tracheal suctioning
Intravenous Therapy	peripheral venous catheter placement crystalloid and colloid fluids pressure infusion blood sampling (mainly blood sugar and typing)
Intracardiac Therapy	usually limited to epinephrine
Drugs for Intravenous Use	alkalizing, vasopressors & cardiotonic drugs, steroids, anticonvulsants, analgesics, vagal blocking, sympatholytic, narcotic reversing drugs
ECG	obtaining, interpretation and transmission of electrocardiographic rhythm strips defibrillation

Monitoring necessary to diagnose properly and to control these therapeutic modalities includes those offered in the basic life support program plus the electrocardiograph (ECG). The ECG leads usually consist of limb leads (commonly lead II), although precordial leads are utilized in some systems. The necessity to disrobe the victim partially prevents some systems from adopting the precordial lead in the prehospital environment. The single lead can be displayed on an oscilloscope or, more recently, viewed on a television-type monitor as a non-fade or moving window display. In the victim with acute collapse, the signal is usually taken from the precordium through the defibrillator electrodes.* As soon as limb or precordial electrodes are placed, the signal immediately can be displayed, written on a single-channel, direct writing oscillograph or transmitted by telephone or radio to a distant control point. The necessity to display this analog (ECG) signal at a remote point deserves explanation.

The paramedic is trained to obtain essential physiologic data and to relay this data accurately to the medical control point. In most systems, medical control is usually hospital-based physicians although there are exceptions (5). Increasingly, these hospital-based physicians are located in the emergency departments or critical care units. They should be conversant with the mission, training, equipment and capability of the paramedic. In addition, they need training in limited diagnosis utilizing data transmitted by voice plus the analog ECG signal (6,7). The system needs two-way voice communication for the physician to direct the diagnostic algorithms and appropriate therapy. Paramedics are trained to recognize patterns, symptom constellations and, when appropriate, to suggest diagnoses to their medical control. The emergency department or intensive care unit nurse, trained in emergency medicine, may be utilized to offer limited medical direction. Some systems utilize standing orders for easily recognized situations such as cardiac arrest or trauma requiring intravenous fluid therapy.

The transmission of analog data to a distant point is seen as necessary to medical diagnosis of the victim. As the data is received by the physician, the victim, in a sense, becomes a patient of an identified physician.

Telemetry of the ECG from the scene was originally carried out to permit defibrillation and definitive care of the arrested victim. The necessity to do this via radio rather than telephone channels related to prehospital environment and lack of telephone channels near all victims. The radio channels originally utilized were in the 150 mHz band and were licensed by the FCC on an experimental basis only. New rules were issued that established a medical services category in the UHF band.

In the urban setting, radio channels tend to be few in number, but the almost ubiquitous telephone has become increasingly important. The modulated ECG signal may be transmitted over conventional telephone channels utilizing relatively inexpensive and simple interfacing equipment. This modality has become so well-accepted that nearly all manufacturers of emergency ECG equipment (i.e. portable radios for field telemetry use, defibrillator-scope monitors) can provide interface devices with 50 feet or more of wire for attachment to the conventional telephone subset. It is now proposed that very low-power telemetry units be provided for transmission of voice and/or ECG from victim location to nearest telephone (i.e. about 1,000 feet). This would permit overcrowded urban units to increase their useage of telephone

*Physio Control Quick Look[R]

channels, and thus preserve the fixed number of radio channels for essential radio use.

In rural areas, crowding is not the main problem. New UHF systems for large regions previously serviced by VHF means completely new equipment and, usually, a different communication system. These new systems are expensive and may not be utilized frequently enough to present an attractive cost/benefit effectiveness.

The selection of monitoring equipment appropriate to the field becomes a question of optimizing a series of conflicting requirements. The data should be relatively essential and the equipment should be light, rugged, dependable, battery-operated, easy to apply, self-correcting, fail-safe, and have appropriate limit alarms. Equipment designed to process the ECG can probably be improved upon to be more reliable than present methods and to furnish heart rates with alarm limits. Simultaneous transmission of three channels of data by multiplex techniques suitable for computer readout might be an advantage. The furnishing of limited biochemical data in the field and the monitoring of ventilation functions could be useful additions. The technology now available is probably adequate to meet many of these needs. The proper adaptation of this technology to the present and future needs of the prehospital systems will require further study of present systems, outcomes and needs.

REFERENCES

1. Department of Health, Education and Welfare, PHS-HSA/HRA: Annual Report Fiscal Year Ending June 30, 1975, Emergency Medical Services Systems Act (PL 93-154)

2. The Emergency Care Research Institute: Evaluation: oxygen-powered resuscitators. Health Devices 3(9): 207, 1974

3. Kaplan BC, Civetta JM, Nagel EL, et al: The military anti-shock trouser in civilian prehospital emergency care. J of Trauma 13(10): 843, 1973

4. Burch LH, Kenney LJ, Doornbos F, et al: A study of external cardiac compression. J Mich State Med Soc 61, Nov, 1962

5. Hirschman JC, Nussenfeld SR, Nagel EL: Mobile physician command: a new dimension in civilian telemetry-rescue systems. JAMA 230: 255, 1974

6. Nagel EL, Hirschman JC, Nussenfeld SR, et al: Telemetry-medical command in coronary and other mobile emergency care systems. JAMA 214: 332, 1970.

7. Lewis AJ, Ailshie G, Criley JM: Prehospital cardiac care in a paramedical mobile intensive care unit. Calif Med 117: 1, 1972

AN OVERVIEW OF PARAMEDIC COMMUNICATIONS IN LOS ANGELES COUNTY

W. D. Hobbs

Department of Communications
County of Los Angeles

Paramedical service is a new and rapidly growing part of pre-hospital care in the nation today. Communication between the paramedic in the field and the physician or nurses at the base hospital constitutes a vital link in providing successful pre-hospital care. Such communication includes the transmission of patient data to the hospital and medical instruction to the paramedic. The role of biotelemetry is of particular significance in the transmission of patient EKG data to the base hospital for evaluation by qualified medical personnel.

BACKGROUND

The Los Angeles County paramedic program commenced in 1969 with two pilot efforts, one based at Daniel Freeman Hospital and the second at Harbor General Hospital. Nurses were sent on emergency runs until the passage of the Wedworth-Townsend Act in July of 1970. Under this Act, paramedics were permitted by law to carry out specific medical procedures, provided they maintained two-way voice communications with a physician or specially trained nurse. Although both pilot efforts were successful, the paramedic program grew rather slowly until 1972, at which time there were eight paramedic units operating within the County. The paramedic program expanded rapidly over the next four-year period to its present size. There are 25 base hospitals and 115 paramedic units operating in Los Angeles County today, serving over 7,000,000 people.

The rapid growth of the paramedic program placed heavy demands upon communicators, among others, involved in engineering, planning, and providing communications equipment. In 1972, the Federal Communications Commission allocated five channels for biomedical communications and telemetry. For a short period of time, these five channels proved to be adequate. Realizing that some sort of frequency coordination was necessary to avoid the impending problems of co-channel interference, the County, together with several participating agencies, formed an ad-hoc committee to resolve communications problems and coordinate channel assignments. This action, fortunately, helped to reduce interference problems. Later the same year, as more base station hospitals were added, incidents of co-channel interference began to increase.

By early 1974 it became apparent that the "system" could not tolerate the addition of more hospital radio base stations. The

Los Angeles County Department of Communications, together with the Department of Health Services, recommended that a moritorium be placed on the addition of hospital radio base stations, until such time a coordinated plan and system could be designed. There were, at this time, approximately 25 hospital base stations and 55 paramedic units operating in the field. About the same time, the FCC allocated three additional channels to the medical services category, thereby expanding available channel resources.

Later in 1974 funds were made available for the design of a Countywide coordinated paramedic communications system. Subsequently, a consulting firm was selected and contracted to provide such a design. Contractor effort commenced in mid-1975 and continued through April 1976.

PRESENT EQUIPMENT CONFIGURATION

The typical paramedic base hospital operates a single channel (two frequency) transceiver in the 463-468 MHz band with 50-60 watts of transmitter RF power. An omnidirectional, gain antenna is used in nearly all locations. This effective power level is substantially more than deemed necessary and contributes, together with the single channel limitation, to the co-channel interference problem between base stations. The base station is usually located in an equipment room atop the hospital and near the antenna, and is remotely controlled from a console(s) located in the hospital emergency department. In addition to the radio remote control console, there is a console which contains demodulating and display devices that are used in the receive mode. Such devices include an EKG demodulator, CRT display and strip-chart recorder for viewing EKGs, and an audio tape recorder for recording both voice and EKG data. The console also has provision for interfacing with the standard telephone, permitting communications with the paramedic via commercial telephone systems.

The typical paramedic radio device or "orange box" contains a transceiver with an RF power output of 12 watts. The unit is approximately the size of a portable typewriter case and weighs 25 lbs. In addition to the transceiver, the device contains a duplexer, which permits full-duplex operation by the paramedic, and an EKG amplifier/modulator with connecting patient leads. Powered by rechargeable NiCad batteries, the unit is capable of continuous RF transmission for approximately one hour. A handset, equipped with push-to-talk (PTT) capability, is employed for voice operation. A selector switch on the front panel permits voice PTT, voice-only, EKG-only modes of trasmission. Multiplexing of voice and EKG is not a feature currently employed in the "orange box" used in Los Angeles County. Nearly all portable devices have eight-channel capability, while only a few devices include Continuous Tone Controlled Squelch Systems (CTCSS). The newer eight-channel devices are capable of functioning as repeaters, when used with the handie-talkies that are provided to the paramedics.

As an adjunct to the "orange box", each paramedic unit carries an acoustical telephone coupler which interfaces the portable device, via a 50-foot cable, with an ordinary telephone. This feature allows the paramedic to transmit voice and EKG data, and receive voice instructions via commercial landline telephone. Unfortunately, this coupler is inconvenient and time-consuming to utilize, and does not provide adequate audio output to compensate for excessive telephone line loss and noise experienced in rural areas. It is estimated that the telephone coupler device is used less than 20% of the time for voice and EKG communications.

TRAFFIC CONSIDERATIONS

A recent tally indicates that nearly 20,000 paramedic runs are dispatched monthly; however, many are false alarms, or deemed minor and not reported, or reported via another agency. It has been estimated that 14,000 paramedic runs per month is a more realistic figure. A communications survey of County records for the month of November 1975 shows that 54% of the runs reported required some form of communications. Twenty-two percent used voice only, while the remaining 32% required both voice and EKG communications. Of the total runs reported, 45% used radio as the medium, while 9% utilized the telephone. Based upon these figures, an estimated 150 paramedic runs requiring EKG communications, using both radio and telephone, are conducted daily.

Radio frequency channel loading is rapidly becoming a matter of serious concern. At present there are an estimated 200 "radio" runs conducted daily on the existing eight radio channels. Recent traffic studies indicate that the number of paramedic runs will double by the 1980-82 time period. Time-sharing of radio channels will become more commonplace, unless additional channels are allocated by the FCC or an improved telephone coupler device is used as a primary means of communications.

COUNTYWIDE COORDINATED COMMUNICATION SYSTEM

The recently completed system design calls for a Communications Coordination Center (CCC), capable of dynamically assigning channels to each base station. Rather than call the base hospital directly on a preassigned channel as he does now, the paramedic will contact the CCC on Channel 9 (paramedic coordination voice network), request a channel and a paramedic base hospital. The CCC will, in turn, assign a channel and alert the hospital as requested. Although somewhat more time-consuming, this method will assure the paramedic of an interference-free channel, if one exists. Coordination between the CCC and the base hospital will take place by landline, or via radio on Channel 10 in case of landline failure. Redundant and extended radio frequency coverage will be provided by mountaintop base stations operating on Channel 7, with Channel 8 as a back-up channel. Paramedics in airborne helicopters will communicate on Channel 8, using Channel 7 as a back-up channel. All paramedic-to-base hospital calls will be attempted on Channels 1 through 6. Base hospitals will be configured in a small-cell arrangement, transmitting with approximately 25 watts (ERP) on all eight channels. Channels 7 and 8 may be used by hospitals in the small-cell configuration only when these channels are not in use by a mountaintop base station.

Bench testing of typical base station and portable devices was included as part of the overall systems design effort. Tests were conducted to determine: a) the minimum acceptable received carrier-to-noise ratio for both voice and EKG, and b) the minimum RF signal-to-interference ratio acceptable for both voice and EKG. The test data were reviewed by engineers, as well as certified base station nurses experienced in reading EKG charts. The test results showed that the EKG data were more critical and would tolerate less noise. Specifically, the minimum acceptable carrier-to-noise ratio was determined to be 11 dB. The minimum acceptable signal-to-interference ratio was determined to be 8 dB, providing a low-pass noise filter was used at the output of the EKG demodulator. These data were later used to establish talk-out and talk-back criteria, as well as co-channel interference zones.

BIOTELEMETRY SYSTEMS

Two major vendors supply the vast majority of paramedic radio and biotelemetry equipment in Los Angeles County. With but few exceptions, both vendors follow the same schema and specifications: subcarrier: 1400 Hz or 1800 Hz - FM modulated; deviation: 50 Hz/mv ± 250 Hz maximum or 150 Hz/mv ± 450 Hz maximum; multiplex: not employed; demodulator sensitivity: -38 dBm to -8 dBm or 0 dBm; frequency response: 0.1 to 100 Hz; RF carrier deviation: ± 4 KHz. Subsequent to the release of the IRAC biotelemetry standard, the vendor supplying the 1800 Hz subcarrier equipment changed over to a 1400 Hz subcarrier.

CONCLUSION

Biotelemetry has played a key role during the growth of the paramedic program in Los Angeles County. The initial pilot programs placed strong emphasis on the telemetering of EKG data, setting the tenor for the years of growth that followed. The Countywide coordinated communications system design supports this sustaining requirement for EKG data transmission. For the foreseeable future, biotelemetry will continue to play an important role in the Los Angeles County paramedic program.

REFERENCES

1. System Applications, Inc: Operational Requirements for Paramedic Communications in Los Angeles County, October 1974.
2. ITT Federal Electric Corporation: Design of a Basin Paramedic Communications System, April 1976.

DEVELOPMENT AND IMPLEMENTATION OF A STANDARD FOR EMERGENCY MEDICAL TELEMETRY

Gordon W. James

Motorola, Inc., Plantation, Florida

Introduction The Office of Telecommunications Policy recently published a proposed standard for emergency medical telemetry (1). Motorola participated in the meetings held to develop this standard and made a number of recomendations which were ultimately incorporated in the final draft. This paper reviews some of the material presented in support of our recomendations and discusses the implementation of a multiplex telemetry system.

Basic Guidelines A standard for Emergency Medical System (EMS) telemetry must be consistent with a number of basic requirements. The telemetered ECG must be of sufficient quality to allow diagnosis of life threatening arrhythmias. This quality should be maintained when the signal is transmitted over a noisy channel which provides usable voice communications. Telemetry should be accomplished by a method which uses simple low cost hardware. It must be compatable with the 300 - 3000 Hz frequency response of voice grade radio and telephone equipment and it must meet FCC and telephone operating company rules and regulations. The method used for telemetry should allow multiplexing voice and ECG for simultaneous transmission. Multiplex equipment should be compatable with non-multiplex equipment.

Method of Telemetry While a number of methods can be used for ECG telemetry (2), the current method of choice because of hardware simplicity and compatability with standard radio and telephone equipment is FM modulation of an audio subcarrier. All of the major EMS telemetry systems in operation at this time use this method of telemetry. It gives good quality over noisy channels and can be adapted to multiplex operation. In order for any other method to be adopted as a standard it would have to be justified by clearly superior operation and competitive cost.

Having determined to use an FM modulated subcarrier for telemetry, it is necessary to choose a center frequency and deviation sensitivity. Deviation sensitivity is determined by the required ECG quality. Choice of a subcarrier is dictated by prior uses of frequencies in the 300 - 3000 Hz band as well as the multiplexing requirement.

ECG Quality The required ECG frequency response and dynamic range for EMS telemetry is subject to a number of qualitative judgements. It is clearly desirable to limit the frequency response so as to suppress artifacts likely to be present when an ECG is taken under nonideal 'emergency' conditions. The limited response must not however interfere with diagnosis. A consensus of medical opinions presented to the standards committee (1) led to accepting a .1 to 40 Hz frequency response and a dynamic range of ± 5 mV. The ECG signal to noise ratio was specified by requiring less than 10 mV RMS of noise when an unmodulated subcarrier having a 20 db signal to noise ratio is being received. Assuming the noise amplitude of the recovered ECG to have a Gaussian distribution the baseline noise would then be within a range of ± 20mV for 95% of the time (3). In terms of an ECG trace having a 1 mV deflection equal to 1 cm, the peak to peak noise would be less than .5 mm.

Subcarrier Deviation Sensitivity The theoretical subcarrier deviation required to yield the desired 10 mV RMS baseline noise can be calculated using the formula (3):

$$(S/N)s = 3\beta^2 (S/N)c \qquad (1)$$

In this formula (S/N)s is the recovered signal to noise ratio, (S/N)c is the carrier to noise ratio of the unmodulated subcarrier to noise contained in a band equal to twice the highest modulating frequency, and β is the modulation index which is given by;

$$\beta = (\text{subcarrier deviation})/(\text{Modulating frequency}) \qquad (2)$$

For this calculation the signal is a 5 mV peak sine wave at 40 Hz. (S/N)s is thus 51 db. Assuming the noise from 300 to 3000 Hz to have a uniform frequency distribution, the carrier to noise ratio is 20 db plus a 15 db improvement in going to the 80 Hz bandwidth. (S/N)c is thus 35 db. The factor 3 must then yield an improvement of 16 db. Since the theoretical improvement is about 3 db more than what is usually measured β will be calculated from:

$$10 \log(3\beta^2) = 19 \text{ db} \qquad (3)$$

This yields a value for β of 5.1. Using equation 2 the deviation is 205 Hz for 5 mV or 41 Hz for 1 mV. Allowing for some margin of safety the final recomended deviation is 50 Hz/mV.

Since radio receivers are often specified in terms of the signal required to yield a 12 db ratio of signal + noise + distortion to noise + distortion (12 db SINAD) it is interesting to calculate the noise for this condition using the 50 Hz/mV figure. In terms of the 1 cm deflection the peak to peak noise is .6 mm which is still acceptable.

The bandwidth required for telemetry can be calculated in two ways. Because of the nature of the ECG signal there is very little amplitude at 40 Hz. The maximum deviation produced by 40 Hz components can be taken as 50 Hz. This yields a modulation index of slightly more than 1. There will be three significant sidebands in the FM signal and the total bandwidth required will be 6 x 40 = 240 Hz. In contrast, low frequency components around 10 Hz may deviate the subcarrier by 250 Hz. The modulation index is 25 which is very large and the required bandwidth is 2 x 250 or 500Hz. In this case the maximum bandwidth is determined by the low frequency components of the signal.

Subcarrier Frequency Much of the discussion regarding a telemetry standard has been concerned with the proper choice of a subcarrier for multiplex . Because a multiplex telemetry system requires removing a portion of the audio spectrum equal to the bandwidth required for telemetry, voice quality will be degraded. Both subjective quality and intelligibility may be effected. Subjective quality is closely associated with the peak energy spectrum of speech which lies almost entirely below 1000 Hz. Intelligibility is spread over the spectrum from 200 to 6000 Hz. Table 1 is a tabulation of 20 frequency bands which contribute equally to speech intelligibility(4). Because speech is highly redundant a number of these bands can be removed without seriously effecting intelligibility. A standard voice channel removes approximately five bands yet the error rate for typical messages is less than 2% under good signal conditions and remains low even in the presence of noise.

From the table it can be seen that removing speech from 2500 to 3000 Hz would remove two bands, Removing speech from 300 to 800 Hz would remove four bands. Removing speech from 1150 to 1650 removes three bands. It is apparent that a high frequency would be best but that the difference is not great. The difference in expected error rates for a system removing four bands as opposed to a system removing two is only 2% (5). For this reason more consideration should be given to other factors.

The band of frequencies from 2450 to 2750 cannot be used for telemetry because telephone regulations require that the power in this band not exceed power in the range from 800 to 2450 Hz. Frequencies below 1000 Hz should not be used in order to preserve subjective quality. Frequencies from 1800 to 2300 Hz are used for signaling and control in many radio systems and should therefore be avoided if possible. Because of these considerations the recomended standard is 1400 Hz. The bandwidth occupied by telemetry is from 1150 to 1650 Hz.

TABLE 1

Band	Limits	Width	Band	Limits	Width
1	200-330	130	11	1660-1830	170
2	330-430	100	12	1830-2020	190
3	430-560	130	13	2020-2240	220
4	560-700	140	14	2240-2500	260
5	700-840	140	15	2500-2820	320
6	840-1000	160	16	2820-3200	380
7	1000-1150	150	17	3200-3650	450
8	1150-1310	160	18	3650-4250	600
9	1310-1480	170	19	4250-5050	800
10	1480-1660	180	20	5050-6100	1050

Implementation of Multiplex System Figure 1 shows a block diagram of a complete telemetry system which can operate in either a multiplex or non-multiplex mode. In this diagram the microphone includes pre-emphasis and limiting as required by FCC rules. Microphone audio is routed directly to the transmitter for non-multiplex operation, Gain of this path is set so that a full 5kHz deviation of the radio carrier is obtained. For multiplex operation the audio is passed through a notch filter which has 20db of attenuation between 1150 and 1650 Hz. This filter prevents interference between the audio and the telemetry subcarrier. Audio from the notch filter is summed with the telemetry subcarrier . The two signals are adjusted so that each deviates the radio carrier by 2.5 kHz.

At the receiving station the subcarrier is detected and demodulated to yield the ECG signal for display. When a subcarrier is detected the speaker audio is routed through a notch filter which has 40 db of attenuation between 1150 and 1650 Hz. This removes the subcarrier from the voice on a multiplex transmission. If no subcarrier is transmitted the speaker audio bypasses the filter.

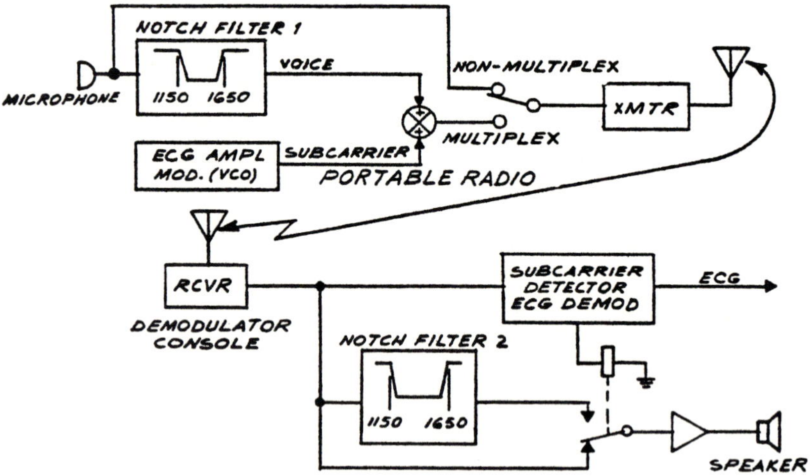

Figure 1

References

(1)A Report on a Biomedical Telemetry Standard for Emergency Medical Communications, Available from Office of Telecommunications Policy.

(2)Watts, R.N., "Some Design Considerations for Narrowband Medical Telemetry Over the Switched Message Network", IEEE Trans on Communications Technology, Vol COM-19 No. 3, June, 1971.

(3)Schwartz, M., "Information Transmission, Modulation , and Noise" McGraw-Hill, 1970.

(4)Beranek, L.L., "The Design of Speech Communications Systems", Proc. of the IRE, Vol 35, 1947

(5)"Methods for the Calculation of the Articulation Index", American National Standards Institute, ANSI S3.5-1969.

DISCUSSION

PHILLIPS: Dr. Zerzaway, what component of isometric exercise is the main contributor to increased arterial pressure?

ZERZAWAY: Cardiovascular reaction to isometric exercise was first described in the early sixties. This was assumed to be a special vascular reflex, the cause of which is not known, but increased potassium outflow from the muscles was thought to be a factor.

NAGEL: Sudden death in athletes is maximum during isometric exercise. It has to do with decreased coronary artery blood flow and increased myocardial oxygen demand.

UNIDENTIFIED: Dr. Neuman, to what do you attribute the higher reliability of radiotelemetry relative to wired systems?

NEUMAN: We attribute it primarily to the reduction in 60-cycle interference.

UNIDENTIFIED: Have you given any consideration to using noninvasive transducers so that labor contractions in pregnant women could be monitored without rupturing the amniotic bag?

NEUMAN: Yes, we have filed an application to develop means to do that very thing. Presently, we introduce the transducer directly into the anmiotic sac.

SESSION IV

NEUROLOGICAL APPLICATIONS

Chairmen

J.A.J. KLIJN, Ph.D.

A.M. DYMOND, Ph.D.

EDITOR'S OVERVIEW

This session represents a continuation of the clinical applications theme, with the difference that it concentrates on neurophysiological data - primarily intracranial pressure, electromyography and electroencephalography. The papers deal with signal measurements, interference, and data-handling and analysis. The benefits afforded by miniaturization are again evident, especially in long-term monitoring of ambulatory psychiatric and epileptic patients and in particular intracranial pressure measurements over long periods of time.

Klijn's paper on powerline interference presents a novel analysis of this problem and suggests the use of a counter-acting voltage to create a substitute "Faraday Cage" with a potential equal to that of the ground of the recorder. The substitute potential might serve to minimize the powerline interference.

The two data-handling papers both consider the use of computers to speed the process of EEG data scanning and epileptic-seizure identification. A paper by Ives deals with the dual use of a computer to present a quick tracing of the EEG data for immediate (within 4 seconds) inspection and an IBM-compatible digital tape which creates a permanent library record of the seizures. Weaver discusses a computerized scheme by which EEG recorder tapes may be played back at 50 times the recording speeds for rapid analysis.

MULTICHANNEL BIOTELEMETRY OF ELECTROPHYSIOLOGICAL PARAMETERS FROM UNRESTRAINED PSYCHIATRIC PATIENTS

A.M. Dymond, R.W. Coger, J.P. Martin, E.A. Serafetinides

Veterans Administration Hospital Brentwood
Los Angeles, California
AND
Department of Psychiatry and BRI, UCLA

Multichannel biotelemetry provides a number of advantages over traditional hard-wired monitoring techniques (1,2). It provides greater flexibility for patient monitoring, while substantially reducing 60Hz and movement artifacts. Telemetry's portability allows monitoring of unrestrained patients in their natural environment for extended periods, making it possible to record episodic behavioral and physiologic events which might not be manifest in the laboratory. It can also provide the long, continuous record required for clinical EEG sleep analysis. Telemetry is also being considered for applications involving restrained patients such as are found in the OR and ICU, since telemetry can reduce the tangle of wires connecting patients to monitoring equipment and can minimize the hazard of electrical shock (3).

Except for single channel EKG monitoring, telemetry has not yet found wide clinical application. This is due in part to such considerations as expense, technical complexity, and reliability. Most use of multichannel biotelemetry has been associated with research projects which have the necessary resources to support the equipment. We have now had five years of experience operating telemetry in a clinical setting with the goal of obtaining EEG recordings during spontaneous seizures from epileptics with implanted electrodes (4). This experience has served as a background for the development of a new multichannel biotelemetry system intended for routine clinical use, as well as for research. In addition to the usual technical specifications, the design goals for this system included simplicity of construction, use, and maintenance, as well as safety, reliability, ruggedness, cost, and size. This system is now complete and has seen its first year of field testing (5).

The equipment is expandable up to fourteen channels and employs multiplexing followed by pulse width modulation (PWM) and crystal stabilized phase modulation (pM) in the TV band (174-216 MHz). Each channel has differential amplification provided by an instrumentation amplifier configuration, the gain of which can be controlled by a resistor. The inputs are capacitor coupled and also include a series resistor at the input connector bulkhead for additional patient safety and for RF protection. The amplifier outputs are sampled by a CMOS multiplexer, whose output is connected to a regulated PWM encoder circuit. The encoder output is assembled into a composite signal and sent to the PM transmitter. The first time interval in the transmitted signal contains a synchronizing pulse, the second time interval contains a PWM encoded zero reference signal, and subsequent time intervals sequentially contain the PWM encoded amplifier signals. Multiplexer clock rate is set to provide time intervals of 140 microseconds. For a fourteen channel system, this results in sampling at about 416 Hz per data channel. The first and last 10% of each of the intervals containing data are digitally locked to limit the transmitted amplitude in the event the input on that channel is overdriven. The pM transmission is accomplished in a three transistor circuit with the phase modulation being impressed on a crystal stabilized carrier which is then multiplied to the final transmission frequency.

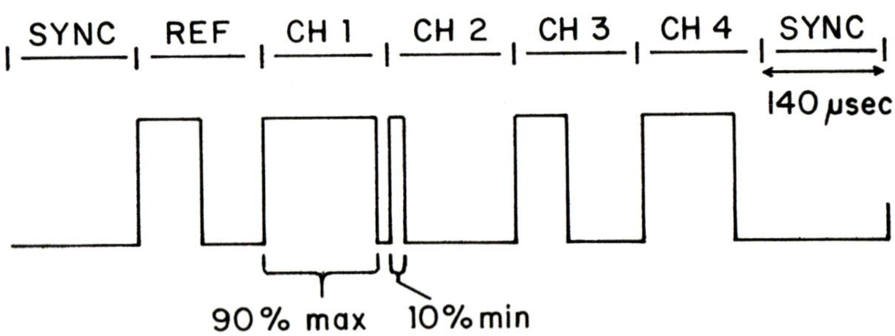

Fig. 1: Composite PWM waveform of a four channel system showing arrangement of synchronizing, zero reference, and four data intervals. The first and second data channels show the maximum and mimimum modulation limits.

The entire telemetry pack is constructed with standard miniature components soldered to printed circuit boards and then sprayed with a thin coating of adhesive. The transmitter and the rest of the electronics are enclosed in individual copper foil shields for RF protection, and interconnections are made using EMI filters. The entire unit is packed in polyurethane foam with an outer coating of fiberglass. This configuration makes the telemeter immune to RF interference, watertight, and resistant to mechanical damage. The final telemeter pack has a polarized input connector for the patient leads, a jack for the antenna, a jack to obtain the PWM prior to the RF stage, and a polarized connector for the battery. Consequently, the telemeter pack is very simple to use. The unit is powered by a single 9 volt battery such as is used in many commercial portable radios (Mallory MN1604). The battery will operate the telemeter pack for over twenty-four hours. Most of the power is consumed by the RF transmitter which has a range of about 100 meters.

Problems in tuning radio receivers are involved in a significant part of telemetry failures in the hands of nontechnical personnel. For transmission in the FM band (88-108 MHz), commercial receivers can generally be found which do not require unusual operator expertise. At TV frequencies this is not the case. Consequently, it was necessary to develop a special receiver as part of this telemetry system. The receiver is based around a commercially available modular system. Internal modifications were made to the RF tuning stage to center it at the band of interest. A crystal controlled local oscillator was also installed, along with a 100 kHz ceramic IF filter. Although the usual receiver adjustments are present inside the unit, the only operator controls on the front panel are an on-off switch and an audio volume control for a small speaker in the panel. Rear panel connections include a 50 ohm antenna input jack and a receiver output jack.

The receiver output is taken to a PWM demodulator where the composite signal is demultiplexed and decoded. Internal adjustments of amplifier gain and baseline are provided, and there is provision for DC or AC coupled outputs. Precision gain control for each channel is also available from switches on the front panel of the demodulator. Additional circuitry inside the demodulator provides for blanking and pulse recognition of the incoming composite waveform, and enhances system performance in situations where there is marginal RF field strength.

An unusual feature of this system is its ability to monitor its own performance. Since the zero reference potential is internally generated in the telemeter pact, monitoring its stability in the demodulator provides an indication of proper systems operation. This information is available as a voltage level coupled to a warning light on the front panel and to a pin on the rear panel of the PWM demodulator. This feature has proven to be very useful in allowing positive identifications of system related artifacts which occur when the patient moves out of range of the receiver.

Some of the specifications for the entire system, including operation of RF link, are: frequency response, 0.8 Hz to 125 Hz (half-amplitude); noise, 3 microvolts maximum through this bandwidth; input impedance, 10 megohms differential; and cross talk, 0.1% typical. In practice, the telemeter is mounted on a patient's head in a pouch inside a soft cloth hat which is affixed to the head with hair clips and adhesive tape. This is both cosmetic and sufficiently comfortable to be worn continuously for hours. For brief recordings, metal disk electrodes and paste are used to make contact with the patient. These electrodes can be further secured with collodion. For long term recordings, conductive silver 'paint-on' electrodes have proven effective, although they require more time to apply. The radio receiver and PWM demodulator are located in an area remote from the patient, and the records are usually stored continuously on an FM magnetic tape recorder.

When an observer is with the patient, it is convenient to provide him with a small pocket-sized FM transmitter, similar to a wireless lapel mike. This pocket transmitter can be used with a commercial FM receiver and the observer's comments can then be put on a voice track of the tape recorder. This system has proven efficient for monitoring EEG and other electrophysiological variables from psychiatric patients, both in the laboratory and on the ward.

ACKNOWLEDGMENT

This work was supported by the Veterans Administration.

REFERENCES

1. Porter RJ, Wolf AA, Penry JK: Human electroencephalographic telemetry. A review of systems and their applications and a new receiving system. Amer J EEG Technol 11: 145, 1971

2. Bruner JMR: Telemetry in a large hospital - A look at the future. IEEE Trans Biomed Engng 18: 325, 1971

3. Ferrer-Allado T, Brechner VL, Dymond A, et al: Ketamine induced electroconvulsive phenomena in the human limbic and thalamic regions. Anesthesiology 38: 333, 1973

4. Lieb JP, Dymond AM, Walsh GO, et al: Clinical and research applications of a multichannel EEG radiotelemetry system in patients with partial epilepsy. Proceedings, 3rd International Symposium on Biotelemetry, Asilomar, Calif., 1976

5. Dymond AM, Fryer TB, Coger RW, et al: A multichannel radio-radiotelemetry system for clinical and research use. Med Instrum 9: 41, 1975

INTRACRANIAL PRESSURE MONITORING USING AN IMPLANTED EPIDURAL CAPACITANCE TRANSDUCER AND TELEMETRY

Silverberg[1] G.D., Ream[2] A.K., Corbin[3] S.D., Fryer[4] T.B., Becker[1] D.H., and Hopkins[2] W.C.

Stanford University Departments of 1: Neurosurgery, 2: Anesthesia, 3: Cardiovascular Surgery. Stanford, California

4: Ames Research Center, NASA, Moffett Field, California

INTRODUCTION

Progress in monitoring intracranial pressure (ICP) using an implantable epidural transducer and telemetered data output has been impeded by the unavailability of a reliable, drift-free transducer (2). In an effort to obtain such a monitor, we have examined two kinds of transducers over the past several years. We have previously reported our experience with a miniature strain gauge bonded to a titanium diaphragm (3). We now are reporting our experience with an induction powered capacitance transducer linked to a variable frequency oscillating circuit.

TRANSDUCER

The transducer used in this work is described in a paper by T. B. Fryer (4).

MATERIAL AND METHOD

Four capacitance units were examined in vivo and in vitro for stability, sensitivity, accuracy, temperature stability, hysteresis, and durability.

In vitro measurements were done in a pressure chamber against a water manometer. The frequency output of the transducer in MH_z was plotted against pressure. Long term stability was examined by periodic measurement of the frequency output at atmospheric pressure and with an intermittent pressure load over three months in two transducers.

In vivo measurements were performed acutely in five dogs. The transducer was fixed epidurally into a parietal burr hole using a flanged housing and self-tapping screws (Figure 1). ICP was measured independently via an 18 gauge spinal needle in the cisterna magna or an 18 gauge catheter in the contralateral lateral ventricle connected to a Statham P23-BC transducer. ICP was varied by inflating or deflating a small epidural balloon positioned contralateral to the pressure transducer and by adding or removing cerebrospinal fluid (CSF). Systemic blood pressure was monitored by a catheter in the femoral artery connected to a Statham P23-AC transducer. All transducers were calibrated before and after each experiment. We found it very important to control the following parameters carefully in order to obtain reproducible results:

Mechanical Placement

Positioning the transducer housing coplanar with the dura is mandatory to get an accurate reflection of CSF pressure (1,5). The transducer must also be rigidly fixed without edge loading or stressing during implantation.

Pressure

Because these cells are all sealed pressure monitors, a barometer accurate to 0.1 torr is necessary to correct for atmospheric pressure.

Temperature

It is necessary to calibrate the transducers in a temperature controlled bath at 37° C before and after use.

RESULTS

In vitro measurements showed roughly a linear correlation between the frequency output of the transducer in MH_Z and pressure (Figure 2). Uncompensated thermal drift was found to be 0.75 torr/degree C° for temperatures in the physiologic range. Long term drift figures show 0.06 torr/day zero point drift over a three month period in two of the four transducers. Two newer units have not been sufficiently bench tested as yet. Although these long term drift figures look very promising, we have seen significant fluctuations of the zero point around this stable baseline in a random manner suggesting electronics instability rather than changes in physical equilibrium (range: ±5.35 torr).

The acute in vivo implantations showed excellent correlation between ICP measured from the epidural space and the CSF pressure measured from the lateral ventricle and the cisterna magna (Figures 3 and 4). The correlation coefficients for the transducers varied from 0.97 to 0.998. The standard deviation of the epidural measurement about the regression line of slope 1 varied from 3.2 torr to less than 1 torr. There was no evidence of hysteresis. Sensitivity was excellent to less than 1 torr.

Both in vitro and in vivo the transducers seemed quite durable. Rough handling did not appreciably affect the zero point or gain.

CONCLUSIONS

Accurate measurement of ICP is practical utilizing an epidural pressure transducer. The long term stability seen with these capacitor transducers suggests that complete implantation with telemetered output accurately and reliably reflects both acute and chronic changes in ICP. The absence of plastic diaphragms and pre-stressed lever systems, as well as the very small diaphragm displacement per torr pressure applied, will hopefully be reflected in stable recordings over prolonged time sequences (6,7,8).

The random fluctuations of output about the otherwise stable baseline should be correctable with a more stable electrical circuit. Chronic implantations will be carried out to assess the tissue response to the housing material and effects on transducer stability.

REFERENCES

1. Dorsch WWC and Symon LJ: A practical technique for monitoring extradural pressure. J Neurosurg 42: 249-257, 1975
2. O'Brien MD and Waltz AG: Intracranial pressure gradients caused by experimental cerebral ischaemia and edema. Stroke 4: 694-698, 1973
3. Ream AK, Silverberg GD, Rolfe J, et al: Accurate measurement of intracranial cerebrospinal fluid pressure by an epidural technique. 28th Annual Conference on Engineering in Medicine and Biology, New Orleans, La., 1975
4. Fryer TB: A pressure telemetry system utilizing a capacitance-type transducer. Proceedings, 3rd International Symposium on Biotelemetry, Asilomar, Calif., 1976
5. Schettini A, McKay L, Majors R, et al: Experimental approach for monitoring surface brain pressure. J Neurosurg 34: 38-47, 1971
6. Rylander HG, Taylor HL, Wissinger JP, et al: Chronic measurement of epidural pressure with an induction powered oscillator transducer. J Neurosurg 44: 465-478, 1976
7. Timm GW, List JS, Bradley WE, et al: Frequency stability of an implantable pressure telemetry capsule. IEEE Trans Biomed Eng, BME-22:3, 250-251, 1975
8. Walker AE, Viernstein L, and Chubbuck JG: Intracranial pressure monitoring. Amer Assoc of Neurological Surgeons. Bal Harbour, Fla, 1975

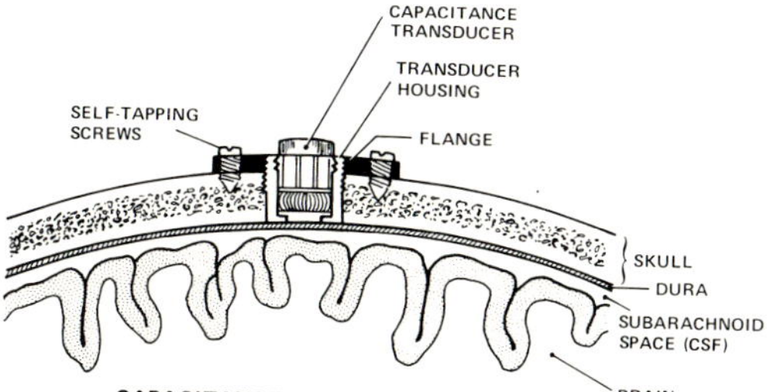

CAPACITANCE TRANSDUCER AND HOUSING POSITIONED IN SKULL

Figure 1.

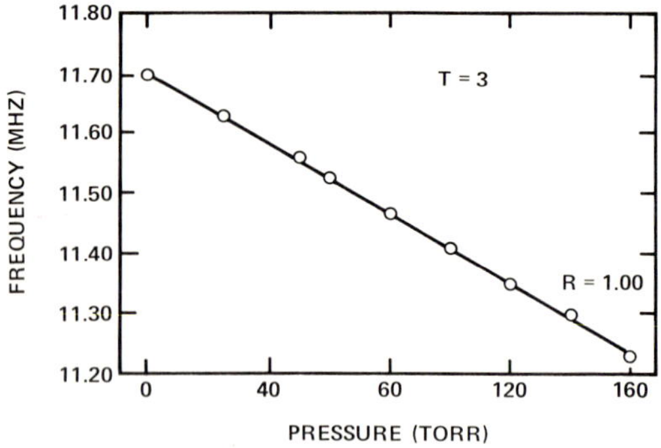

Figure 2. In vitro calibration. The frequency output of the transducer is plotted against the applied pressure.

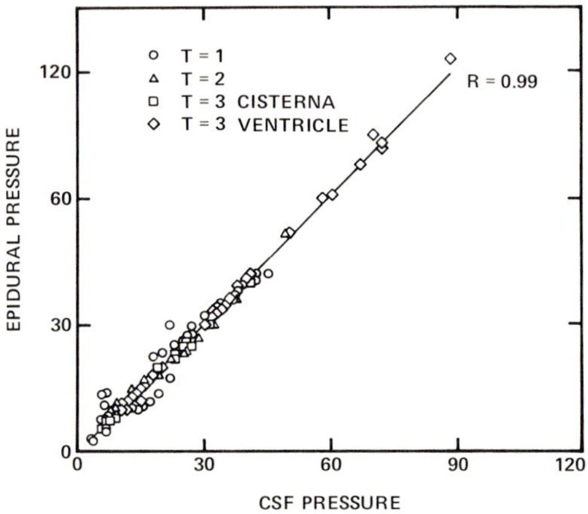

Figure 3. In vivo correlation between the epidural pressure as measured by the capacitance transducer and CSF pressure as recorded from the cisterna magna and the lateral ventricle.

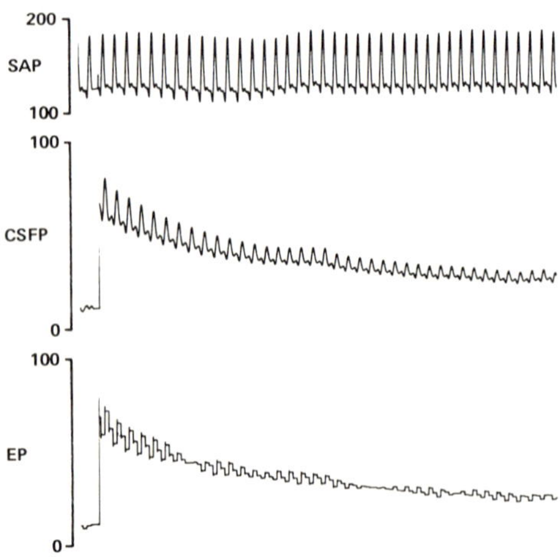

Figure 4. Record of one experiment. Systemic arterial pressure (SAP) is monitored from a catheter in the femoral artery. CSF pressure (CSFP) is measured by a catheter in the left lateral ventricle. Epidural pressure (EP) is measured with the capacitance transducer. The intracranial pressure was elevated by inflating the epidural balloon.

TWO-CHANNEL INTRACRANIAL PRESSURE AND TEMPERATURE MONITORING SYSTEM

Albert Leung, Ronald Lorig, Wen H. Ko, Eddie Cheng

Engineering Design Center
Case Western Reserve University
Cleveland, Ohio

In long-term intracranial pressure (ICP) measurements, the use of transcutaneous connections may induce infection. This hazard can be eliminated by a totally implantable ICP monitoring transmitter. The design and construction of such a special purpose telemetry system is described here.

Although pressure measurement is the major objective of this instrumentation system, a temperature channel is added. This provides information to the receiver for the correction of error in the pressure reading due to the thermal drift of the pressure transducer. The implant is powered inductively by an external 3.5MHz oscillator (see Figure 1).

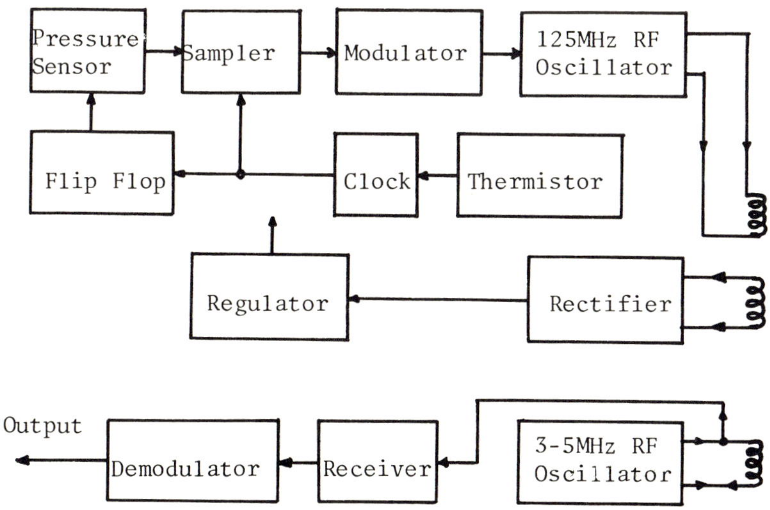

Figure 1. Block Diagram

By placing the coil of the oscillator within four inches of the implant, the detector coil can obtain enough power (400μW) for proper operation.

Pulse rate modulation is used in the temperature channel. A thermistor is used as the temperature sensor and the period, T_θ, is modulated by its resistance (see Figure 2).

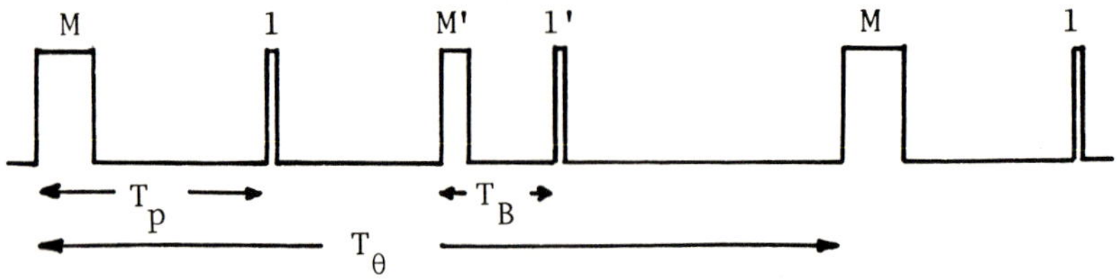

Figure 2. Signal Format

The pressure sensor is pulse powered at a 5% duty cycle in order to reduce the implant's power consumption. When the pressure sensor is powered, its output is amplified and sampled. This sampled value is used to modulate the position of pulse 1 relative to the marker pulse, M, (T_p). However, it should be noted that the pressure sensor is only powered at every other sampling period. The sensor's output is constant at zero when it is not powered. This null voltage is fed through the same amplifier, sampler and modulator as the pressure voltage, and it modulates the relative position of pulse 1' relative to M'. This chopper type telemetry system provides a zero reference to the receiver to cancel the effect of the thermal drift of the electronic components.

The pressure sensor used is a piezoresistive effect pressure transducer bridge (1). However, special care is required to mount the transducer in order to minimize thermal drift and the effect of mechanical disturbance (see Figure 3).

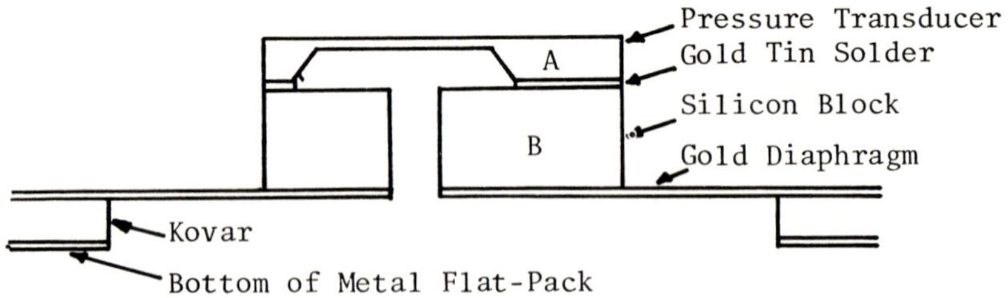

Figure 3. Pressure Transducer Mounting

In such an assembly, the inside of a metal flat-pack is used as a pressure reference chamber. The pressure transducer bridge, A, and the silicon block, B, form a mechanically rigid unit which sits on the gold diaphragm at the bottom of the flat-pack. This diaphragm acts as a strain relief and the silicon block, B, further reduces the undesirable environmental stress applied to the pressure transducer bridge, A. Since A and B are both of the same material, thermal drift, due to differential thermal expansion, is minimized. This assembled unit has been tested and the following results have been obtained:

Excitation voltage:	4V
Pressure sensitivity (differential pressure):	4.5μV/V/mmHg
Temperature sensitivity:	-3.6μV/V/mmHg
Temperature coefficient:	-0.8mmHg/°C

Thermal effect still exists; however, this can be corrected at the receiver, as described above.

The implant is sealed inside a 5/8" x 5/8" x 1/8" metal flat-pack. The only external components are the power detector coil and the R.F. oscillator coil. Such a package is resistive to body fluid invasion, and after coating with polyurethane, the implant can have a long life expectancy inside a patient or an animal.

ACKNOWLEDGMENTS

This work, supported by NIH grant GM 14267-10, was done in collaboration with the Neurosurgery Department at University Hospitals with Dr. Frank E. Nulsen and Dr. Gerald S. Brodkey.

REFERENCE

1. Ko WH, Hynecek J, Conrad D, Yon ET, "Miniature Pressure Sensor with Possibility of Intravascular Application" 10th International Conference on Medical and Biological Engineering, Dresden, Germany, 1973.

CLINICAL TELEMETRY OF EMG AND TEMPORAL INFORMATION DURING GAIT

A. O. Quanbury, C. D. Foley, D. A. Winter*,
R. M. Letts, T. Steinke

Shriners Hospital for Crippled Children
Winnipeg, Manitoba
*Department of Kinesiology,
University of Waterloo
Waterloo, Ontario

INTRODUCTION

Over the past 2 years we have developed and refined our locomotion lab facilities from an experimental operation to a point where the equipment can be used virtually at a moment's notice to perform routine clinical gait studies. (1,2,3). We have, in this time, conducted gait studies on 100 patients with walking problems and are now doing studies not only on patients in our own hospital and from other hospitals within the city but also on patients from another major rehabilitation hospital outside the Province of Manitoba. The use of a multichannel telemetry system forms a major component for making these studies. (4,5). Muscle activity is recorded from surface electrodes and temporal information from small electrical switches mounted in the heel and sole of a specially instrumented shoe and then transmitted to a multi-channel pen recorder. The lightweight system is worn easily and comfortably on a belt by all but the youngest children and allows free ranging over our gait study area without the encumberance of trailing or overhead cables. In addition the input electrical characteristics of the system and the relatively short interconnecting cables require only a minimum of skin preparation to yield a clean artifact free signal from surface electrodes. In general the multi-channel telemetry system provides a near ideal method of obtaining the necessary data.

METHODS AND EQUIPMENT

Although the telemetry system itself does not limit the region where the subject must walk, the simultaneous video recordings for kinematic assessment do dictate that the walking be done along a 5 m. walkway in front of a black background. (2). The telemetry system is a 6 channel FM-FM unit manufactured by Bio-Sentry and modified to allow transmission of the envelope of the EMG signal. (4,5). Figure 1 shows a 4 year old patient wearing the telemetry belt. The customary procedure is to use 4 channels for EMG signals and 2 channels for the footswitch signal. The footswitches used are of 2 types. If the subjects can wear normal footwear they are fitted with "Hush Puppy" shoes which have small switches mounted in the heels and soles and resistor networks on the sides (5) as shown in Figure 2. If the subjects must wear their own shoes, then flat switches made from two layers of phosphor bronze are taped to the bottom of the shoes. Permanent records are obtained on an eight channel Brush recorder.

Surface electrodes, and occasionally indwelling fine wire electrodes, are used to pick up the signals from the various muscles of interest. Depending on the patient's particular problem these might include tibialis anterior and posterior, gastrocnemius, quadriceps, hamstrings, hip adductors and abductors and the paraspinal muscles. The simultaneous recording of the foot-switch signals enables the determination of the phases of the walking cycle during which the activity of the various muscle groups occurs.

The 100 patients that have undergone gait studies can be broken down as follows: hemiplegia (adult) 22; cerebral palsy 33; muscular dystrophy 7; scoliosis 11; amputees 7; other 20.

The cerebral palsy group included various levels of dysfunction from mild hemiplegia to quadriplegia where all four limbs were affected. Most of the patients had received or were about to receive orthopaedic surgery which included Achilles tendon lengthening, adductor tenotomy, hip flexor release, hamstring release and muscle transfer.

RESULTS

Figure 3 shows the type of output obtained. From this data basic temporal information such as pace period, swing phase/stance phase ratio, double weight bearing period may be obtained as well as information on muscle activity patterns.

DISCUSSION

The information obtained from these gait studies has been applied clinically in a variety of situations. It has proven most useful in the improved assessment of the severely involved cerebral palsy patient. In these situations surgery to improve one area of concern may generate an additional problem unless the status of all the major muscles in the leg is known. The success of a tendon transfer depends on the strength and condition of the muscle whose action is being transferred and an increased knowledge in this area will improve the chances of success. All the information gained is of a quantitative nature and so the measurements of changes over a period of time or as a result of surgery are much more positive than those available from a qualitative assessment.

CONCLUSION

Over the past 3 years the walking disabilities in 100 patients have been assessed with the aid of a multi-channel telemetry system to obtain information on the muscle activity and on the temporal aspects of gait. This information has been proven valuable from a clinical point of view in improving the medical team's assessment of the patient's particular problem.

REFERENCES

1. Dinn FD, Winter DA, Trenholm BG: CINTEL-computer interface for television. IEEE Trans on Computers C-19: 1091, 1970

2. Winter DA, Greenlaw RK, Hobson DA: Television-computer analysis of kinematics of human gait. Comput Biomed Res 5: 498, 1972

3. Letts RM, Winter DA, Quanbury AO: Locomotion studies as an aid in clinical assessment of childhood gait. Can Med Assoc J 112: 1091, 1975

4. Winter DA, Quanbury AO: Multichannel biotelemetry systems for use in EMG studies, particularly in locomotion. Am J Phys Med 54: 142, 1975

5. Winter DA, Hobson DA, Greenlaw RK: A microswitch shoe for use in locomotion studies. J Biomech 5: 553, 1972

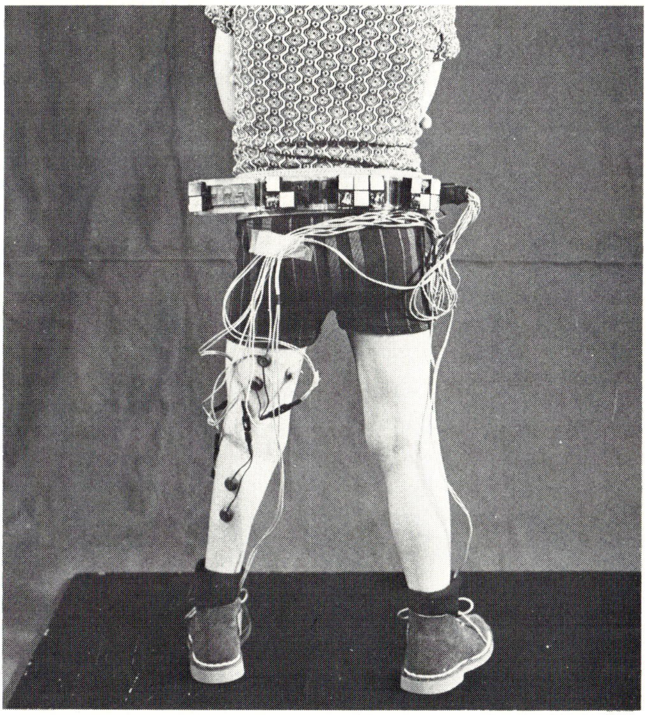

Figure 1. Four year old patient wearing the telemetry belt.

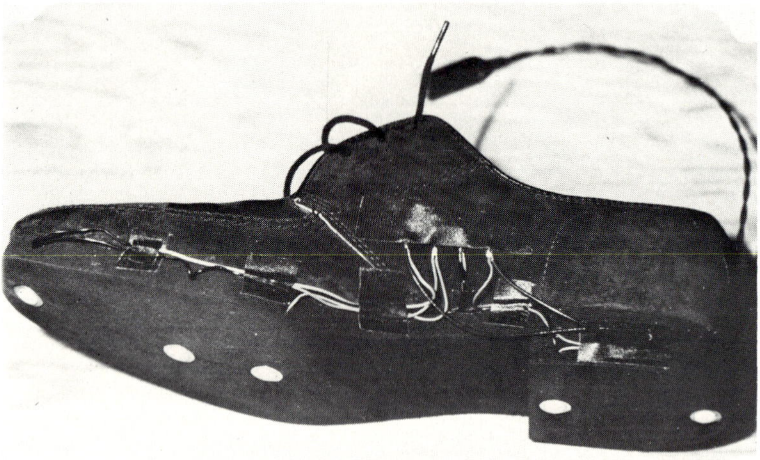

Figure 2. Microswitch shoe.

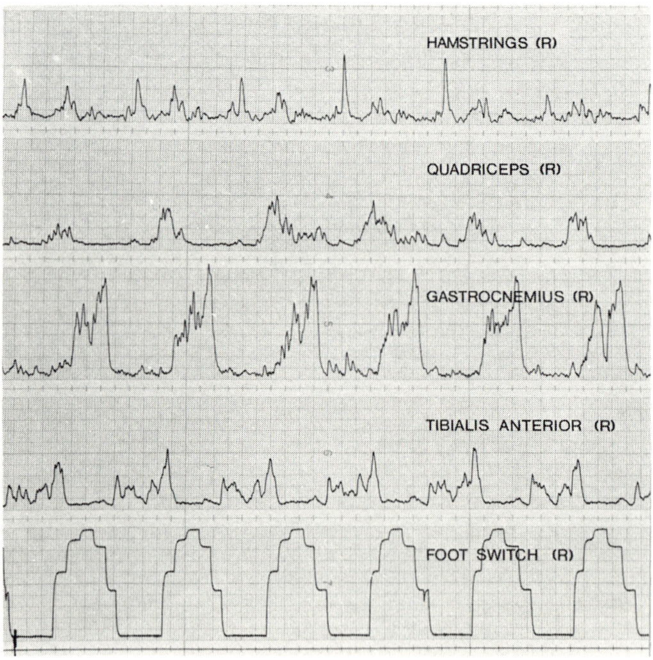

Figure 3. Record of E.M.G. and footswitch information.

POWERLINE INTERFERENCE IN ELECTROPHYSIOLOGICAL RECORDINGS

J.A.J. Klijn

Laboratory of General Neurophysiology
University of Nijmegen
Nijmegen (The Netherlands)

INTRODUCTION

Due to its simplicity, cabled telemetry is often used in acute electrophysiological experiments on the brain of restrained anesthetized animals. The capacitances of the cables change the effective input impedance, however, and this results in inequality with that of the amplifiers. The inequality causes a tremendous loss in common mode rejection ratio due to the high impedance of the specific electrodes (cortex and needle electrodes) used in those experiments. If we only take into account the inequalities between electrode impedances which, in spite of identical manufacture and processing, deviate 50 percent from the nominal value, the CMRR is importantly degraded below the value specified by the manufacturer. This reduction is shown in figure 1.

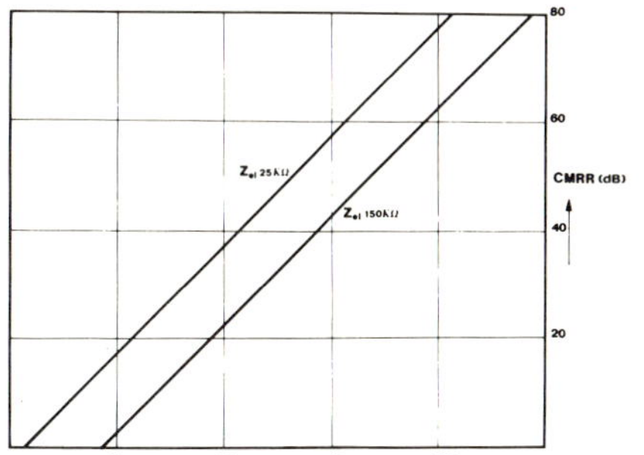

Figure 1. Maximum CMRR when the impedance of one electrode is Z_{el}-50% and the other Z_{el}+50% in relation to the input impedance for cortex electrodes (25 kΩ) and needle electrodes (150 kΩ at 50 Hz). ←

DISCUSSION

The frequency dependence of the input impedance makes a high CMRR hard to realize and lowers the CMRR below the levels indicated in figure 1. Apart from the powerline interference susceptibility, the effective CMRR affects the reproducibility of experiments, certainly when response enhancement by computers is used. To measure the effective CMRR the subject is enclosed in a cage which is subjected to a 50 Hz signal with respect to the ground of the recorder. The resulting U_{body} is measured between the ground and an electrode not drawn in figure 2 on one channel of the recorder. This U_{body} serves as a common mode signal to another channel connected to EEG electrodes in the usual way.

Figure 2. Measurement of the effective CMRR under practical circumstances. →

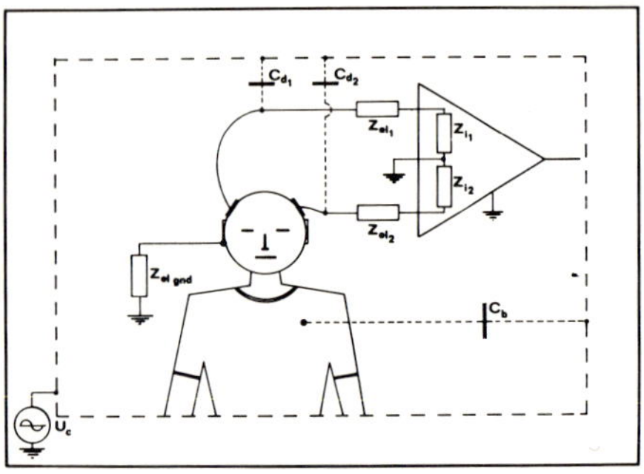

Although these measurements can be made simpler by insertion of a signal generator in the subject's ground, this method was preferred, because it includes all aspects of powerline interference and is simpler to relate to the powerline interference measurements discussed below. An effective CMRR of 25 dB with a variance of 7 dB was found by the method indicated.

The body potential can be calculated from

$$\frac{U_{body}}{U_{cage}} = \frac{R_{el.\ ground}}{1/\omega C_{body} + R_{el.\ ground}} = R_{el.\ ground} \cdot \omega C_{body}$$

Here $R_{el.\ ground}$ can be omitted in the denominator due to its relatively low value. When the subject is grounded by a conventional ear electrode, U_{body} is, in accordance with this formula, 60 dB below U_{cage}. If we now assume that 50 µV/cm is a common sensitivity in EEG recording and that an interference level which doubles the line width is acceptable, then the maximum differential mode signal amounts to U_{DM} = 2.5 µV. The U_{body} that causes this U_{DM} in relation to an effective CMRR of 25 dB is also 35 µV, which gives a voltage of 35 mV for U_{cage}. Figure 3 explains this calculation once again, the instrumental CMRR being mentioned to indicate the possibilities for further improvement.

Figure 3. The values of interference, U_{DM}, U_{body} and U_{cage} in relation to the instrumental CMRR. →

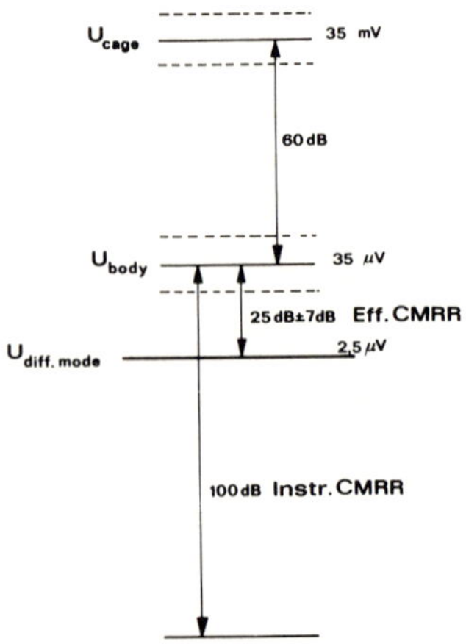

To measure a level of interference comparable to U_{cage}, a capacitative probe was previously developed. Details of the construction, which can be seen in figure 4, and a justification of this statement were the subject of a previous paper.

Figure 4. The capacitative probe and associated circuitry for measurement of the equivalent of U_{cage} in areas with discontinuous distributed sources of interference. →

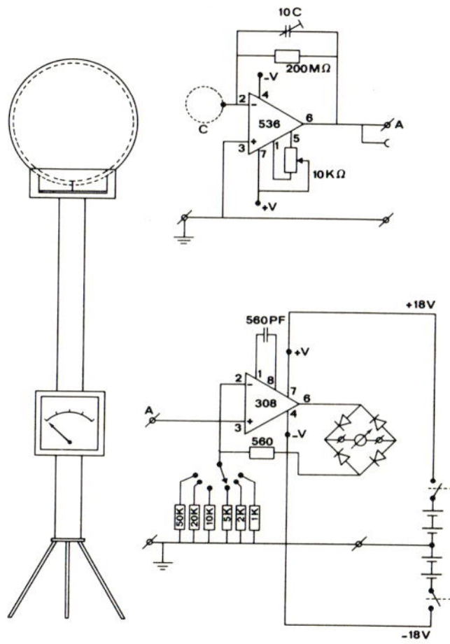

One advantage of this scalar, which we have called U_{ce}, over the electric field strength used up to now in interference analysis, is its direct applicability as a driving force in substitution diagrams.

Another advantage is that the spatial behaviour has been studied by Carslaw and Jaeger (1) on comparable thermal problems. From the solution they derived, the following aspects should be mentioned. Starting from the radiating wall U_{ce} remains constant within a few dB up to 1 meter. From that point on U_{ce} decreases by approximately 35 dB per room height until the opposite wall begins to interfere. The valley that is generated in this way proves to be very wide and U_{ce} only varies 5 dB over half the room width provided it is greater than 4 meters. It can be proven that the location of the minimum is affected only slightly by differences in radiation strength between opposite walls. From this theoretical evidence and its experimental verification, it was concluded that the U_{ce} measured in the geometrical center of a room is characteristic for the lowest interference level obtainable in that room. This location, however, has little experimental relevance. It has therefore been modified to 1 meter above the geometrical center of the floor. Results obtained in different rooms, some of them in use for electrophysiological experiments, are given in table 1.

Table 1. The Equivalent of U_{cage} as Measured in Different Rooms with Single-phase 50 Hz 220 V_{RMS} Powerlines

Specifications	U_{cage} Indicated in RMS
Common electric wiring	8 V
Screened electric wiring	0.6 V
Excessive electric walls	14.5 V
Metal-plated walls	16 mV
Self-supporting gauze cage	10 mV

From this table it can be concluded that none of these rooms is suited for electrophysiological experiments at the EEG level unless it is screened. What can we do now to improve the situation? As indicated already in figure 3, we have to restore the effective CMRR as closely as possible to the instrumental one. There are two possibilities: first, we can provide the electrode wiring with a screen that follows, by means of feedback, the electrode potential. This method also reduces the effective cable and input capacitance and the inherent CMRR degeneration, but it does not reduce cable artifacts. As shown by the work of A.G. Ratz (2), cable artifacts due to friction between core and insulating sleeve can be seen as a piezoelectric transduction of the bending of cables. The related substitution diagram shows that only a low impedance of the conductor on which the signal is induced can suppress this and comparable artifacts such as static. This is why some work has also been done on electrode amplifiers in our laboratory. These amplifiers offer a high impedance to the electrode necessary for a high CMRR and a low impedance to the electrode wiring. From the point of view of thermal noise, an amplification of 10 times would be desirable, but the cumbersome alignment procedures made us decide to choose unity gain, which can be realized within less than 0.1 percent from DC to 2kHz without such procedures. This means that the maximum obtainable effective CMRR amounts to 60dB, which is an improvement upon the effective CMRR of figure 3. The reduction of cable artifacts equal to the ratio of the output impedance of the amplifier and the electrode impedance while skipping with the electrode wiring is clear in recording b of figure 5.

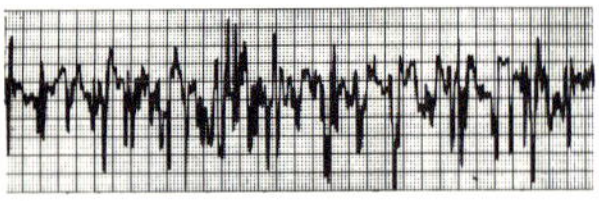

a) Normal ECG recording.

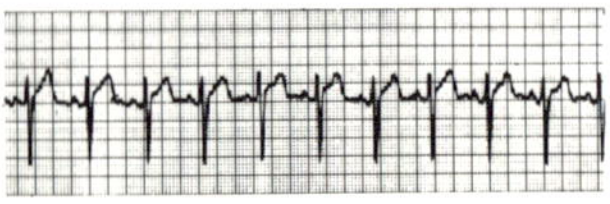

b) Recording with impedance converter.

← *Figure 5*. Recording of an EEG without and with electrode amplifiers during skipping with the electrode wiring.

↓ *Figure 6*. Miniature amplifier mounted on needle. The main problem encountered during several years of use has been the moisture-tight encapsulation.

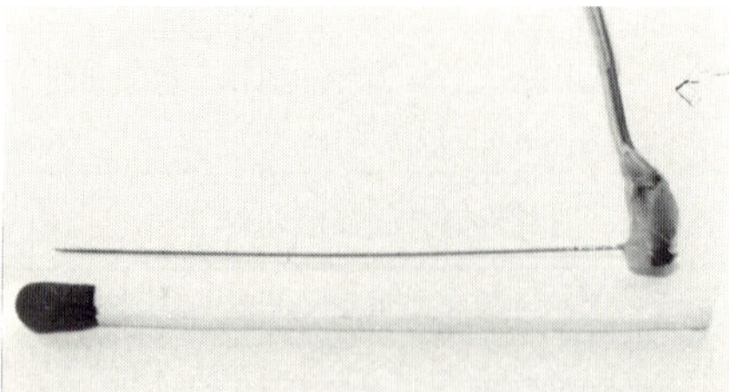

REFERENCES

1. Carslaw HS, Jaeger JC: In Conduction of Heat in Solids. London, Oxford University Press, 1950, pp 154-158
2. Ratz AG: Triboelectric noise. ISA Transactions 9: 154-158, 1970

Twenty-Seven Days in the Life of an Epileptic Patient

J.R. IVES, A. WILKINS, M. JONES, F. ANDERMANN, and J. WOODS

Montreal Neurological Institute
Montreal, Quebec, Canada

INTRODUCTION

Until recently (3) it has been technically impossible to monitor the EEGs of ambulatory epileptic patients outside the confines of a hospital.

Recordings were usually limited at most to a few hours because the requirements of traditional EEG recordings restricted the patient to lie as still as possible in bed in a shielded room since any movement and/or 60 cycle artifact would completely obscure the record and make its reading uninterpretable.

With the introduction of EEG radio telemetry (1), its increasing involvement in clinical applications (6) and its technical refinement into a multichannel seizure monitoring system for surgical candidates (2,4), telemetry has been useful in specific areas. However, the monitoring of ambulatory outpatients is not technically feasible with radio telemetry at this time, while a 4 channel 24 hour EKG cassette (5) has had some success (3) in this area.

AN APPLICATION

One area of application is the use of the cassette to provide continuous long term monitoring of a patient's EEG in his or her natural environment and evaluate the effectiveness of antiepileptic medication or other forms of treatment on the number of clinical and electrographic seizures as well as interictal epileptogenic discharges over a significant period of time (2 to 5 days).

Recording Equipment

The equipment has been previously described (3) and consists basically of recording electrodes, EEG preamplifiers and the cassette recorder applied and arranged in a cosmetically acceptable manner to allow the patient to venture back into his real world without embarrassment.

Patient

The patient in this report is a 14 year old moderately to severely retarded girl living with her parents and has been reported elsewhere (8). Her parents were given instructions and a schedule as to how and when to re-jelly the electrodes and change the cassette and batteries.

Investigation

One aim of the investigation was to study the effect of

monocular occlusion by frosted glasses on this girl's extremely frequent pattern sensitive spike and wave seizure discharges.

The second study was designed to assess the effect of the oral administration of Taurine (7) on the epileptic activity.

The patient was monitored for a total of 675 hours (27 days) during a 21 month period as outlined below:

Session	Reason	Start Date	Length
1	base-line, pre glasses	Mar. 20, 1974	2 days
2	on and off glasses, trial I	April 3, 1974	5 days
3	on and off glasses, trial II	June 12, 1974	5 days
4	on and off glasses, trial III	June 18, 1975	5 days
5	base-line, pre-Taurine	Sept 30, 1975	4 days
6	on 100 mg Taurine B.I.D. P.O.	Nov. 3, 1975	4 days
7	on 200 mg Taurine B.I.D. P.O.	Nov. 21, 1975	4 days

During this period only a total of 2-3 hours was lost due mainly to one incident of the patient unknowingly disconnecting herself and another of premature battery failure (fig. 3, dotted line).

Playback, Data Reduction, and Analysis

The C-120 cassettes were replayed at 60 times their original recording speed onto a Mingograf ink jet EEG machine so that a 24 hour tape took 24 minutes to replay while the paper speed was 15 mm/sec to obtain a write out of 20 minutes/EEG page as illustrated in fig. 1. More conventional EEG write speeds can be obtained to verify discharges as has been previously reported (3).

Since this patient's discharges consisted of irregular bilaterally synchronous 3 per second spike and wave bursts, the 4 channels were summated in a manner which emphasized the synchronous activity and reduced irregular background; then this resultant signal was bandpass filtered between 2-4 Hz (4 pole Butterworth filter) and written out as channel 5 in figure 1. This signal was then rectified to obtain a DC voltage proportional to the presence of "3 per second spike and wave" activity and used as an indication for the occurrence of a discharge as shown on channel 6 of figure 1.

By now counting the discharges in 3 pages then the number of discharges per hour can be easily and quickly obtained and transferred to a graph as shown in figures 2 and 3.

Results

In figure 2 the continuous 5 day summary of the patient's hourly discharge frequency as occurring in her natural environment, her sleep/wake cycle and the dramatic effect of left or right monocular glasses, no glasses, or "placebo glasses" can be clearly seen (for a full discussion, see (8)).

In figure 3, the 4 day chart of discharge frequently after receiving 200 mg of Taurine (7) B.I.D. P.O. (an amino acid with

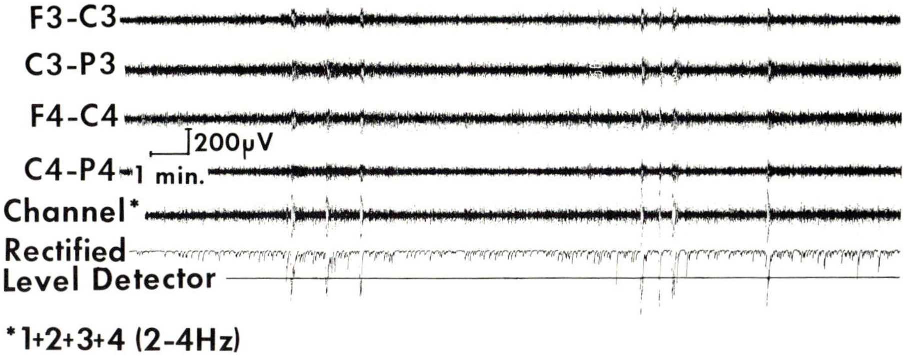

Figure 1

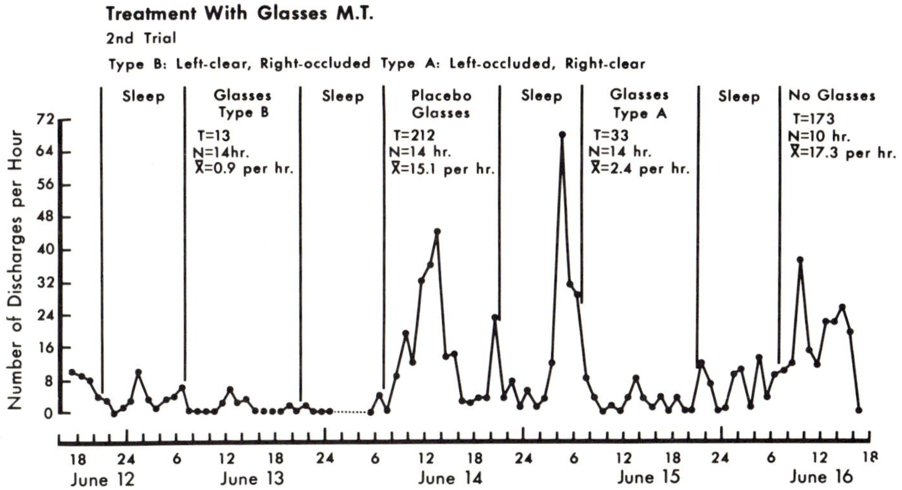

Figure 2

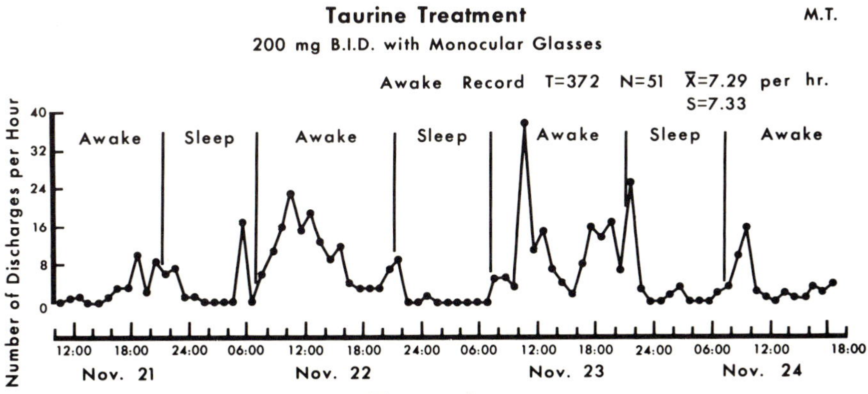

Figure 3

some antiepileptic properties) is seen, and in comparison to the 2 previous sessions (not shown), A) pre-Taurine, and B) 100 mg of Taurine B.I.D. P.O., it was demonstrated with the statistical student t test that there was no significant difference between the three sessions.

DISCUSSION

In this one patient, the advantage of cassette recording of EEG activity in both a clinical and research setting is well exemplified. The information obtained was quantifiable and thus useful to her physicians in decisions regarding her antiepileptic drug regime.

In figure 3, the mean discharge rate per hour in the awake record is X = 7.29 while the standard deviation is S = 7.33. This indicates a tremendous variability in the occurrence of this patient's seizure discharges and is a possible explanation for the general lack of correlation of the number of discharges seen in the traditional EEG studies done for only a short time and the clinical effects of anticonvulsants.

REFERENCES

1. HOLTER, N.J. and GENERELLI, J.A. Remote recording of physiological data by radio. Rocky Mtn. Med. J., 1949, 46: 747-750.
2. IVES, J.R., THOMPSON, C.J., GLOOR, P., OLIVIER, A. and WOODS, J.F. Multi-channel EEG telemetry-computer monitoring of epileptic patients. In Neukomn, P.A. (Ed.), Biotelemetry II, Karger, Basel, 1974, 216-218.
3. IVES, J.R. and WOODS, J.F. 4-channel 24 hour cassette recorder for long term EEG monitoring of ambulatory patients. Electroenceph. clin. Neurophysiol., 1975, 39: 88-92.
4. IVES, J.R., THOMPSON, C.J. and GLOOR, P. Seizure monitoring: a new tool in electroencephalography. Electroenceph. clin. Neurophysiol., in press.
5. MARSON, G.B. and McKINNON, J.B. A miniature tape recorder for many applications. Control and Instrumentation, Dec., 1972, 46-47.
6. PORTER, R.J., WOLF, A.A. Jr., and PENRY, J.K. Human electroencephalographic telemetry. Amer. J. EEG Technol., 1971, 11: 145-159.
7. VAN GELDER, N.M., SHERWIN, A.L., SACKS, C. and ANDERMANN, F. Biochemical observations following administration of Taurine to patients with epilepsy. Brain Res., 1975, 94: 297-306.
8. WILKINS, A.J., ANDERMANN, F. and IVES, J. Stripes, complex cells and seizures. Brain, 1975, 98: 365-380.

A WEARABLE DIGITAL BIOLOGICAL RECORDER

C.S. Weaver, J.M. Yarborough, J.H. Chadwick, and F.A. Brown

Stanford Research Institute
Menlo Park, California

INTRODUCTION

Frequently a very low power cut-off frequency and high dynamic range are desirable when biological signals (e.g., ECG's) are recorded.

There are three ways in which very low frequency data can be stored on magnetic tape: baseband (or modulated carrier) AM, FM, and digital recording. Each of these techniques has obvious strengths and weaknesses. Baseband amplitude modulation can handle extremely low frequencies, and for a 0.05 Hz lower cut-off frequency (the maximum lower cut-off frequency that should be used with an ECG waveform), it is a possibility. The problem with AM is its sensitivity to the tape system's inherent signal-to-noise ratios and tape speed variations. A good tape recorder and playback system has a signal-to-noise ratio of 35 to 40 dB with this modulation. Frequency modulation, while overcoming the problem of tape noise to a certain extent, is still limited by speed variations and frequency response, variation to a dynamic range in voltage of about 200 to 1, about 46 dB under ideal conditions.

Digital recording has the distinct advantage of requiring only one zero-type decision from the recording medium. Since the data are in a digital format, the signal-to-noise ratio and dynamic range are set by the quantizer, not the tape recorder. A 10-bit A-D conversion, for example, would be sufficient to cover three decades. The converter can handle DC signals and, therefore, the problem of low-frequency response is eliminated, and the upper end cutoff can be set by selection of a suitable sampling rate. There is no drift in the linearity of a digital recording system in contrast to an FM recording system, and the change in gain and DC level of a digital system is small compared to an FM system.

An important advantage of digital recording, when compared to FM recording, is the relative lack of sensitivity of the digital recorder to changes in tape speed (wow and flutter). Since the bits are recorded under control of an accurate clock, the time stability at the output depends only on the accuracy of the clock. The wow and flutter in a well engineered system cause virtually no distortion of the recorded data. This is not true for an FM system, since these speed variations cause not only time-base changes but also distortions in the waveform shape and signal-to-noise ratios of 40 dB and 2 percent linearity typically can be obtained.

The digital recording technology now available is greatly superior to the other methods. Digital recorders can record for a duration that is two or three times longer than that available from FM recorders using the same length of tape. Signal-to-noise ratios of 60 dB and virtually perfect linearity can be obtained. There is no change in gain and no baseline drift, and dc response is available. Since essentially no distortion is introduced by the recording process, the high-precision measurements made by several modern ECG computer processing programs, and manually by a cardiologist, are practical when digital recorders are used.

A SYSTEM DESCRIPTION

The system that has been developed at SRI is the first lightweight, high-capacity, wearable digital cassette recorder that is suitable for recording biological signals. The specifications are given in Table 1. There are four 40 Hz channels suitable for recording, for example, EECs, and two 40 Hz channels may be combined into one 80 Hz channel that is suitable for ECG recording. The digital playback speed is at 45 times real time, and there are analog outputs from each of the four channels that may be played back at different speeds between real time and 45 times real time.

Table 1. Specification for NINCDS Wearable 4-Channel EEG Recording System

Recording medium	C-120 cromium oxide Phillips cassette
Recording format	Digital-Miller encoded
Recording density	7000 BPI
Recording speed	.156 inches/second
Recording capacity	180×10^6 bits
Recording time	12 hours
Battery type	Lithium
Battery life	40 hours
Weight	2.5 lbs
Input dynamic range	60 dB
Analog input noise	1 μV rms
Analog bandwidth	40 Hz
Analog-to-digital conversion rate	100/second/channel

Figure 1 is a simplified system diagram. The waveforms are filtered by small active filters to form the input to the A-D converter. The output of the A-D converter is prepared by the formatter for recording on the four tracks in the record transport, where a parity check bit is added to each of the ten-bit words in the A-D converter output. The ten-bit words that represent the analog waveforms are placed on the four tracks.

A recording code called the Miller Code is used to place the bits on the tape. This code minimizes the recorder bandwidth requirement and is self-clocking, e.g., a phase-locked loop may be locked to the recorded data to generate the signal that is used in the phase comparison during playback to detect the presence of a 0 or a 1. A bias signal consisting of a 22-kHz square wave is added to the Miller-encoded bit train.

In the playback unit, the cassette tape output is processed at high speeds through an equilizer, a decoder, an error checker, and a formatter. The decoder consists of the previously mentioned phase locked loop, a demodulator, and the error checker that determines the presence of errors and allows word synchronization.

The double buffer is used during the reformatting process to allow running of the Phillips cassette transport at a constant speed while the IBM-compatible transport runs in a start-stop fashion as required by its blocked format. In the quick-look playback mode, the double buffer allows a continuous playback, while the transport again has a discontinuous flow of data.

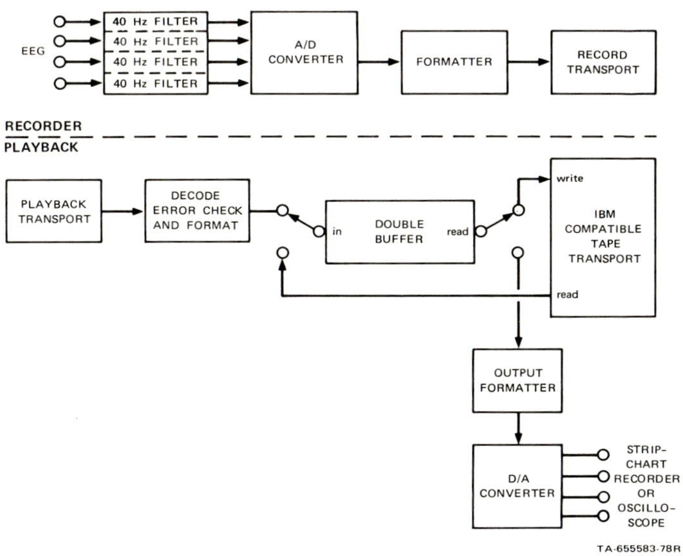

FIGURE 1 SIMPLIFIED BLOCK DIAGRAM OF RECORDER AND PLAYBACK

The reformatting logic is shown in Figure 2. It can be seen that each of the four data channels is treated independently, thus eliminating any potential skew problems. The recovered signal is first equalized in both frequency and phase. The output of the equalizer is then fed to the Miller decoder. This decoder is based on a special phase-locked loop circuit developed by SRI. The characteristics of this circuit allow both a large offset in static frequency and a large amount of jitter to be present without causing decoding errors.

After the decoder, the data stream is processed by the parity lock-and-check circuits. Finally, the data are formatted and sent to the double buffer.

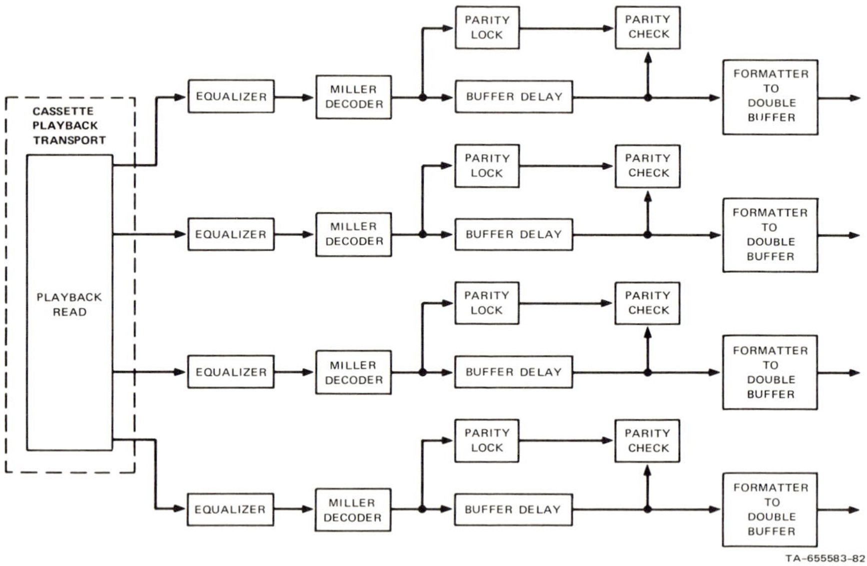

FIGURE 2 DIAGRAM OF THE REFORMATTING LOGIC

ACKNOWLEDGEMENTS

This recorder was developed under a contract from the National Institute of Neurological and Communicative Diseases and Stroke (Contract No. N01-NS-3-2322).

DISCUSSION

ENGSTROM: Dr. Dymond, to what do you attribute the observed reduction in movement artifact with the wireless system relative to the wired system?

DYMOND: The main factor is that all the wiring connectors are secured. With wires, any movement results in cable sway and accentuates noise. With the wireless system the amplifiers are close to the electrodes and the wiring is secured. Implanted systems are particularly free of movement artifacts.

GRAYSTONE: It has been our experience that approximately 99 percent of measurement interference is due to static electricity. Nylon garments are great offenders. That is, they create thousands of volts of noise in the realm of microvolts levels of signal.

SESSION V

MEASUREMENTS FROM UNRESTRAINED ANIMALS

Chairmen

J.A. GESSAMAN, Ph.D.

B.A. WILLIAMS

EDITOR'S OVERVIEW

Several applications of biotelemetry to measurements of the free-ranging activities of animals are presented in this session. The use of telemetry permits measurement of physiological parameters without distortion from the influences of handling and restraint. Miniature systems and components now permit such measurements to be made on small animals (rats, squirrels, rabbits, birds, and fish) as well as large animals (sheep, cattle, elk).

Measurements have been made with both implanted and surface-mounted sensors and have included temperatures, heart rate, pressures, EMG, EEG, and ECG. Studies have involved the effects of hibernation and natural environment on the health and physical condition of caged and free animals. The final paper in the Session deals with a basic study in neurophysiology. In this study EEG data were telemetered from the isolated olfactory bulb in rabbits to determine whether the olfactory response was dependent upon the neural connection to the rest of the brain.

USE OF REMOTE EAR CANAL TEMPERATURE MEASUREMENTS FOR EVALUATING VIRAL VACCINES IN CATTLE

Gary L. Seawright, William Mort Sanders,* and Richard Van Deusen

Veterinary Services Laboratories, Ames, Iowa, and the
*Los Alamos Scientific Laboratory, Los Alamos, New Mexico

INTRODUCTION

The use of telemetry for continuous acquisition of physiological data is an attractive but little-used resource for the study of infectious diseases of food animals. The febrile response is a well-known corollary to infectious disease processes and is highly amenable to continuous data acquisition by telemetric methods. We were therefore interested in the potential of telemetry of temperature data as a quantitative tool for evaluating the protective efficacy of animal vaccines. We report here a pilot experiment in which remote-continuous monitoring of ear canal temperatures was used to evaluate a viral vaccine in cattle whose immunity was experimentally challenged with known virulant virus.

MATERIALS AND METHODS

The instrumentation developed and used in these experiments (1) is shown schematically in Fig. 1. A thermister type probe is inserted about 70 mm into the animal's ear canal. Changes in the ear canal temperature alter the electrical resistance of the thermister which changes the resonant frequency of an oscillator circuit. The frequency signal from the oscillator circuit is used to pulse on a small 150MHZ transmitter for about 200 ms during each cycle of the signal from the oscillator circuit. The time interval between bursts from the transmitter (0.3-1.5 s) relays temperature data from the animal to a remote receiver.

The RF burst from the transmitter is converted to an audio output signal. The audio signal is rectified and integrated to minimize spurious noise pulses. This signal is then converted to a square pulse for input to the trigger circuit of a time interval meter. This meter digitizes the time interval between subsequent pulses and supplies a digital output to the recording system. An ancillary analog signal from the time interval meter is used to record the data on an analog recorder.

The time interval meter and a keyboard terminal with dual tape cassettes is interfaced to a microprocessor type computer (microcomputer). The digital output from the time interval meter is strobed onto the digital tape cassettes by the microcomputer. This terminal has an acoustical coupler and can be used to

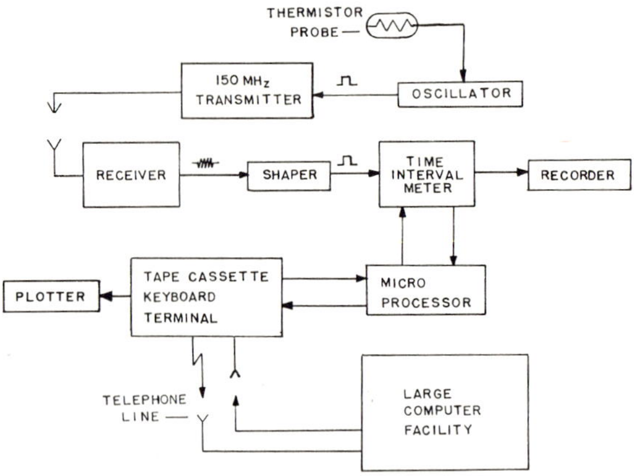

Fig. 1: Block diagram of the temperature monitoring system. The oscillator and transmitter are mounted on the animal's halter.

transmit the recorded data over telephone lines to a large computer facility for conversion of time-intervals to temperatures.

The results from the analysis can be transmitted back to the terminal from the computer facility, printed for evaluation, and recorded on cassette tape for driving a plotter. The computer facility may also be used, via the terminal, for writing, compiling, and executing other computer programs for data analyses. Programs prepared for the microcomputer on the large computer can be transmitted back to the keyboard terminal tape cassette for subsequent loading into the microcomputer for execution.

RESULTS

A steer weighing 650 lbs. was intranasally vaccinated with a candidate vaccine for infectious bovine rhinotracheitis (IBR). A second steer of the same size was held as a nonvaccinated control. Twenty-eight days after vaccination, both steers were intranasally challenged with $10^{7.0}$ infectious units of a standard virulent IBR challenge virus (Cooper strain). Remote ear canal temperature measurements were taken at 1 minute intervals and mean temperatures were computed for 5 min or 60 min blocks. Rectal temperatures were also taken with conventional glass thermometers. The steers were housed in a high security building maintained at 18.5 ± 2.5^{o} C.

Figure 2 shows the normal continuous temperature record of the unvaccinated steer for a period of 7 days before challenge. A diurnal variation of about 0.5^{o} C was observed that was independent of ambient temperature. Temperature minima and maxima usually occurred at 0500-0700 hrs and 1800-1900 hrs, respectively. Figure 3 shows that within the diurnal cycles, there were also variations of higher frequency but lower amplitude.

Figure 4 shows the temperature histories of the steers upon

Fig. 2: Pre-challenge ear canal temperatures over a 7-day interval. Points are hourly means of measurements taken at 60 s intervals.

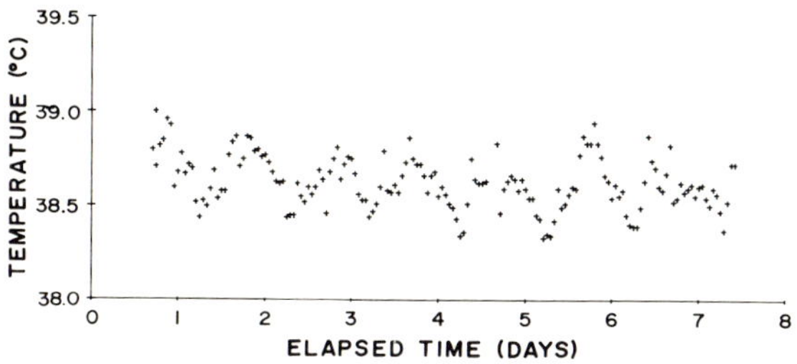

Fig. 3: Pre-challenge ear canal temperatures over a 24-hr period. Points are 5 min means of measurements taken at 60 intervals.

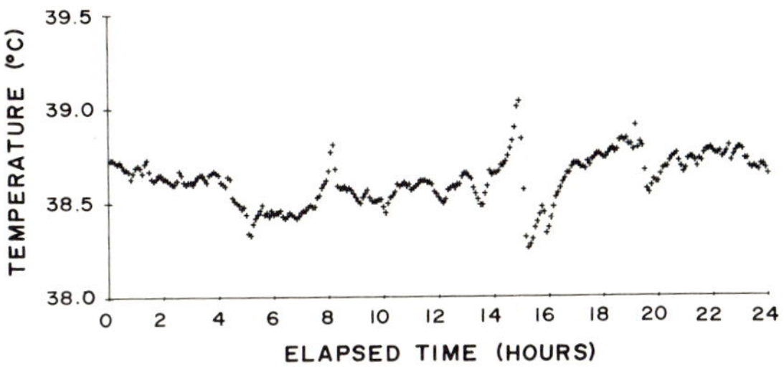

Fig.4: Post-challenge temperatures of vaccinated and unvaccinated steers. Points are hourly means of measurements taken at 60 s intervals.

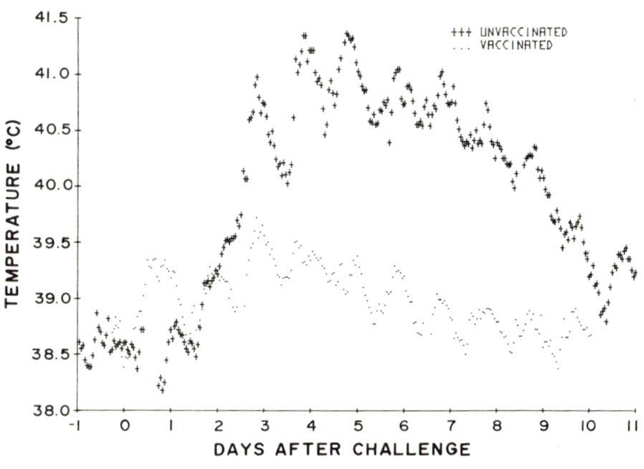

challenge with virulent IBR virus. Temperatures of the unvaccinated steer rose sharply from 48-72 hours after challenge and peaked at 96 and 120 hours. Except for the second day of infection, the diurnal variation continued during the febrile response. However, compared to the prechallenge variations, the amplitude of postchallenge diurnal cycles increased during the ascending phase of the febrile response. The short-term variations also continued during the febrile response.

The difference between temperature records of the vaccinated and unvaccinated steers (Fig. 4) demonstrates the protective effects of prior vaccination. The vaccinated steer developed a low grade and short-lived febrile response with only minor changes in diurnal variations.

Mean ear canal and rectal temperatures over the course of the experiment were 39.45° C and 39.28° C, respectively. Analysis of variance revealed the difference of 0.17° C to be significant ($p < 0.02$).

DISCUSSION

We believe that temperature monitoring by remote methods is a useful tool for studying infectious diseases and for evaluating vaccine performance in cattle. Continuous ear canal temperature measurements greatly increased the quantity of data obtained but also permitted the measurement of parameters not possible by conventional methods of temperature measurement. For example, diurnal variations, which are well-known in normal cattle (2), were perturbed in our experiment as a result of IBR infection. If this proves to be a consistent finding with IBR or other viral infections, it will be a useful quantitative parameter.

Our data indicate that ear canal temperatures were slightly higher than rectal temperatures. This finding is consistent with a similar difference between deep-body and hypothalamic temperatures in the ox (2). Remote ear canal temperature measurement may, therefore, be a tool for monitoring the hypothalamic temperature regulating center in cattle.

Although our experiments were done in animals that were experimentally exposed and housed in security facilities, remote temperature monitoring would enable studies to be made in a typical environment where subjects are naturally exposed, e.g., in feed lots.

Our immediate goal is to include automatically scanning multichannel capability in our system. This improvement would provide automatic acquisition and computer analysis of large amounts of data from several animals studied in parallel.

ACKNOWLEDGEMENTS

We thank Dr. MR Swanson for assistance in data analysis and SK Hanson for technical assistance. This work was supported by the US Department of Agriculture and the Energy Research and Development Administration under ERDA/USDA interagency agreement.

REFERENCES

1. Holm DM, Sanders WM, Payne RJ: Probe activities: Los Alamos Scientific Laboratory report LA6110PR, 1975

2. Whittow GC: Ungulates. In Comparative Physiology of Thermoregulation, Vol 2 (ed. GC Whittow). New York and London, Academic Press, 1971, pp 191-281

A SYSTEM FOR CONTINUOUS MEASUREMENT OF METABOLISM, BODY TEMPERATURE, FEEDING, DRINKING, AND ACTIVITY IN UNRESTRAINED RATS

G. N. McEwen, Jr., R. Miranda, T. B. Fryer,
E. Satinoff and B. A. Williams

NASA-Ames Research Center
Moffett Field, CA 94035

INTRODUCTION

A major reason for using biotelemetry is to record data from unrestrained subjects. When studying animals, this becomes extremely important especially when investigating central nervous control systems which might be influenced by restraint or handling. The effects of restraint and handling have become increasingly troublesome when studying circadian rhythms. Studies with biotelemetry devices have shown marked differences in the measured rhythms as opposed to studies using attached probes. In this paper, we describe a system for measurement of several physiological and behavioral parameters of rats, while the animals are neither restrained nor handled for long periods. Information on the components of the system is commercially available, or will be available from the Technological Utilization Office of NASA-Ames Research Center.

APPARATUS DESCRIPTION

Figure 1 is a block diagram of the system.

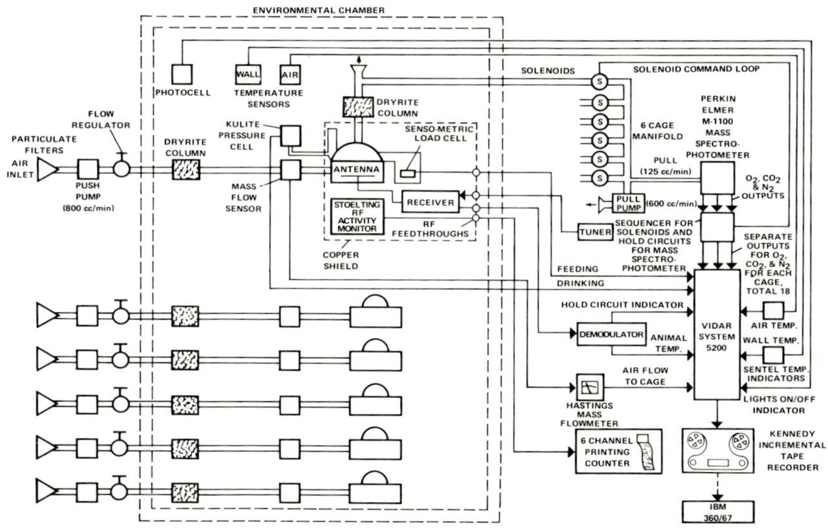

Air is pulled through a particulate filter by an air pump with flow to the cage regulated by needle valve flow regulators. The air stream then passes through a quick change dryrite absorbent column, then through a Hastings mass flow sensor, and then through the cage. This assures that the air volume through the cage is continuously measured. Any portion of the air stream passing through the cage may be monitored for percent oxygen and carbon dioxide to determine metabolic rate. The air stream then passes through another dryrite absorbent column and out an exhaust tube. The flow through the system is maintained at 800 cc/min for rats but can be adjusted for animals of different sizes. A portion of the cage exhaust air is pulled through a time-sequenced solenoid and mixing manifold at 600 cc/min. A mass spectrometer samples 125 cc/min of air from the manifold. The delay time for this system from introduction of known calibration gas into cages to 90% response on the mass spectrometer is 11 sec. A clock-sequencer and sample-hold circuits are provided which program the solenoids and the outputs from the mass spectrometer to buffer circuits at the rate of 20 seconds/channel. This ensures that the buffer circuits are updated every 2 minutes for each cage. A separate buffer circuit exists for each parameter (O_2, CO_2, and N_2) for each cage. Drinking is sensed by Kulite 0-0.5 psig pressure cells which monitor the height of a column of water in the drinking tube. Food consumption is monitored by Senso-Metric dual arm load cells (0-100g), which measure the weight of a food cup containing 45 mg. Noyes food pellets. The environmental test chamber has sensors for lights on/off, wall temperature and air temperature. All of the above analog data is directly sampled by a Vidar 5200 data acquisition system at 2 minute intervals, and recorded on a Kennedy incremental tape.

Animal activity is sensed by a Stoelting RF activity monitor which has an adjustable threshold for differing activity levels. The output of this system is recorded on a six channel printing counter. Body temperature is monitored by Konigsberg temperature telemetry transmitters. The receiver is placed inside a copper mesh shielding cage which isolates each test cage. All leads through the shield are made through RF connectors. This practice minimizes RF noise pickup and intertransmitter cross-talk at the receiver. The demodulators for the temperature telemetry signal are designed to retain their latest valid voltage output even during signal drop-out. This is accomplished by means of a capacitive hold circuit. The voltage output and a hold circuit indicator are recorded as discreet data channels on the Vidar data system. Data from the Kennedy tape recorder are then analyzed on an IBM 360/67.

SUMMARY

A system is described which continuously monitors several physiological and behavioral parameters using telemetry of body temperature and automated systems for controlling and measuring feeding, drinking, activity and metabolism in unrestrained rats. The system is particularly applicable to studies of circadian rhythms and other long duration experiments in small animals.

HEAT FLUX AND TEMPERATURE DATA TELEMETERING SYSTEM FOR MAMMALIAN ENERGY FLOW STUDIES

J. L. Stuart and S. M. Zervanos

J. Stuart Enterprises, Grass Valley, CA 95945
Pennsylvania State University, Reading, PA 19608

The collared peccary's (Tayasu tajacu) interaction with the desert environment was investigated by placing instrumented animals in an unused enclosure in the Phoenix Zoo. The telemetered data monitored the physiological responses of the free roaming animals.

During each study period one male and one female or one male and two females were released into the enclosure. One female was equipped with the heat flux-skin temperature telemetry unit, in addition body core temperature and heart rate were measured with other telemetry units. The heat flux-skin temperature unit was sutured mid-dorsally on the skin of the peccary. The temperature probe and the heat flux transducer (HFT) were inserted subcutaneously in the shoulder on each side of the mid-dorsal line, 20 cm anterior to the transmitter. No discomfort to the subject was apparent from this type of implantation. The unit was calibrated before and after use, with no corrections required. All units were encapsulated with an Elvax^R-paraffine compound (1). The transmitter and its battery pack weighed 85 grams and were designed for surgical implantation.

The telemetry signals were received on a modified commercial FM receiver and recorded on a portable strip chart recorder. The telemetering transmitter consisted of the sensors, the encoding and transmitting electronics and the battery pack. Power was taken from four mercury cells providing 0.5 amp-hours of energy at 5.4 volts. Operating lifetime was approximately 10-11 days. To provide an accurate indication of the HFT output, calibration signals were also transmitted along with the HFT data. The calibration points selected were +1.0 mV, 0.0 mV and -1.0 mV signals against which the HFT signal was compared.

The HFT signal was amplified and multiplexed with the calibration signals as shown in figure 1. The MPX clock was varied by the skin temperature sensor (2). The resulting multiplexed output signal controlled the variable frequency oscillator (VFO). The VFO amplitude was scaled to set the transmitter deviation ratio. The FM-FM signal was recovered and decoded into a suitable data format. Since this system was used on free ranging animals in desert conditions, it was completely portable.

The FM receiver used a four element Yagi antenna located at the edge of the enclosure. The Schmitt trigger shown in figure 2 was toggeled by the demodulated FM signal and divided by two to provide a symmetrical square wave. Its frequency was proportional to the transmitted modulation frequency. The variable pulse width one-shot

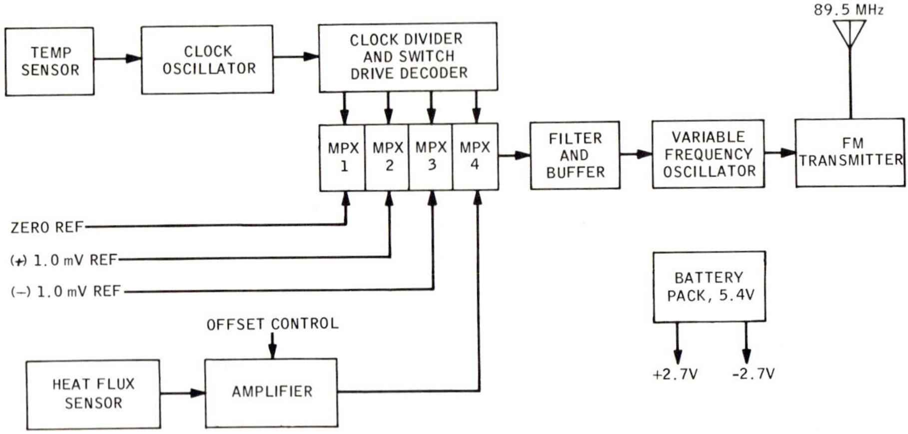

Figure 1 Transmitter Block Diagram

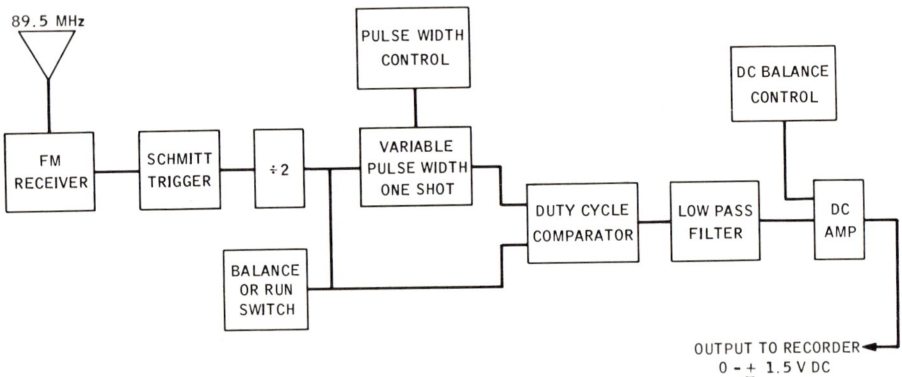

Figure 2 Receiver Block Diagram

was triggered by the square wave from the divider. A pulse width control set the one-shot pulse width to match that of the square wave when the square wave frequency represented the "0.0mV" reference voltage in the transmitter multiplexer. The duty cycle of both the square wave and the one-shot were then exactly the same and the comparator output was zero. When the transmitter multiplexer sampled any other voltage, the square wave frequency was either higher or lower in frequency. When compared to the reference duty cycle of the one-shot, the comparator output would change from the zero reference level. The filter and DC amplifier processed and buffered this signal to provide a recorder output that faithfully reproduced the multiplexed signal present in the transmitter. The data signal was taken ahead of the FM deemphasis network.

In the "balance" position, the balance or run switch inhibited the one-shot and allowed the operator to balance the DC amplifier to set the output level to zero. The recorder pen was set to zero input at the midpoint of the chart. In the "run" position the pulse width control was adjusted to make the "0.0 mV" reference signal coincide with the midpoint of the strip chart. By adjusting the recorder gain the ±1.0 mV reference points were set at the opposite edges of the chart. With the HFT in an isothermal medium, the sensor should show an output of almost zero mV. The offset pot for the HFT amplifier was then adjusted for coincidence of the data and zero reference signals on the strip chart recorder. This adjustment removed the small input offset voltage present at the HFT amplifier input terminals. The recorder now indicated the HFT sensor directly in mV. Multiplying this voltage by the transducer constant (0.00433 cal $cm^{-2}min^{-1}mV^{-1}$) established the heat flux in Ly/minute. A sample of the data recording is shown in figure 3.

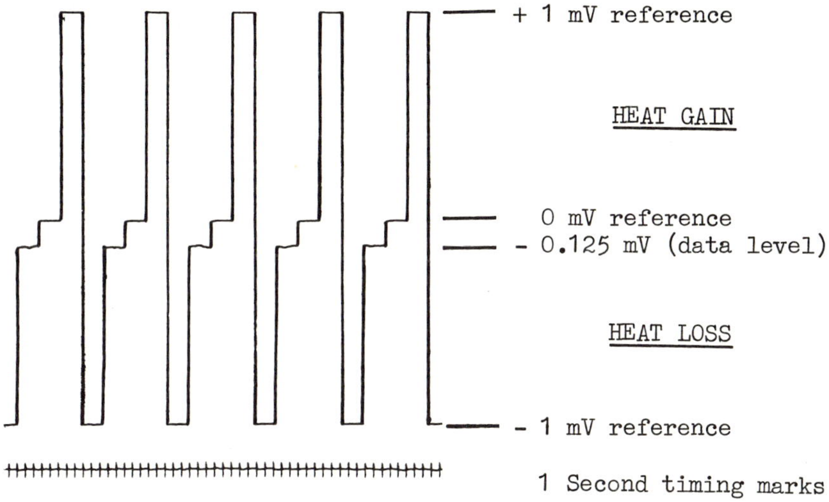

Figure 3 Sample of Recovered Data

Skin temperature was determined by counting the number of data cycles per minute and using equation 1 to calculate the temperature.

$$\text{Log } T = 0.32438(D - 4.05) + 1.0000 \qquad (1)$$

T = Calculated temperature in °C
D = Number of data cycles per minute

The above equation was derived from the empirical data taken from the temperature sensor calibration data.

The peccaries were allowed to roam freely within the enclosure. Activity, location, feeding and orientation data were collected each hour throughout the day and night. Respiration rates were determined visually, using a stopwatch. Environmental data were also collected simultaneously with the physiological data from the peccaries.

The skin temperatures during the winter remained above ambient at all times, with a significant increase to 42.3°C during the basking periods or when the animals were active in the sun (3). This rapidly decreased to 29.7°C as the solar radiation decreased. Heat flux during daytime basking showed a net heat flow to the animal at a rate of 0.4 Ly/min.. This rate was maintained during periods when solar radiation was high (10 am to 4 pm). Heat loss, which began soon after sunset, averaged 0.20 Ly/min. throughout most of the night. Since the heat flux probe was in an area of the body exposed to the environment during huddling behavior, the data may not be indicative of actual heat loss from most of the animals' surface, especially those areas making contact with other peccaries.

In addition to free roaming animals, the physiological responses of peccaries placed in restraining cages were studied and the data compared with that received from the telemetering units. Body and skin temperatures from both tethered (caged) and telemetered animals were basically the same during comparable study periods (2). The greatest difference was in the winter minimum body temperature, where the tethered temperature was 1.1°C lower than that monitored by telemetry. Skin temperature differences obtained from the two methods of measurement ranged from 0.4°C for the winter maximum value to 4.9°C for the summer daytime average.

Telemetered heat flux values differed from those measured in tethered animals, especially winter nighttime heat loss. Observed heat flux patterns were quite similar for both tethered and free roaming animals and justified tethered animal positioning in the environment to simulate the behavior of free roaming animals.

REFERENCES

1. Stuart JL: Biomedical Transducers. Proceedings of a Symposium on Biotelemetry (1971), pp343-362
2. Stuart JL: Avian Biotelemetry. IEEE 1975 Region Six (Western USA) Conf., pp 21-24
3. Zervanos SM, Hadley NF: Adaptational Biology and Energy Relationships of the Collared Peccary (tayasu tajacu). Ecology 54:4 pp 759-774, 1973

This work was supported by the National Science Foundation Grant (GB-27706).

RF TELEMETRY SYSTEM FOR PRESSURE AND TEMPERATURE MONITORING OF FREE SWIMMING FRESHWATER FISH

H. Guckel A. K. Scidmore
J. B. Beyer S. Chitre
Department of Electrical Engineering
University of Wisconsin-Madison
Madison, Wisconsin 53706

Biologists from the Limnology Laboratory at the University of Wisconsin-Madison are conducting studies of salmon migration in Lake Michigan. One phase of their work involves the determination of modifications in migration habits due to thermal fallout near nuclear power plants. In particular, questions are asked which involve the depth and temperature of the water at the location of the fish in the thermal plume area. The instrumentation package which is discussed here was designed and constructed to provide water pressure and temperature data and to serve as a prototype for future transmitters.

The transmitted signal is a pulse tone modulated 52 MHz RF-carrier. The choice of an RF rather than an acoustic carrier was based on the fact that in shallow water near the shoreline, reflections of acoustic signals cause problems with data recovery for the chosen type of modulation scheme. The disadvantage of using an RF scheme is primarily that a RF-antenna is more difficult to handle than an acoustic transducer. Furthermore, adequate range can be obtained only as long as water depth is not too large and water conductivity is not too high. Thus, half-power depth is roughly 2 meters for a water conductivity of 260 μmhos/cm. It should be pointed out that since the transmission angle of a radio wave emerging from the high dielectric constant water medium is nearly 90°, the resulting signal path is essentially vertically upward from the fish to the surface and then along the surface to the receiver.

The pressure signal is represented by the pulse repetition rate. At zero pressure a typical value is 175 pulses/minute. This pulse rate decreases by 4 pulses/minute for every 1 psi increase in pressure. A single pulse per minute represents, therefore, roughly 17 cm of water which is the water depth resolution. The water depth signal is readily decoded by computer processing of the taped experimental data. Decoding in the field is possible by utilizing the human ear and a stop watch. This property of the tag was considered important and led to the choice of the present coding scheme.

The temperature signal is encoded as a tone burst. The upper frequency for the tone modulation is 500 Hz which occurs typically at 30°C. The modulation frequency decreases to 100 Hz at 0°C. Resolution is therefore roughly 0.1°C. Functioning of the transmitter in the field is readily checked by ear, which is a highly desirable property of the device.

A summary of the data format is shown in Figure 1.

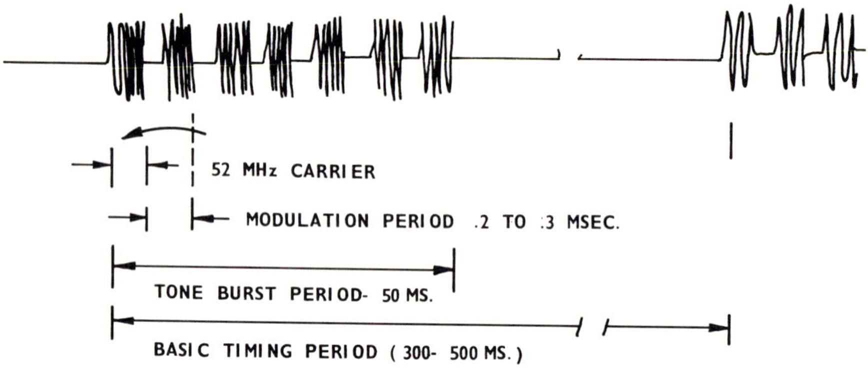

Figure 1. Data Format

Electronic design of the transmitter hinges on the type of pressure sensor which is used. The choice for practical reasons involved a silicon pressure sensor. A bridge impedance of 5000 Ω was chosen. The device is constructed on a 6 mm × 6 mm silicon substrate of 0.008" thickness. The device is mounted so that two elements are in compression and two in tension. Measured output voltages are 1 mV/psi at 4 volt bridge excitation. The device may be used successfully for the present application only if the off-set voltage due to bridge unbalance is small and if the bridge power supply is stable. The off-set problem was cured by proper processing. The bridge power supply was stabilized by utilizing a well designed voltage regulator. An additional requirement, namely, low current drain was satisfied by utilizing a switched power supply for the bridge. The resultant block diagram appears in Figure 2.

The detailed circuit design considerations will not be covered here. However, some unusual aspects will be pointed out. Figure 3 details the voltage regulator. This circuit has an internal power dissipation of 1.2 mW, a temperature stability of 4 mV/°C and a required input-output differential of only 0.2 volts. The reference diode is operated near 50 μA, which makes device selection somewhat difficult. The pressure circuitry is shown in Figure 4. This circuitry provides the basic timing cycle for the transmitter. Here G_1 acts as a level sensitive comparator. C_1 is charged negatively by the OTA and positively by the OTA-R_1 combination. The OTA current is determined by the voltage across C_2 which provides the holding effect needed. V_2 controls

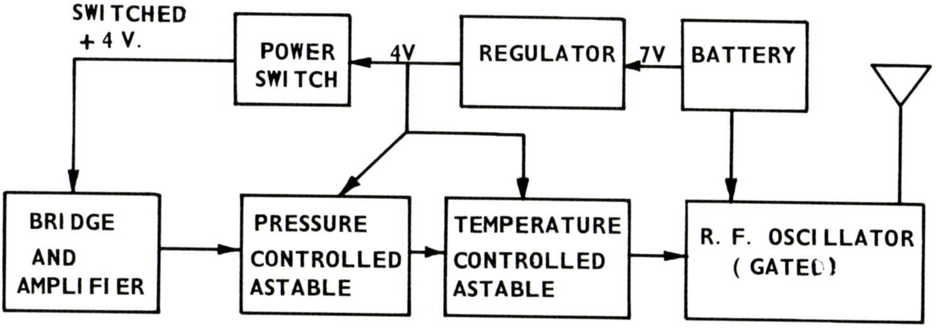

Figure 2. Block Diagram

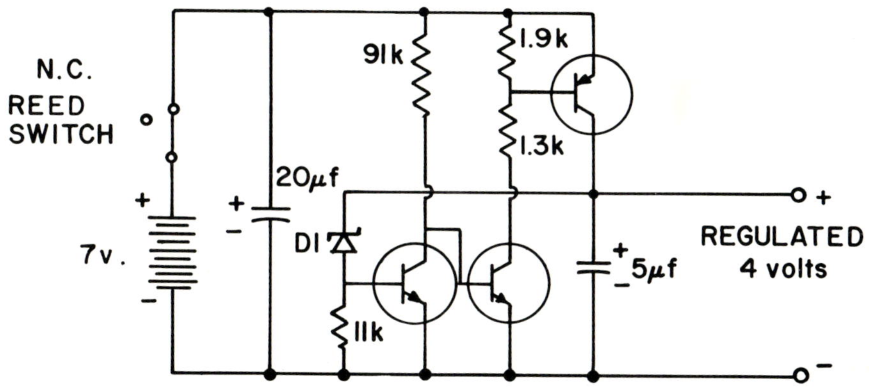

Figure 3. Low Voltage Regulator

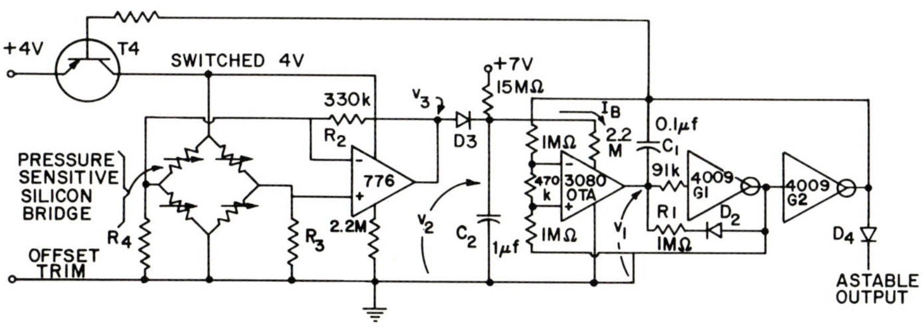

Figure 4. Pressure Controlled Astable

the basic data format. The pressure transducer is switched into the circuit when C_1 is charging. Thus V_3 is a pulsed voltage with amplitude proportional to the sensed pressure. The pressure sensitivity is controlled by R_2. R_3 or R_4 is used to provide the proper zero pressure offset.

The temperature signal is generated by a two-inverter CMOS-thermistor type oscillator of standard design.

The 52 MHz circuitry is a buffered crystal oscillator. Antenna matching is achieved with a transformer coupled power amplifier stage.

The transmitter was constructed from discrete components. The usual packaging difficulties were experienced. Shielding problems between the RF section and the high impedance pulse circuitry were solved by proper layout. The final result was a prototype which was field tested successfully.

The engineering conclusion which one might draw from the collected data is simply that P-T transmitters are feasible but complicated. The reason for this is basically the pressure transducer. The resistive device is quite simple. However, the interconnecting circuitry is cumbersome and expensive. The reason for this is in part due to the long lifetime - low battery voltage requirement which is needed to construct a transmitter which is acceptable to biologists. These problems are essentially eliminated by utilizing a capacitive pressure sensor. Such a device which consists of a silicon-silicon dioxide-silicon assembly with a 2 micron air gap has been designed, constructed and tested. Results indicate adequate pressure sensitivity in a CMOS oscillator. The circuitry for the transmitter becomes very simple, the efficiency is increased considerably, and the transmitter size is reduced. This leads one to the conclusion that future designs of P-T tags will utilize capacitive sensors.

TELEMETRY OF BODY TEMPERATURE AND HEART RATE OF HIBERNATING AND NORMOTHERMIC GROUND SQUIRRELS

M.J. Devaney, D.R. Deavers, A.M. Coria,
G.E. Tempel, and X.J. Musacchia

Dalton Research Center, University of Missouri,
Columbia, Missouri 65201

INTRODUCTION

In our studies of water turnover rate (WTR) in 13-lined ground squirrels (Spermophilus tridecemlineatus) it was considered essential to monitor body temperature (T_b). During physiological investigations such as water turnover studies, which require repeated blood sampling over periods of several days, it is advantageous to know T_b throughout the data collection period. This is particularly true when one is studying animals, such as ground squirrels, that are capable of torpidity in which T_b may be lowered to various levels for variable durations. Through continuous recordings from implanted heart rate (HR) and T_b transmitters, it was our goal to characterize the physiological state of ground squirrels during water turnover studies. It is well established that metabolic rate of torpid animals is much lower than that of normothermic animals (7), but there is little information on WTR during torpor (4,5,6).

METHODS

The study of water turnover in hibernating and normothermic ground squirrels required a temperature and heart rate telemetry system which could accomodate a large number of caged squirrels. Barr (1) considered a somewhat similar system though restricted to T_b. Ground squirrels implanted with transmitters were placed in copper mesh Faraday cages arranged side by side in an environmental chamber. The transmitter was a variation of the Barrows (2) squegging oscillator circuit in which the EKG signal is used to modulate the squeg frequency about its mean value which is determined by T_b. The transmitter was varied to limit the minimum squeg frequency at reduced temperatures to facilitate the detection of heart rate. The squeg rate varied from approximately 200 Hz at 7°C to 700 Hz at 37°C and the RF was in the AM Broadcast Band.

The receiver subsystem of Fig. 1 was designed to accomodate up to ten cages. The receiving antennas consisting of 40 turns of #26 enamel clad copper wire on a 9" x 11" form were placed between the animal and Faraday cages. The receiving antennas were time multiplexed using CMOS CD 4016 quad transmission gates driven from a decade counter. Channel one, monitoring a reference transmitter, served as a cycle marker on the polygraph recording and permitted continuous verification of receiver performance.

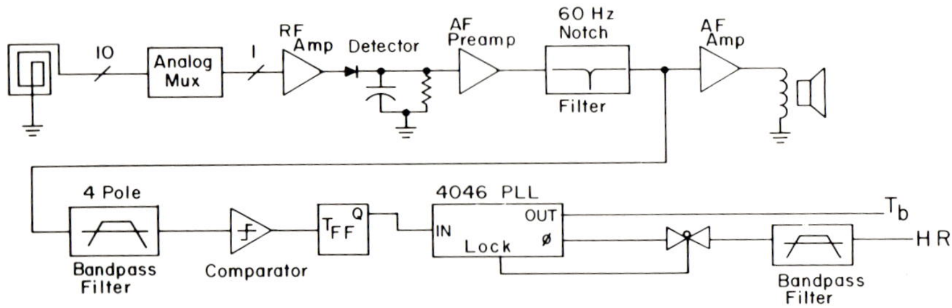

Fig. 1. Telemetry receiver subsystem.

The signal from the multiplexor was amplified prior to envelope detection. The detected audio was further amplified and passed through a 60 Hz notch filter. The filter output was monitored via an additional audio stage and loud speaker to provide aural verification of the system performance. The notch filter output was also passed through a 200-700 Hz 4 pole active bandpass filter to remove frequency components outside the squegging range and converted to a constant amplitude symmetrical square wave by passing the signal through a comparator driving a T-Flip-flop. The square wave from the flip-flop drove the input to a CD 4046 phase locked loop which performed the FM detection. The average error voltage from the loop was proportional to the average input frequency which was a function of T_b. This level was supplied to one channel of the stripchart recorder, while the EKG was obtained by AC coupling the variations in the loop error voltage. The phase locked loop was equipped with a loop lock detector which drove a front panel LED indicator when the loop was locked on the input signal,and when the loop was not in the locked state, this signal served to decouple the EKG signal from the second channel of the recorder by disabling an analog switch. This served to eliminate the multiplexor switching transient from the recorder output.

Each squirrel was anesthetized with nembutal, the transmitter placed midventrally in the peritoneal cavity, and the HR electrodes were threaded subcutaneously to the lower-left and upper-right ventral thoracic region, and sutured in place. The right jugular vein was cannulated, and 0.50 mCi tritiated water (^3H_2O) was injected i.v. The left carotid artery was cannulated and arterial blood was sampled (0.05 ml) after ^3H_2O had equilibrated with body water (2 1/2 hr pers. observ.) (8). The animals were kept at 20°C for a few days, then transferred to 7°C with the cage placed over the antenna for reception of T_b and HR. Daily blood samples were taken in heparinized hematocrit tubes and centrifuged. Then 0.015 ml plasma was added to 10 ml Bray's solution (3), and radioactivity was determined with a liquid scintillation counter (Beckman, LS-250). Counting efficiency was determined by adding a known amount of ^3H_2O and 0.015 ml nonradioactive plasma to 10 ml Bray's. Total body water (TBW) was determined from the first blood sample by the dye dilution principle. Log ^3H_2O concentration of plasma of cold exposed animals was plotted against time of sample collection. Slope (K), or rate of decrease of plasma radioactivity, was determined by least squares regression. WTR is the product: K x TBW.

RESULTS

Fig. 2 shows examples of the rate of decrease in plasma radioactivity and the regression equations describing these relationships. The three squirrels in Fig. 1 are known to have been normothermic (T_b 35-37°C) during the experimental period. Water turnover calculated from such plots of 12 squirrels averaged 14.5 ml/day.

Fig. 3 is a similar plot for 2 squirrels. One was normothermic for 2 days and then hibernated while the other exhibited nightly lowering of T_b to about 20°C for 2 days and was normothermic the third day. The slope for animal #15 (Fig. 2) during the hibernation period, is zero, indicating water loss is so low that it cannot be detected with the ^3H_2O method. The effect of

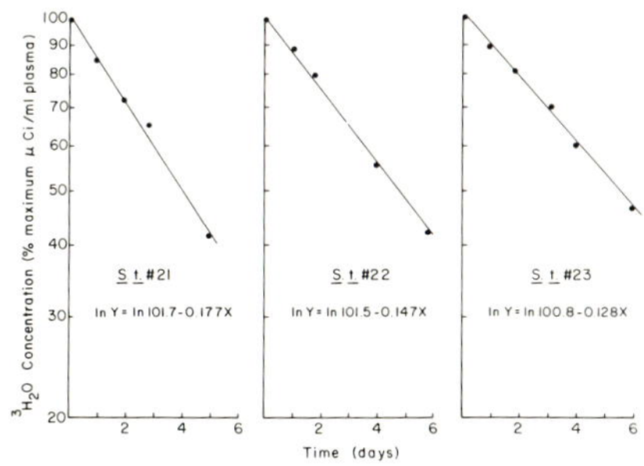

Fig. 2. Decrease of plasma ^3H_2O concentration of normothermic cold exposed ground squirrels. Equations are of the form: ln Y = ln a + KX, where a is the Y-intercept and K is the slope and is negative.

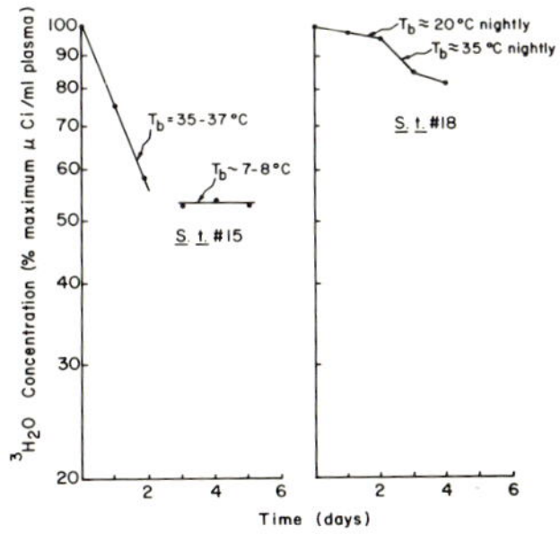

Fig. 3. Effect of lowered T_b on rate of decrease in plasma ^3H_2O concentration.

lowering T_b to about 20°C on water loss is also seen in the low slope for animal #18 as compared to the normal slope when the same animal was at $T_b \sim 35°C$ (Fig. 3).

DISCUSSION

It is clear that T_b has a marked effect on WTR. Hibernation at low ambient and body temperatures results in very low water loss. It is possible that water conservation as well as energy conservation may have been a factor in the evolution of hibernation in this species.

The use of T_b and HR transmitters during water turnover studies of animals capable of torpidity is useful in that the investigator knows the state of the animal during the study. In the absence of nocturnal T_b measurements, data from animals like #18 (Fig. 3), which exhibited nightly lowering of T_b, would be anomalous and might incorrectly be interpreted as a very low WTR in a normothermic animal.

REFERENCES

1. Barr RE: A telemetry system for recording body temperature on large numbers of caged rodents. Med. and Biol. Eng., Vol. 10, pp. 677-684, 1972.

2. Barrows WF: Implanted telemeter for electrocardiogram and body temperature. NASA Tech Brief B72-10035, 1972.

3. Bray GA: Determination of radioactivity in aqueous samples. In The Current Status of Liquid Scintillation Counting (ed ED Bransome, Jr). New York, Grune and Stratton, 1970, pp. 170-180.

4. Deavers DR, Musacchia XJ: Water turnover, total body water, and metabolism of hibernating and cold acclimated normothermic 13-lined ground squirrels. Fed Proc 35: 639, 1976.

5. Fisher KC, Manery JF: Water and electrolyte metabolism in heterotherms. In Mammalian Hibernation III. London, Oliver and Boyd, 1967.

6. Hudson JW: Torpidity in mammals. In Comparative Physiology of Thermoregulation, Vol. 3, Special Aspects of Thermoregulation (ed GC Whitton). New York, Academic Press, 1973, pp. 97-165.

7. Kayser C: The Physiology of Natural Hibernation. New York, Pergamon Press, 1961.

8. Richmond CR, Langham WH, Trujillo TT: Comparative metabolism of tritiated water by mammals. J Cell Comp Physiol 59: 45-53, 1962.

AN EXTRACORPOREAL MULTICHANNEL TELEMETRY SYSTEM FOR PHYSIOLOGICAL APPLICATIONS

R. Miranda, G. N. McEwen, T. B. Fryer, and B. A. Williams

Ames Research Center, NASA
Moffett Field, California 94035

INTRODUCTION

A small multichannel transmitter (2 x 3 x 4 cm), designed to be plugged into a skull-mounted connector in rabbit-sized animals, is described in this paper. The system provides for the measurement of ECG, EEG, EMG, and up to six body temperatures. Multichannel telemetry systems for measuring ECG, EEG, and EMG (1,2), as well as multiple temperatures (3,4) have been reported in the literature (1,2). The temperature measurements have normally been made using thermistors (3,4). The system described here employs copper-constantan thermocouples for temperature measurements. Thermistors have been used in the past because of their large signal output for small temperature changes, their low cost, and their availability. However, thermistors also have disadvantages; for example, they are often inaccurate when exposed to moisture or high humidity. Sealing implanted devices from moisture is one of the more difficult instrumentation design problems, especially when transducers must be remotely located from the electronic components. Thermocouples are widely used in temperature measurements, but they provide extremely low signals, of the order of tens of microvolts, for physiological applications where temperature changes are normally only a few degrees or less. The salient feature of the system described here is that dc voltages as low as 4 μV (equivalent to 0.1°C) can be accurately measured. By overcoming the low signal difficulty, the many advantages of thermocouples can be exploited. For multiple measurements at diverse locations, the thermocouples offer the advantages of small size, easy adjustment to length during surgery, linear output, low cost, and low impedance values which minimize their susceptibility to moisture.

CIRCUIT DESCRIPTION

A block diagram of the telemetry transmitter is shown in figure 1.

All thermocouple output readings are sampled by CMOS subcommutated electronic switches and are amplified to suitable levels by a properly compensated amplifier. Commutation with CMOS switches eliminates the need for bulky dc amplifiers and achieves

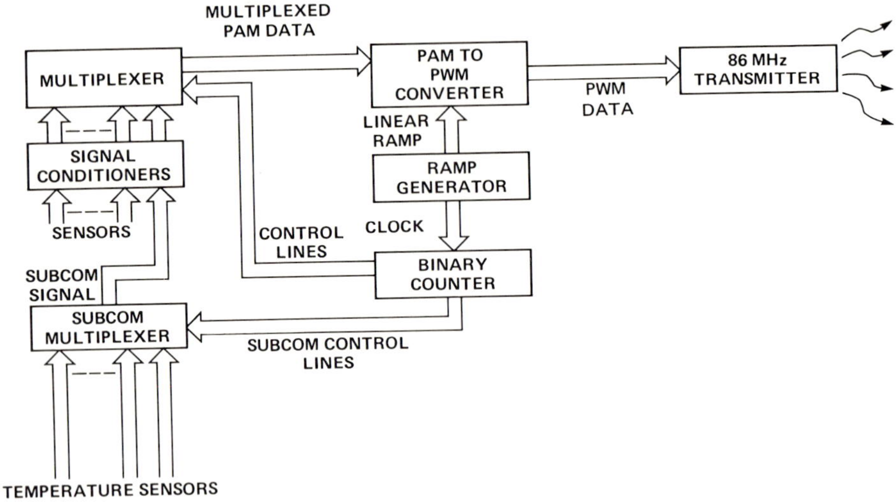

Figure 1. Telemetry transmitter block diagram.

the required static (dc) accuracy by use of an ac amplifier. The low zero offset of the CMOS switches allows measurements accurate to within a few microvolts. Overall system accuracy is maintained by ratio detection in the demodulator to compensate for possible telemetry and signal conditioning drift and errors. Since a constant temperature reference junction is not feasible, a thermally insulated junction point in the implanted skull connector is used as a reference for all the thermocouples, and this point is then monitored with a moisture-sealed thermistor. All the electrode signals, including subcommutated temperature signals, are preamplified by differential instrumentation amplifiers with a frequency response to 100 Hz.

The transmitter operated with four mercury batteries (±2.6 V, 1.15 cm diam, 0.33 cm height, and 100 mAhr) which provided continuous operation for approximately 50 hr.

Figure 2 shows the complete unit with both sets of batteries mounted on the side.

The ramp generator is a free-running oscillator that provides the fixed timing for sampling and multiplexing of the input signals by generating a linear sawtooth waveform. Narrow pulses derived from the negative-going transition of the sawtooth ramp serve as the clock pulses to operate the multiplexing unit. Each incoming clock pulse closes one of eight electronic CMOS switches, which sequentially sample each data channel. The next pulse, which arrives approximately 180 μsec later, opens the switch and closes the next and so on for a total of eight channels (one frame). Thus, every channel is sampled at a rate of approximately 0.7 kHz. The amplitude modulated series of pulses out of the

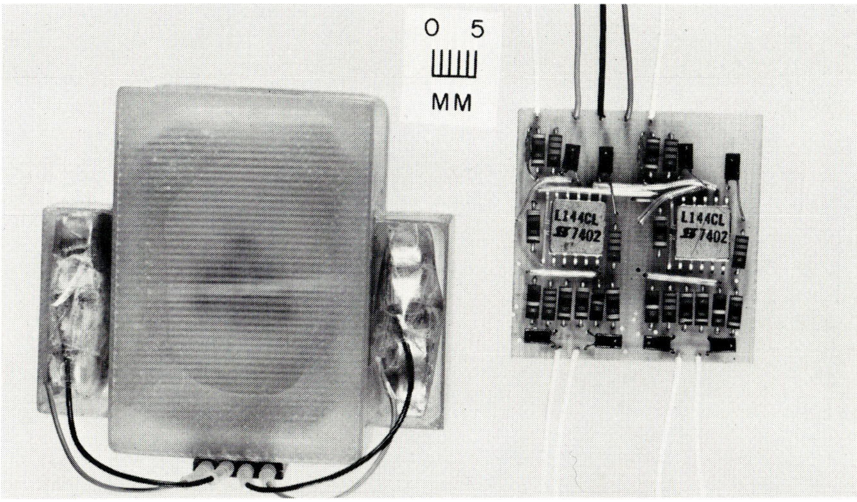

Figure 2. Complete unit with both sets of batteries and signal conditioning board showing assembled components.

multiplexer is then converted into a pulse width modulated format. Every eighth channel is blanked for frame synchronization, and digital limiting is provided in the PWM converter to prevent broken transducer leads or overdriven signals from generating false synchronizing pulses.

The eight data channels are assigned in the following way: Channel 1 is permanently grounded to provide a zero reference; channels 2 thru 6 are used to monitor EMG, EEG, and ECG signals plus two spare channels that are available for future expansion; and channel 7 subcommutates into eight additional low-speed channels every 0.115 sec. Temperatures such as core, nasal (breathing rate), brain, and back are then measured. The basic PWM and multiplexed telemetry system is the same as has been described previously for use with other transducers, such as pressure and flow transducers (1). A more detailed description of the PWM modulation and demodulation scheme is given therein.

The transmitted rf signal is received using a telemetry or high fidelity type FM receiver. The PWM signal out of the receiver is then demodulated and demultiplexed to obtain the original analog input waveforms. Both the transmitter and demodulator are independent of any linearity adjustments because the data are coded into the pulse duration.

PERFORMANCE

The accuracy of the temperature measurement was tested by various calibration tests. In one test the reference thermocouple junctions, as measured by the sealed thermistor, were varied while the sensor thermocouple temperatures were kept constant. The analog voltages out of the demodulator were converted to temperature; the results, summarized in table 1, demonstrate an accuracy of approximately ±.1°C.

Table 1.

Thermocouple temp., °C	Reference thermistor temp., °C	Demodulation Thermocouple output, °C	Thermistor output, °C	Thermistor + thermocouple, °C
37.6	26.1	11.2	26.1	37.3
37.5	27.4	9.8	27.4	37.23
37.6	28.2	9.0	28.2	37.2
36.9	30.0	6.7	30.0	36.7
37.4	35.0	2.0	35.0	37.0
37.7	36.0	1.3	36.0	37.3

SUMMARY

The unique feature of this multichannel telemetry system is its ability to accurately measure and telemeter microvolt dc signals. This allows the use of thermocouple sensors with their many advantages for such studies as thermal regulation in small laboratory animals where multiple temperature measurements at diverse body locations are required. The system also provides the capability of measuring other significant parameters such as ECG, EEG, and EMG that are required in most experimental studies.

REFERENCES

1. Fryer TB, Sandler H, Freund W, et al.: A multichannel implantable telemetry system for flow, pressure and ECG measurements. J Appl. Physiol 39:2, 318-26, 1975

2. Fryer TB, Pope JM, Winget C: An implantable multichannel temperature transmitter. Proc. 20th An Conf on Eng & Bio, 1967

3. Smith EN: Multichannel temperature and heart rate radio telemetry transmitter. J Appl. Physiol 36: 252-5, 1974

4. Pauley JD, Reite M, Walker SD: An implantable multichannel biotelemetry system. EEG Clin Neurophysiol 37: 153-160, 1974

CHANGES IN HEART RATE AND RESPIRATORY FREQUENCY ASSOCIATED WITH SPONTANEOUS SUBMERSION OF DUCKS

P.J.Butler and A.J.Woakes

Department of Zoology and Comparative Physiology, University of Birmingham Birmingham B15 2TT, U.K.

It might be expected that in birds and mammals, the cardiovascular responses to submersion (i.e. vasoconstriction and bradycardia), which serve to reduce drastically the rate at which oxygen is supplied to all tissues except the central nervous system (1), would be initiated at their maximum level at the beginning of the dive (5). This certainly happens in harbour seals (4). Studies on ducks, however, have revealed that heart rate declines slowly during a dive, reaching its lowest level, of approximately 15% the pre-dive value, some 40-60 sec after submersion (2,3). At least this is the situation when the birds are lightly restrained in the laboratory with only their head placed under water (hereafter called laboratory dives). Because of these apparent differences in the cardiac chronotropic response to diving between mammals and birds, and because of the obvious differences in the activity of the animal during laboratory and spontaneous dives, it was decided to telemeter changes in heart rate associated with spontaneous dives of birds on an outside pool (4 x 8m x 1m depth).

Initially, a single channel, commercially-available FM transmitter (SNR 102F, Devices Instruments Ltd., Welwyn Garden City) was used. It was placed in a water-tight box which was held in position on the back of the bird by straps passing underneath the base of the wings. The e.c.g. electrodes were placed in the thoracic cavity close to the heart, run underneath the skin and exteriorized close to the transmitter. The wing straps and the e.c.g. electrodes regularly broke as a result of the pecking they received from the bird, and consequently useful results were only obtained from one animal, an Australian white eye (Aythya australis).

Spontaneous dives often occurred in a series, with 3-15 dives following in quick succession. The duration of each dive was 5-20 sec and the first dive of a series (or a single isolated dive) was heralded by an increase in heart rate to approximately double the value recorded when the animal was swimming. Immediately upon submersion heart rate fell to a level which was only slightly below that recorded during swimming (Fig.1). This level was maintained throughout the whole of the period of submersion. Upon surfacing, there was a clear tachycardia and during the next dive in a series heart

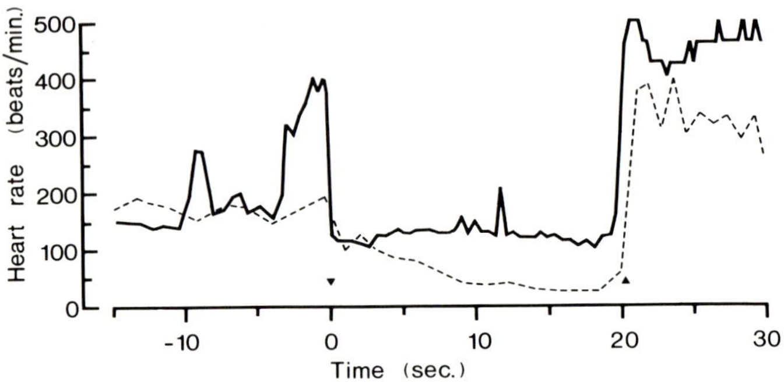

Fig.1 Changes in heart rate during first spontaneous dive of a series (——) and during a laboratory dive (---) of similar duration. Aythya australis using commercial F.M. transmitter. Arrows indicate submersion and surfacing.

rate fell immediately to a level similar to that recorded during the first dive. Thus, although a bradycardia occurred during submersion with respect to the heart rate immediately preceding a dive, it was not really a bradycardia in terms of the heart rate recorded during swimming, nor was it anything like the bradycardia recorded during a laboratory dive of similar duration (Fig.1).

This data indicated that it would be worth continuing with telemetry of information during natural submersion of birds and a two channel transmitter was designed and built for the purpose (7). It monitored respiratory frequency by way of a thermistor

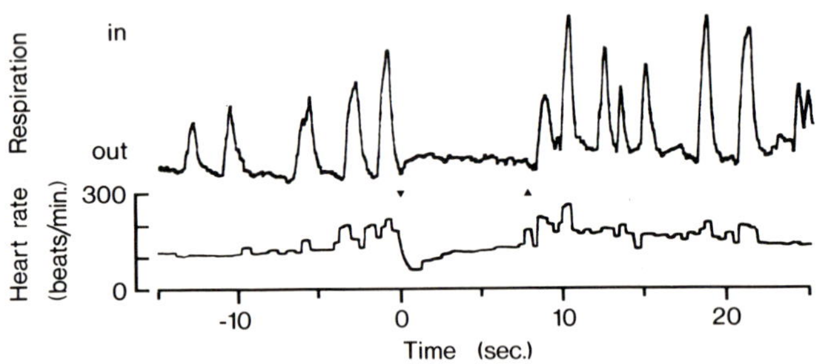

Fig.2 Changes in heart rate and respiratory frequency during first spontaneous dive of a series. Same duck as in Fig.1 using implanted purpose-built transmitter. Arrows indicate submersion and surfacing.

placed in the trachea and e.c.g. by a bipolar electrode placed close to the heart. The whole unit; transmitter, electrode and leads, was implanted inside the animal. Although it was used predominantly with tufted ducks (Aythya fuligula) and pochards (Aythya ferina), Fig.2 shows the first spontaneous dive of a series from the Australian white eye (i.e. the same bird from which the data in Fig.1 was obtained). It can be seen that qualitatively the change in heart rate is similar to that seen in Fig.1, although the absolute rates are lower in Fig.2. Also in Fig.2, the heart rate immediately upon diving is lower than the swimming heart rate, and it increased progressively during the period of submersion. This sequence was quite common when the implanted transmitter was used. Respiratory frequency increased just before the dive.

The recording of ventilatory activity made it possible to relate the onset of apnoea (or at least the cessation of air movement through the trachea) to the onset of the bradycardia. Generally speaking, the birds breathed out upon submersion and,

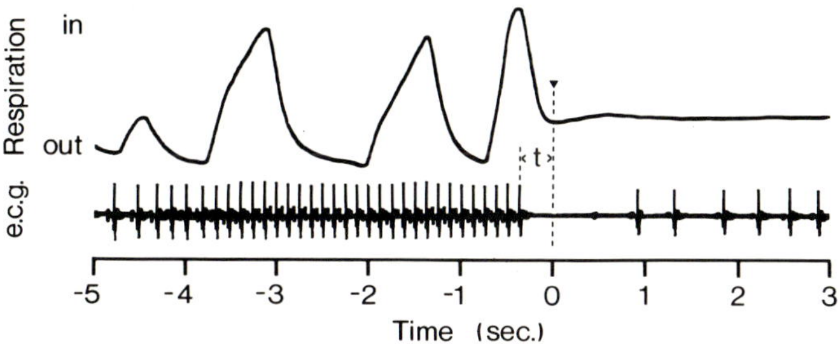

Fig.3 Relationship between cessation of air flow through the trachea (indicated by the long, vertical, dashed line) and the onset of bradycardia at the beginning of a spontaneous dive. Tufted duck (A.fuligula).

on average, heart rate began to decrease before the flow of air through the trachea had stopped and hence before the closure of the glottis (Fig.3). From 19 dives (not necessarily first dives of a series) by 3 birds, the mean interval between the last QRS wave of the e.c.g. before submersion and the cessation of air flow through the trachea ('t' in Fig.3) was 263 $\pm$ 28 msec (S.E. of mean) whereas the mean cardiac interval of the 5 heart beats before submersion was 170 $\pm$ 15 msec (S.E. of mean). As the former is significantly ($P<0.01$) greater than the latter, it is clear that not only is there tachycardia and tachypnoea before the first spontaneous dive of a series, but that the onset of bradycardia precedes apnoea when birds dive spontaneously. Thus it appears that these responses to diving are initiated centrally.

During laboratory dives, the progressive bradycardia seen in birds is thought to result from a progressively more intense stimulation of the arterial chemoreceptors as PaO_2 declines and $PaCO_2$ increases (3). This is not apparently the case during spontaneous dives, as the bradycardia, such as it is, is instantaneous and does not become progressively more intense. Whether or not the peripheral chemoreceptors are involved in the cardiovascular responses to spontaneous submersion of birds remains to be seen. The likelihood of this seems doubtful in view of the fact that in penguins, PaO_2 changes very little during spontaneous dives, although it falls markedly during laboratory dives of similar duration (6). The whole question of the control of the circulation during submersion of birds needs to be re-investigated, this time with respect to spontaneous (or perhaps trained) dives.

ACKNOWLEDGEMENTS This work was supported by the Science Research Council.

REFERENCES

1. Blix A.S. Kjekshus J.K., Enge I. et al: Myocardial blood flow in the diving seal. Acta physiol. scand. 96: 277-280, 1976.

2. Butler P.J. & Jones D.R: The effect of variations in heart rate and regional distribution of blood flow on the normal pressor response to diving in ducks. J.Physiol. 214: 457-479, 1971.

3. Butler P.J. & Taylor E.W: The effect of hypercapnic hypoxia, accompanied by different levels of lung ventilation, on heart rate in the duck. Respir. Physiol. 19: 176-187, 1973.

4. Elsner R.W: Diving mammals. Science Journal 6: 68-74, 1969.

5. Irving L: Elective regulation of the circulation in diving animals. In Whales, Dolphins and Porpoises (ed K.S.Norris). London, Cambridge University Press, 1966, pp 381-395.

6. Millard R.W., Johnansen K. & Milsom W.K: Radiotelemetry of cardiovascular responses to exercise and diving in penguins. Comp. Biochem. Physiol. 46A: 227-240, 1973.

7. Woakes A.J. & Butler P.J: An implantable transmitter for monitoring heart rate and respiratory frequency in diving ducks. Biotelemetry (in the Press) 1976.

A HEART RATE--ACTIVITY BIOTELEMETRY SYSTEM FOR USE ON WILD BIG GAME ANIMALS

Jerry J. Cupal, Rocky Mountain Forest and Range Experiment Station, Laramie, Wyoming.

R. W. Weeks, Electrical Engineering Department, University of Wyoming, Laramie, Wyoming.

C. Kaltenbach, Animal Science Department, University of Wyoming, Laramie, Wyoming.

Knowledge of physiological functions such as heart rate and animal activity are valuable in the management of our wildlife resources. Because earlier studies present strong evidence that heart rate is affected by external disturbances (2), monitoring of this parameter could enhance behavioral studies. In addition, heart rate shows a good correlation with metabolic rate (2,3) and could be an important parameter in energy budget determination on free-roaming animals.

A telemetry system for collecting internal biological information from big game animals requires specialized equipment. Internal electrodes and an implanted transmitter are required to monitor heart rate. Other systems have been developed, but short life and limited transmission range restrict their usefulness. To overcome these problems, repeater-type systems have been developed (1). A similar technique is used to retransmit heart rate information up to 20 kilometers from wild elk (Cervus canadensis). The same system collects animal activity data by monitoring RF signal strength. This paper describes the biotelemetry system and procedures used in the field installation of the heart rate monitor.

TELEMETRY SYSTEM

The telemetry system consists of the following: an implanted heart pulse transmitter, a repeater neck collar, and appropriate decoding circuits for analog recording of heart rate and signal strength.

Implanted transmitter. The function of the implanted transmitter is to monitor the EKG signal and emit a pulse whenever the R wave occurs. The EKG signal is processed through two bandpass amplifying stages (fig. 1) (CA3080 and A1 of L144) and keys on a low-frequency (1Mhz) transmitter (D26E1) whenever a pulse of sufficient amplitude appears on the input. An automatic gain control voltage (AGC) is generated by means of a peak detection circuit (A2 of L144) to insure that the system has sufficient gain so the R wave triggers the pulse detector (A3 of L144). Electrode placement is not critical for operation of this system because of the AGC circuit.

The EKG is detected by means of implanted sternum electrodes. These electrodes are constructed in the form of an "electrode

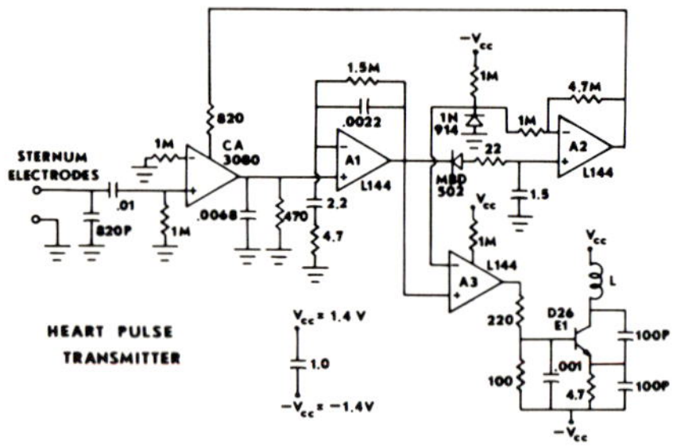

Fig. 1. Schematic diagram of implanted heart pulse transmitter.

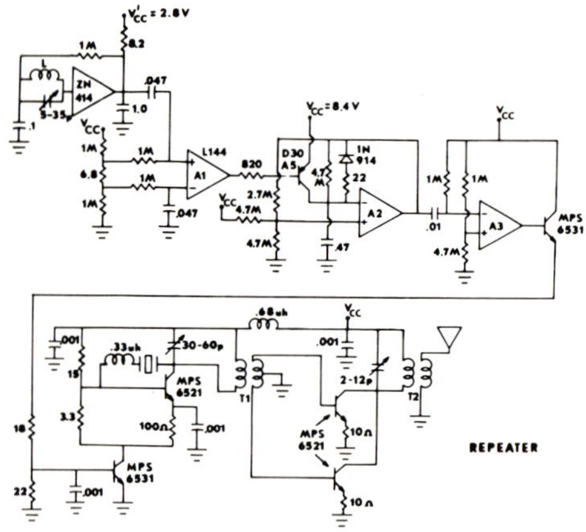

Fig. 2. Schematic diagram of neck collar repeater.

sheath" consisting of stainless steel wire wrapped on the outside of 1.5 mm OD polyvinyl tubing, with the insulated electrode lead-in wires running inside the tubing to the transmitter. The final implant dimensions (excluding the 50 cm electrode leads) are 1.4 cm x 3.5 cm x 4 cm. An operating life of 1 year is expected from the 220 mampHr mercury batteries (±1.4v) used to power this device.

Neck collar repeater. The function of the neck collar repeater (fig. 2) is to retransmit the heart pulse data over a long-range telemetry link. The incoming heart pulse signal is sensed by a low-frequency receiver (ZN144). This pulse is detected (A1 of L144) and used to control a synchronizable astable multivibrator (A2 of L144). The output of this multivibrator keys on a VHF (172MHz) transmitter for long-range data transmission. In this manner, the neck collar system retransmits the heart pulse information over ranges exceeding 20 kilometers. If the heart pulse signal is absent (due to excessive implant-to-repeater separation, for example) the multivibrator runs at its own free running rate of 1 pulse per 3 seconds, allowing for continued tracking of the animal. The system will synchronize with the heart pulse as soon as signal from the implant reappears.

The circuit of figure 2 is packaged to fit within a 1.5 cm x 2.5 cm x 4.5 cm volume. The entire repeater, including circuits, batteries, and ¼ wave radiating antenna, is fabricated into a conventional neck collar form and weighs 0.9 kg. The low-frequency receiver is powered with a 3 ampHr lithium battery (2.8v). The remainder of the circuit is powered with three 8 ampHr lithium batteries in series (8.6v). A 2-year battery life is expected.

Data display. The data display electronics are shown in figure 3. The transmitted heart pulse train is received and

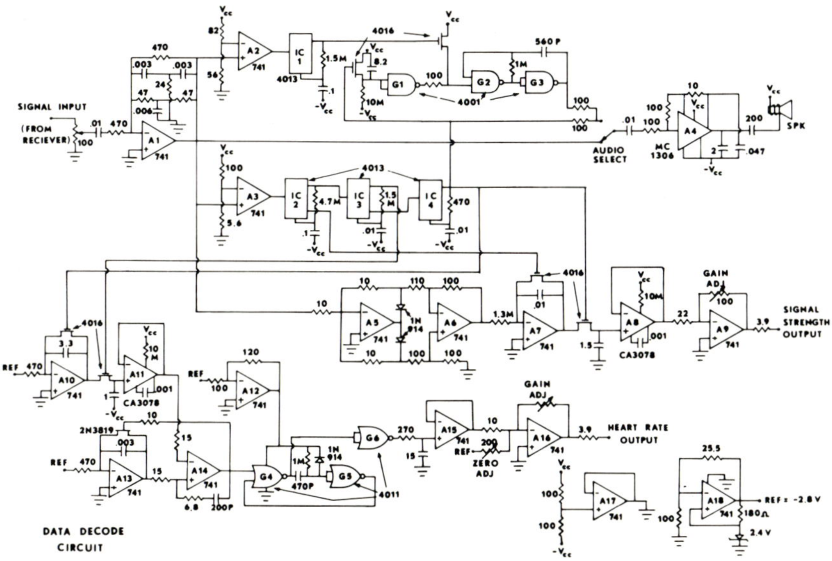

Fig. 3. Schematic diagram of data display circuit.

detected by a commercially available VHF receiver. The pulse frequency is calculated and displayed on the heart rate channel of the analog chart recorder. Signal strength (height of the received signal) is measured and displayed on the second channel. Since the antenna in the neck collar is not an isotropic radiator, the received signal strength is modulated by animal motion. Animal activity can be determined from the resulting pattern of signal strength.

In figure 3, the incoming signal is processed through a band-pass amplifier A1. Amplifier A3 (along with IC2-IC4) acts as a pulse detector and supplies appropriate control voltages to the remainder of the circuit. Amplifiers A5 through A9 act as a signal-strength-to-analog-voltage converter while amplifiers A10 through A16 act as a pulse-frequency-to-analog-voltage converter. Output is made to a dual channel analog chart recorder. The remainder of the circuits of figure 3 act as operator aids and produce audible warnings if signal is either absent or exceed the saturation levels in the receiver.

The entire data display system, including receiver, electronics, chart recorder and a 4.6 ampHr 12 volt battery, measures 15 cm x 19 cm x 30 cm and weighs 2.3 kg. The data recording system can be easily carried to faciliate animal monitoring. Portability is a desirable feature since the test animals are extremely mobile.

FIELD APPLICATION

For the system to be functional on wild animals, the method of installation must be quick, safe, and capable of being performed under unpredictable field conditions. The technique described in this section was perfected during many trials with domestic sheep and penned elk, and then performed on elk captured in the wild.

Captured test animals are immobilized with M-99 (Etorphine) and the hair along the lower neck and sternum is shaved. The area is disinfected and three incisions are made: one at the base of the neck where the transmitter is implanted, and one at the anterior and posterior ends of the sternum. The electrode sheath is implanted subcutaneously with the aid of a catheter (Number 10, 45 cm long). The incisions are closed in a mattress stitch fashion with wound clips and/or stainless steel sutures. The neck collar repeater is then slipped over the animal's head. The antagonist M50-50 (Diprenorphine) is administered and the animal is released. The entire operation takes approximately 15 minutes, and the animal apparently suffers no ill effects.

SUMMARY

The telemetry system described has some unique features that enhance its usefulness for wildlife biotelemetry applications. The automatic gain control circuit within the implanted transmitter allows for non-critical electrode placement. The repeater neck collar permits long-range data transmission of an internally sensed parameter. The pulsed operation of both the implant and the repeater extends battery life. And finally, the anisotropic radiating properties of the neck collar antenna allows animal activity to be monitored without the need of a more sophisticated data channel.

Techniques have been developed for rapid field installation of this system on big-game animals captured in the wild. The system has operated satisfactorily on two wild elk. Data are being used to document reactions of elk to man-caused stress-producing stimuli.

REFERENCES

1. Cupal JJ, Ward AL, Weeks RW: A repeater type biotelemetry system for use on wild big game animals. Instrum Soc Am Trans, 14, 1975, pp 101-108.
2. Gessaman JA: Ecological energetics of homotherms. Utah State Univ Press, Logan, Utah, Vol. 20, 1973.
3. Holter JB, Urban WE, Hayes HH, et al: Ambient temperature effects on physiological traits of white-tailed deer. Can J Zool, 53, 1975, pp 679-685.

TELEMETERED EEG ACTIVITY FROM THE ISOLATED OLFACTORY BULB OF RABBITS

Patricia Black-Cleworth* and Gerda Verberne**

Space Biology Laboratory, Brain Research Institute
University of California
Los Angeles, California

The electrical activity of the mammalian olfactory bulb, during regular breathing consists of slow potentials with superimposed regular sinusoidal bursts of 30-60 Hz (Adrian's induced waves, 1). The slow potentials coincide with the respiratory rate and both slow and induced waves depend on stimulation of the olfactory mucosa (3, 6). Similar activity occurs in prepyriform cortex and amygdala (5, 6, 8). Although most authors suggest that induced waves are initiated in the olfactory bulb, McLennan and Graystone consider them to be generated in the amygdala (8). There are now several papers showing that basic features of the bulbar EEG persist in the isolated olfactory bulb (2, 9).

The purpose of the present experiments was to verify the presence of a normal EEG in isolated olfactory bulbs of rabbits over long periods and to investigate changes in activity produced by sniffing substances of interest to the rabbit, viz.their own scent-marking products. The rabbits were released into an enclosure where they could deposit urine and faecal pellets and sniff their own and other rabbits' products while the electrical activity of olfactory bulb and amygdala was recorded by biotelemetry.

MATERIALS AND METHODS

Eight rabbits were used (7 males,1 female). Four were subjected only to electrode implantation in olfactory bulb and amygdala; in the other four, both olfactory bulbs were separated from the brain by subpial aspiration of the peduncle prior to electrode implantation. The bulbar circulation was seen to be intact after section. Seven rabbits survived the 6 month recording period (4 intact and 3 bulb-sectioned). The enclosure, methods of observation, electrodes and telemetry system have been described previously (3, 4). After the experiments, the animals were sacrificed and perfused. The olfactory bulbs and anterior brains of the lesioned animals were embedded in paraffin and sectioned sagitally or horizontally to check completeness of the sections.

*Present address: Laboratory of Neurophysiology, MRRC, NPI, University of California, Los Angeles, Ca. 90024

**Present address: Dierfysiologisch Laboratorium, Kruislaan 320, Amsterdam, The Netherlands

RESULTS

Only one bulb in one of the 3 rabbits examined histologically was completely separated (No. 6, LOB). The amount of tissue remaining in other bulb sections varied from a minimum of a few central fibers in one slide (No. 7 ROB) to a maximum of part of the lateral olfactory tract remaining in one bulb (No. 5, LOB). Continuities between bulb and brain were ventral in three bulbs, lateral in one, central in one, and non-existent in the last.

The spontaneous EEG of all disconnected bulbs showed the same features as in intact rabbits, viz. slow waves that followed the respiratory rate and superimposed bursts of induced waves. In order to characterize changes in slow and induced waves during sniffing (Sn) of various substances we tabulated all records of Sn of one sec or longer duration and arranged them according to rabbit, lesion, slow wave frequency and amplitude, and changes in amplitude of induced waves during and following Sn (Table I). Changes were similar in both intact and lesioned rabbits. Generally, the frequency of slow waves increased while the amplitude decreased. The induced waves either decreased or remained the same during Sn; increases were rare. Large increases in induced waves (rebounds) following Sn were common. Induced wave changes were most frequent when urine was sniffed and least frequent when objects were sniffed, but differences were not significant (χ^2 test); hence categories were combined in the table. Relations between induced wave changes and duration of Sn were also not significant. A typical record from a lesioned rabbit is shown in Fig. 1 where simultaneous recordings from both separated olfactory bulbs are shown. Slow potentials clearly decrease in amplitude and increase in frequency during Sn of urine; induced waves decrease during Sn and show rebounds following Sn. Note also similarities and differences between the two bulbs.

DISCUSSION

There is little doubt that induced waves can originate in the olfactory bulb itself, since these waves persist in isolated bulbs (2, 9). Tonic changes in amplitude of induced waves, as well as increases (5, 6) and decreases (5,7) during Sn or exposure to olfactory stimuli have previously been attributed to centrifugal influences (2, 7, 8, 10). McLennan and Graystone noted decreases in induced waves during Sn in alert or aroused animals, but increases in inattentive animals (8).

In our experiments, Sn of scent-marking substances caused decreases or no changes in induced wave amplitudes. While our rabbits could be considered as "aroused" in that they were actively exploring, the same changes were observed in bulb-sectioned animals where neural centrifugal influences were largely or totally eliminated. One might think that the effects of arousal could be mediated extraneuronally (i.e. humorally), but suppression of induced waves by RF stimulation did not occur after isolation of

Table 1. Changes in EEG characteristics of bulb during sniffing.

Group	SW Sn Hz	Comp. Hz	Dec Amp SW %	Dec Amp IW %	Inc Amp IW %	Reb. %
Intact						
1	6.1	3.4	92	50	8	45
2	6.1	3.6	88	24	0	33
3	5.8	3.0	84	28	0	34
4	6.3	4.5	100	100	0	100
Lesioned						
5 LOB	5.4	3.9	69	31	0	25
6 LOB	6.4	4.6	61	31	0	46
7 ROB	5.9	4.0	77	62	0	46
LOB	6.5	4.6	89	62	7	55

Abbreviations: SW Sn, slow wave frequency during Sn; Comp., Slow wave frequency before or after; Dec Amp SW, decreased amplitude slow waves during Sn; Dec Amp IW, decreased amplitude induced waves during Sn; Inc Amp IW, increased amplitude induced waves during Sn; Reb., rebound of induced waves after Sn. Individual rabbits are numbered under group. LOB, left olfactory bulb; ROB, right olfactory bulb. Compare bulb designation with description of lesion extent above.

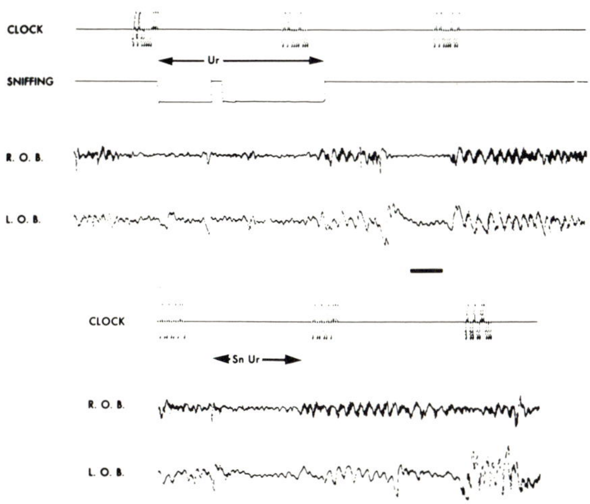

Fig. 1 Telemetered record from olfactory bulbs of lesioned rabbit No. 7 during Sn of urine. Clock shows digital code recorded every 5 sec. Sniffing channel indicates behavioral observation of Sn. Time marker: 1 sec. Ur, urine.

the bulb (10). Alternatively, an active animal may sniff in such a way that induced wave suppression occurs as a result of intrinsic bulbar mechanisms. In this regard the striking changes in frequency and amplitude of slow waves, probably reflecting decreased air intake with each breath during Sn, are noteworthy. The increase in induced waves following Sn, as slow waves become lower in frequency and larger in amplitude, also suggest that volume of air flow is at least one factor involved in induced wave amplitude.

ACKNOWLEDGMENTS

We thank Dr. W.R. Adey for his support and Dr. E.E. Decima for his surgical expertise in performing the pedecular sections. This research was supported by the NIMH Interdisciplinary Training Grant MH-06415, NASA under Grant No. NGR 05-007-195 and Contract No. NSR 05-007-158, and the AF Office of Scientific Research of the Office of Aerospace Research under Contract F 44620-70-C-0017.

REFERENCES

1. Adrian ED: The electrical activity of the mammalian olfactory bulb. EEG Clin Neurophysiol 2:377, 1950
2. Affani JM, Morita E, Garcia Samartino L: Efecto de la seccion de los pedunculos olfatorios y de la comisura anterior sobre la actividad del bulbo olfatorio del marsupial *Didelphis azarae*. Revta Soc Argent Biol 44:183, 1968
3. Black-Cleworth P, Verberne G: Telemetered EEG and neuronal spike activity in olfactory bulb and amygdala in free moving rabbits. In Biotelemetry (eds. HP Kimmich, JA Vos). Leiden, Meander NV, 1972, pp 317-325
4. Black-Cleworth P, Verberne G: Scent-marking, dominance, and flehmen behavior in domestic rabbits in an artificial laboratory territory. Chem Sen Flav 1:465, 1975
5. Freeman, WF: Correlation of electrical activity of prepyriform cortex and behavior in cat. J Neurophysiol 23:111, 1960
6. Gault, FP, Leaton RN: Electrical activity of the olfactory system. EEG Clin Neurophysiol 15:299, 1963
7. Hernandex-Peon R, Lavin A, Alcocer-Cuaron C, et al.: Electrical activity of the olfactory bulb during wakefulness and sleep. EEG Clin Neurophysiol 12:41, 1960
8. McLennan H, Graystone P: The electrical activity of the amygdala, and its relationship to that of the olfactory bulb. Can J Physiol Pharmacol 43:1009, 1965
9. Vacarezza OL, Fachinelli CC: Evolucion temporal de la actividad bioelectrica del bulbo olfatorio privado de sus connexiones centrales y perifericas. Acta Physiol Latinoam 24:365, 1974
10. Yamamoto C, Iwama K: Arousal reaction of the olfactory bulb. Jap J Physiol 11:335, 1961

EEG MONITOR FOR FREE SWIMMING FISH

A. K. Scidmore J. B. Beyer
D. Weiss H. Guckel
Department of Electrical Engineering
University of Wisconsin-Madison
Madison, Wisconsin 53706

INTRODUCTION

Extensive studies have been made of the hypothesis that adult salmon use olfactory information in relocating their home stream. Fingerling salmon have been "imprinted" with synthetic chemical "odors" and their behavior studied during spawning migration 18 months later. EEG responses from the olfactory bulb of captive salmon have been recorded and analyzed. The telemetry system discussed in this paper was produced as part of an effort to correlate the physiological response (EEG and ECG) with the behavior activity of the free swimming salmon.

Some of the constraints on the fish-borne package or "tag" will be revealing. Once activated the design life for the power supply for the tag was to be one week. The tag was to function in fresh water of conductivity as large as 600 micro mho/cm. The range of the transmitter was to be a kilometer if possible. In addition the usual goals for such equipment were to be met as far as possible: infinitesimally small, neutrally buoyant, waterproof, shockproof, etc. The usual simple FM telemetry system would not simultaneously meet the range and life requirements. A sampling system was felt to be necessary to reduce the average power drain on the battery. Previous work indicated that the signals of most importance in this application were in frequency range of 4 to 25 Hz and signal amplitudes of 10 to 100 μV would be delivered by coaxial micro-electrodes chronically implanted in the olfactory bulb.

THE FISH-BORNE TAG

The EEG sampling system and transmitter block diagram are shown in Figure 1. The electrodes are capacity coupled to a differential amplifier, filtered by a third order Bessel filter, and the amplitude of the EEG signal sampled. A train of periodic pulses whose period is proportional to EEG amplitude is generated.

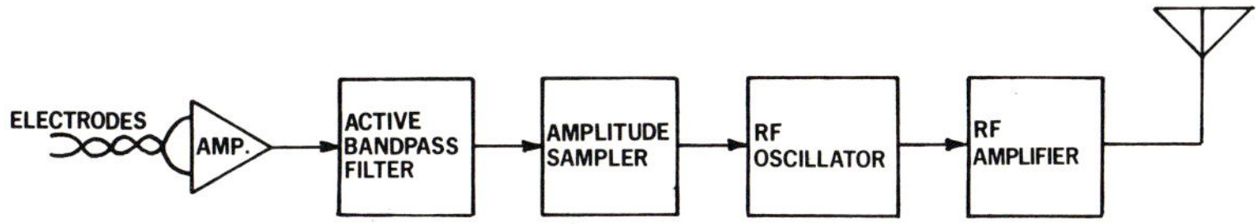

Figure 1. Block diagram of fish-borne tag.

The sampling process is detailed in Figure 2.

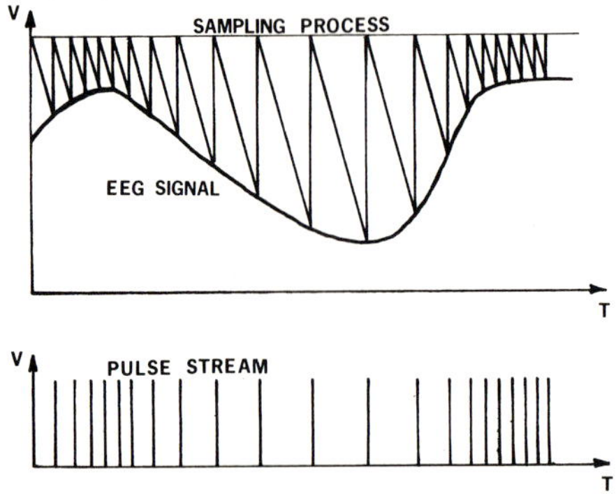

Figure 2. The sampling process.

The power consumed by the RF oscillator and amplifier when gated on is very much greater than for the remainder of the tag. Specifically, the RF oscillator and amplifier draw almost 20 mA from the battery when on, but the remainder of the tag draws only 200 μA. Because of the sampling system used a 24 to 1 reduction in average current drain is obtained so that the average current drain from the battery is less than 1 mA.

The amplifying and filtering part of the electronic circuitry is shown in Figure 3 and the sampling and RF circuitry in Figure 4. Considerable shielding and by-passing is necessary to keep the RF out of the very sensitive EEG differential amplifier. In water a half wavelength for the 52 MHz signal is only 32.5 cm rather than the 5.77 m found in air.

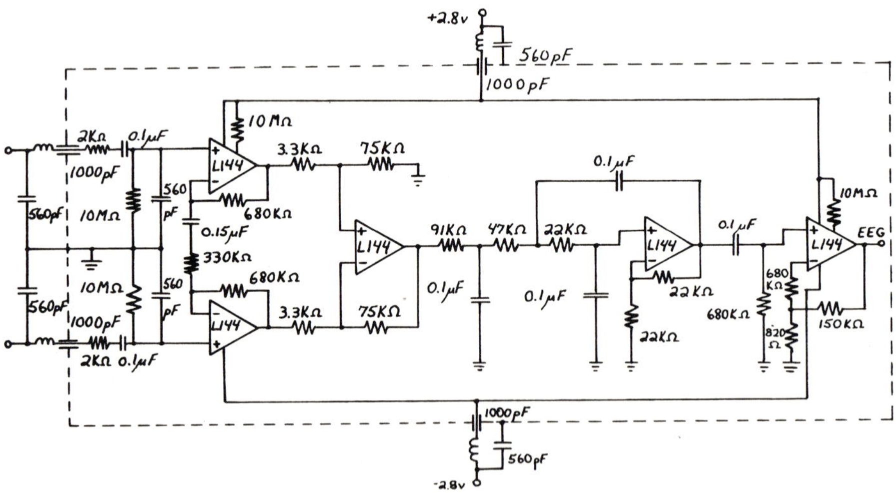

Figure 3. The amplifying and filtering circuitry.

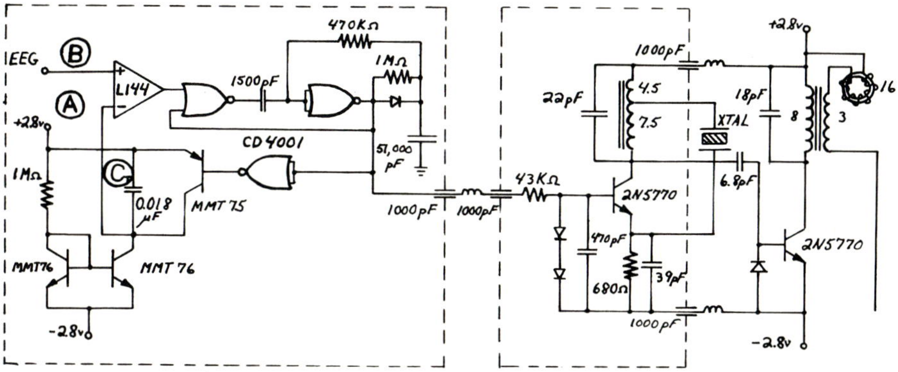

Figure 4. The sampling and RF circuitry.

At 25 Hz, the highest frequency component of the EEG signal, the system samples at least twice per period. The system parameters are such that the mean sampling period is approximately 12 ms. Each sample generates a ½ ms wide pulse, giving an average 24 to 1 off-to-on duty cycle. The entire tag is powered by a center-tapped 5.6 V battery consisting of either a set of four 1.4 volt mercury cells or two lithium cells (2.8 V each).

The detailed electrical characteristics of the fish-borne tag are listed in Table 1.

Table 1. Tag Characteristics

Differential input resistance	20 MΩ
3 dB bandwidth	4 to 25 Hz
Common-mode rejection ratio	> 66 dB
Gain (variable)	≈ 20,000
Average battery drain	1 mA
RF voltage to antenna	9 V rms

EEG RECOVERY

The information leaves the fish as a train of RF pulses. Conventional antennas and communications receivers have been used to recover the pulse information. The receiver output is coupled directly to the demodulator described by the block diagram of Figure 5. The demodulation process is pictured in Figure 6.

The sampling process reconstructs the original signal, but it also introduces some high frequency components not in the original signal. These unwanted components are filtered out by a third order low-pass active filter.

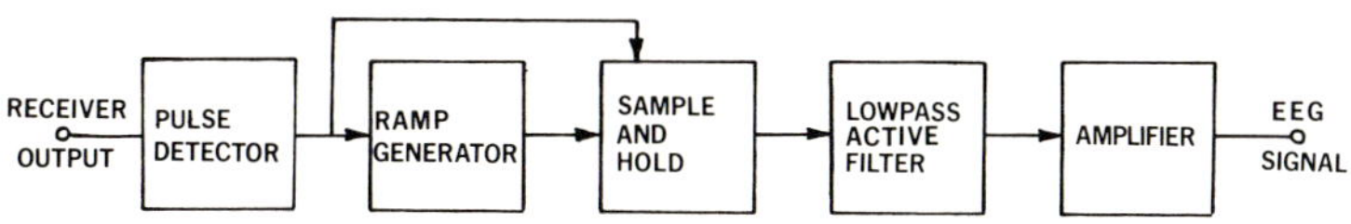

Figure 5. Demodulator block diagram.

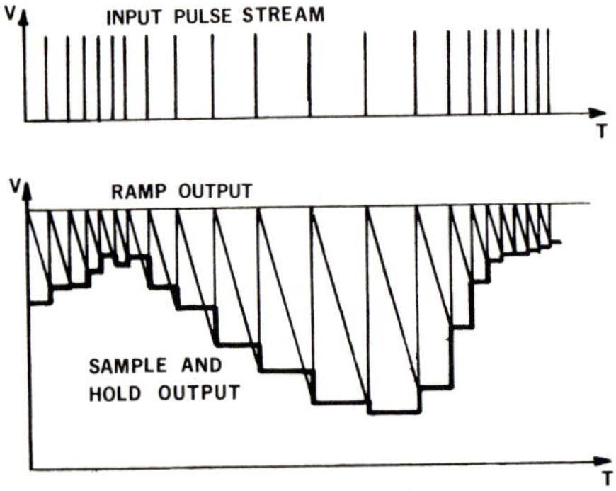

Figure 6. Demodulation process.

The gain of the output amplifier is adjusted to give a 2 volt peak output signal from a 100 μV peak EEG signal. With no input (EEG) signal the noise at the demodulator output is 10 to 20 mV rms.

The first prototype was constructed using conventional discrete circuit components. The circuit board and batteries occupied an approximately cylindrical volume of 11½ cm × 2 cm. These dimensions have been reduced about 25% in the second discrete version. Further reductions in size and weight will be obtained in the hybrid integrated circuit version currently under development.

CONCLUSIONS

The simplest method of telemetering ECG and EEG information is to use the amplified biological signal to frequency modulate an RF carrier. In this application the constraints on battery and signal strength dictated a different approach. This sampling system satisfies these requirements and in addition has excellent linearity and low distortion. Except for upper frequency response this instrumentation conforms to published recommended specifications for EEG telemetry equipment. The sampling system has a limit on the upper end of the frequency response imposed by power supply requirements and receiver bandwidth.

ACKNOWLEDGMENTS

The authors especially thank John C. Cooper of the University of Wisconsin, Laboratory of Limnology for his collaboration in this project. This project was supported in part by NSF Grant GB 35382, the University of Wisconsin University/Industry Program, and the Department of Electrical and Computer Engineering.

A REMOTELY CONTROLLED DEVICE FOR CONTROLLING ELECTRICAL STIMULATION OF THE CENTRAL NERVOUS SYSTEM.

A. A. Perachio, E. Lide†, D. H. Rice, and J. Sikes

Yerkes Regional Primate Research Center, Emory University and Lockheed-Georgia Company†, Atlanta, Georgia

The technique of electrical stimulation has played a significant role in the investigation of neural mechanisms in the central nervous system related to physiological functions and behavior. Responses elicited by stimulation of the brain are markedly altered by anesthesia, tranquilization or restraint (1). Thus, a variety of solutions have been offered to the problem of delivering electrical stimulation through chronically implanted electrodes in freely moving animals. The most successful of these have employed a radio link that in the simplest device functions solely to key a pre-set stimulus (2) and, in more complicated instruments, regulates stimulus parameters, electrode selection and simultaneous control of separate remote units. One of the latter types of devices had been in use in this laboratory for four years (3). We undertook to redesign the system, employing PCM/PDM telemetry techniques and integrated circuits for improved reliability, expanded function and miniaturization.

SYSTEM DESCRIPTION

The system consists of a console unit and four independent animal-worn units (headsets). The console transmits a 16-bit PCM/PDM instruction word at a repetition rate variable from 0.5 to 100 Hz. The carrier frequency is 138.5 MHz variable in broadcast power from 0.1 to 10 watts in 10 db increments. The instruction word is composed of: (1) the address of each of the four units, (2) the codes for each of four electrodes in the selected headset, (3) the pulse current and (4) the pulse width for each stimulus. The frequency of stimulation is determined by the word repetition rate. The stimulus pulse consists of constant current, unidirectional square waves. Pulse current is remotely selectable in two decade ranges: (1) 0.01 mA to 0.1 mA, in 0.01 mA increments, or (2) 0.1 mA to 1.0 mA, in 0.1 mA increments. Stimulus pulse duration is variable from 0.1 msec to 1.0 msec in 0.1 msec increments. Thus, pulse current, pulse width and electrode selection are independent for each headset. Thick-film hybrid microcircuits were used in the headset, employing T^2L logic. (We are presently redesigning this system to use CMOS integrated circuits for simplification of construction and repair and reduction of power consumption.) In addition to the basic system elements, each headset is designed to operate four channels of telemetry to monitor electrophysiological para-

meters during and between stimulation. (The headset illustrated in Fig. 1 has had the telemetry circuit boards and protective cover removed.)

Fig. 1 Head Unit

CONSOLE UNIT

The console depicted in the block diagram, Fig. 2, consists of: four 16-bit recirculating shift registers each associated with one headset, the control for parallel data entry into the registers, A PCM/PDM telemetry transmitter operating at 138.5 MHz and a receiver for a parity pulse transmitted from the headset at 148.5 MHz following each stimulus pulse. Prior to transmission, data are set in parallel in the register for the animal chosen for the experiment. The stored data are shifted out sequentially from each register by a 16-bit clock generator and the serial output from each register is used in sequence to modulate the transmitter. A separate switch on the console panel is associated with each headset. The switch is divided into two segments that are lighted during transmission of the broadcast word and the parity pulse. The function of each bit in the 16-bit instruction word is listed in Table 1.

TABLE 1

MAKE UP OF 16-BIT INSTRUCTION WORD

BIT NO.	BIT FUNCTION
1 & 16	Framing bits
2 & 3	Animal selection
4 & 5	Electrode selection
6	Amplitude decade (1 X or 10 X)
7 - 10	Pulse amplitude selection
11	Undesignated (always "1")
12 - 15	Pulse width selection

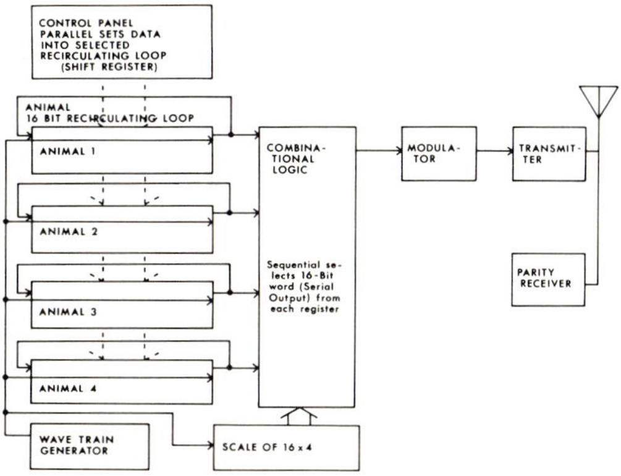

Fig. 2 Block diagram of transmitter console

HEAD UNIT

The head-mounted stimulator is secured through an aluminum interface plate to a Delrin block chronically implanted on the skull of the animal. The principal components of the headset are represented in the block diagram (Fig. 3). The components include a superheterodyne receiver, the logic section for storing and decoding the 16-bit instruction word, a D/A (digital to amplitude) convertor which is controlled by the word and which in turn controls the amplitude of each current pulse, a D/W (digital to width) convertor which, in like manner, controls the width of the constant current pulse, a constant current generator which generates one pulse per each instruction word, a parity transmitter which is modulated by a pulse of the same width as the D/W output, a multiplexer which directs the constant current pulse to the selected one of four outputs tied to four separate implanted electrodes and a power supply consisting of 23 silver oxide cells. The system is mounted on six 1 1/4" x 1 1/2" ceramic substrates in single or multiple layers, mounted back to back in three pairs. The boards are interconnected through a 5-layer printed circuit mother board. The logic circuits are de-energized until the appearance of the first (framing) bit of the word. If the instruction word is not addressed to a particular unit (bits 2 and 3), its logic is turned off after the word has been decoded. Otherwise, the logic remains energized long enough for the constant current pulse to be generated. Since the logic circuit constitutes approximately 80% of the battery drain, the power supply gating is effective in prolonging battery life.

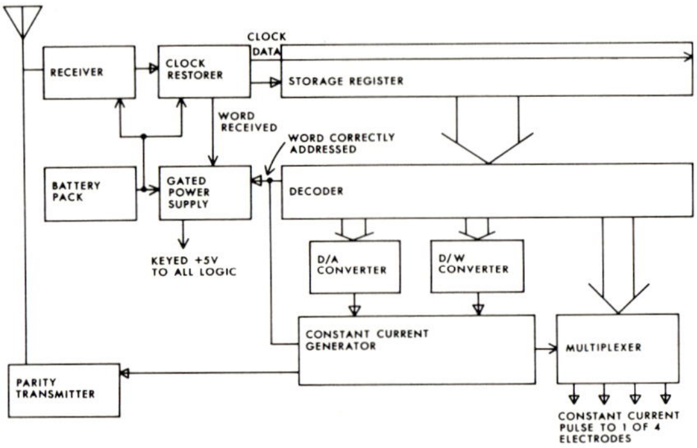

Fig. 3 Block diagram of head unit

SYSTEM PERFORMANCE

This remotely controlled stimulator has a number of advantages over previous designs used in this and other laboratories. The unit is relatively free from interference from other RF sources because of the digital code. The feedback provided by the parity circuit provides verification of the operation of the unit. The range of selection of stimulus parameters is as great or greater than that of any previous design. The unit is compatible with telemetry of physiological parameters as well as with a remotely controlled blood withdrawal system described in a separate paper in this Symposium. The system has been in operation in the laboratory for six years and has proved to be highly reliable in its performance. The size of the unit offers some disadvantages, especially for species smaller than the adult rhesus monkeys used in our research. Redesign of the system with CMOS electronics should help reduce the overall weight and size of the power supply. Although under some experimental circumstances, less flexibility in remote control of stimulus parameters may be acceptable, we have found that each capability described for this unit has been a useful and important feature in experiments involving electrical stimulation of the central nervous system to produce complex behavioral responses.

1. PERACHIO AA, ALEXANDER M: Neurophysiological approaches to the study of aggression. In Neuropsychology of Aggression (ed RE Whalen). New York, Plenum Publishing Co., 1975, pp 65-86.
2. MAXIM PE, SPELLMAN, FA: A radio-controlled constant current biphasic stimulator system for primate studies. Physiol Behav 14: 663, 1975
3. WARNER H, ROBINSON BW, ROSVOLD HE, et al: A remote control brain telestimulator with solar cell power supply. IEEE Trans Biomed Engn BME-15: 94, 1968

This work was supported in part by N.I.H. grants RR-00165 and NSO 9688 and NASA grant NGR 11-001-065.

TELEMETERED NEUROPHYSIOLOGICAL ACTIVITY IN THE OLFACTORY SYSTEM IN RABBITS AS A RESPONSE TO RESPIRATION AND PHEROMONES

Gerda Verberne and Michel Hofman
Laboratory of Animal Physiology, Kruislaan 320, Amsterdam

Information processing in sensory systems can be investigated in several ways. In the present study we focussed our attention on the analysis of field-potentials in the olfactory system in male rabbits.
Three sets of experiments were carried out for this purpose: comparison of simultaneously recorded field-potentials in the first and second projection areas of the olfactory sense organ, the olfactory bulb (olf.b.) and olfactory-prepyriform-cortex (olf.c.) during: 1) rest and normal breathing, 2) in case the natural input of the nose: (unfiltered air) is blocked, 3) when the animal roams around and sniffs scent sources produced by the animals.
It will be clear that the use of telemetry is indispensible for the success of this type of research (2).

METHODS

Three male rabbits were implanted, each with 16 stainless steel (60μ) electrodes. The electrodes were placed in groups in the olf.b. and in the olf.c. The distance between the individual electrodes in a group amounted to 1 mm along the dorso-ventral axis. The behaviour of the rabbits, when producing or responding to the olfactory stimuli, was recorded simultaneously with the telemetered electrophysiological signals.

RESULTS

1) The electrophysiological activity in the olfactory system, in the present paper restricted to the olf.b. and the olf.c., is during rest characterised by spindle shaped fast waves. These waves, with a frequency (Fr.) around 60 Hz in rabbits, are superimposed on slow waves, varying between 2 and 6 Hz (Fig. 1, Fig. 2, upper row, left side). The fast waves are amplitude modulated by the slow waves; in case of electrode 18 (Fig. 2) resulting in a power spectrum with a peak around 65 Hz with side Fr. of about 62 and 68 Hz.
The fast waves in both projections seem to undulate in the same Fr., the olf.c., however, also in a subharmonic component. This could be shown by applying bicoherence techniques. Not all locations in the olf.c. provided such a powerful contribution of subharmonics in the complete power spectrum. Slow and fast waves showed in the olf.b. a strong, in the olf.c. a weak correlation.

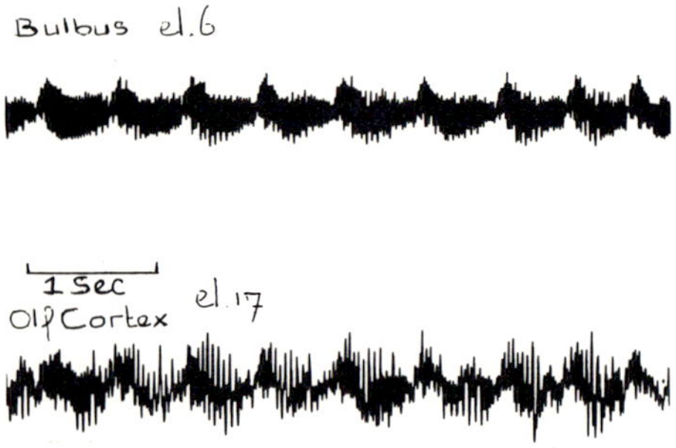

Fig. 1 Example of slow and fast waves in the olfactory bulb (upper row) and olfactory cortex (lower row). The amplitude of the slow waves varies between 200 and 500 μv, that of the fast waves between 200 and 1,000 μv. The registration of the activity in the olfactory cortex suggests the presence of the subharmonic component of the basic frequency. The coherence ratio showed a confidence level below 5%.

2) Blocking the natural input of the olfactory sense organ, namely the inhale of air, by closing the nostrils results in a drastic decrease of the above described slow and fast waves in the olf.b. These effects in rabbits have been published earlier (1). The next question to be answered is how the olf.c. would be affected in this case. The following step: closing only one of the nostrils, presents the opportunity to investigate first the influence of the ipsilateral bulb on the contralateral b. and v.v.; secondly the influence on the ipsi- and contralateral olf.c.
The nostril is closed by gentle pressure of one finger. The effect is a decrease of respiration rate and a more powerful inhalation through the open nostril. The electrophysiological consequences are (Fig.2): an almost complete absence of slow and fast waves in the ipsilateral olf.b. (upper row, right side); b) a drastic decrease of the fast, but only a slight decrease of the slow waves in the ipsilateral olf.c.; c) the fast waves in the contralateral olf.b. (lower row, left side) are still present, while the peak of the Fr. band seems to narrow. The subharmonic component increases clearly. The original peak of about 6 Hz is still present but some new peaks representing lower Fr. emerged. d) Closing both nostrils resulted in a complete lack of slow and fast waves in the olf.b. The olf.c., on the other hand, showed a rest of fast waves and two powerful peaks in the slow components. These results are representative for a number of 25 experiments on different locations in olf.b. and cortex in 3 animals.
3) Pheromones. Almost all rabbits observed in the past, prefer, after having sniffed urine or fecal pellets, strange above own, as expressed a.o. in the duration of sniffing. They can therefore be considered as pheromone carriers. The study in the information processing of

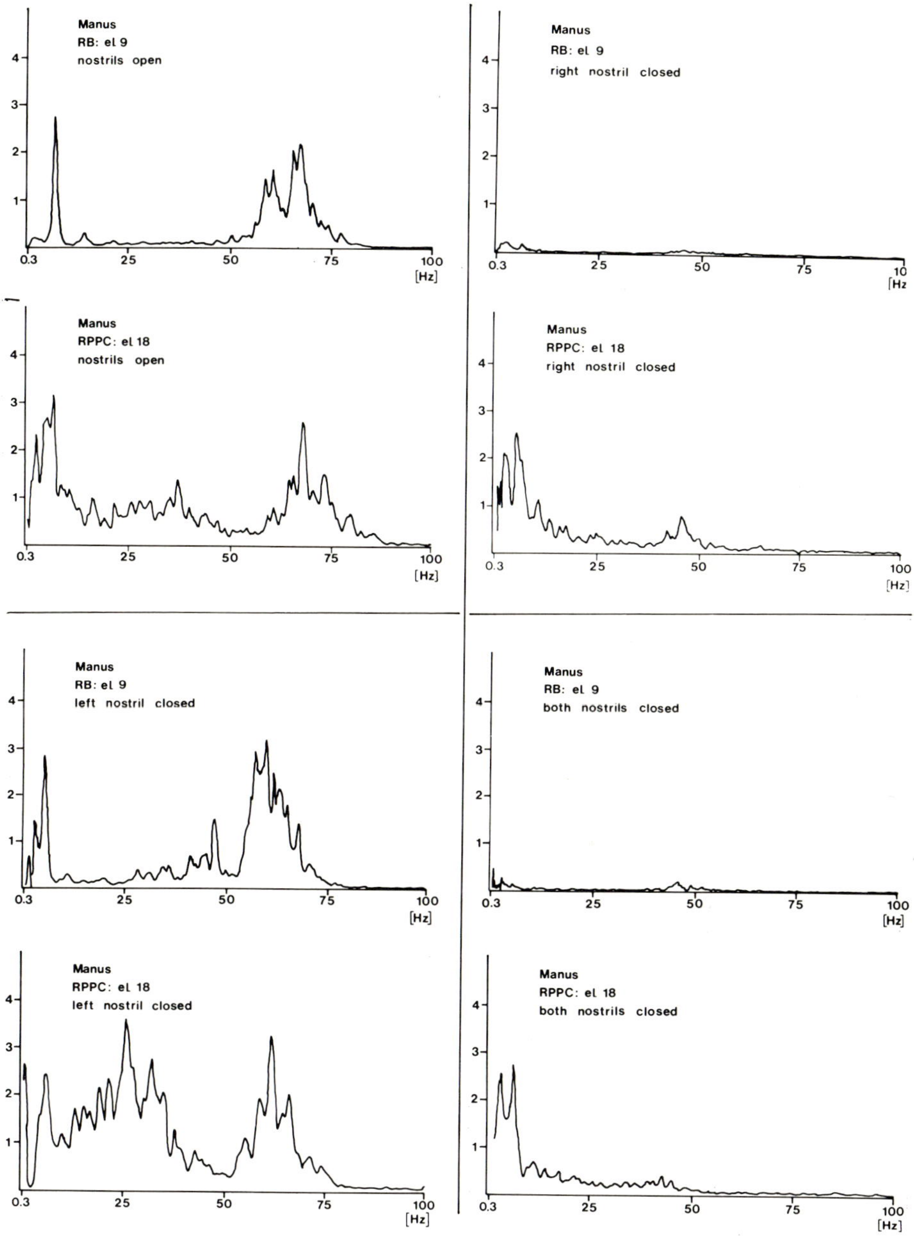

Fig.2 Effects of uni- and bilateral closing of the nostrils on the electrical activity in the olfactory bulb and cortex. Abbreviations:R.B.:right olfactory bulb, L.B.:left olfactory bulb, RPPC:right prepyriform cortex, LPPC:left prepyriform cortex.Further details in text.

of these pheromones was carried out in 60 experiments. a) Sniffing either of these 4 pheromones causes considerable changes in the spindle shaped fast waves: the mean amplitude ($\bar{A}$) decreases, whereas Fr. increases in both projection areas, as compared with that of a reference period; the subharmonic component - if present - fades away. This appeared to be true for the 3 rabbits. b) In case the animal sniffs the preferred scent, the $\bar{A}$ decreases even more. This holds only for two rabbits in the olf.b. for both urine and fecal pellets. Similar results were found for the increase of the Fr., but only in the first rabbit for both types of pheromones. c) The first rabbit produced also in the olf.c. incidental effects; the third rabbit showed only a higher Fr. in the olf.c. when it sniffed the preferred fecal pellet.

DISCUSSION and CONCLUSIONS

The basic spindle shaped waves, as described in the olf.b. and cortex under exp. 1 cannot be transferred to contralateral sides as has been demonstrated in exp. 2. Transfer from the olf.b. to the ipsilateral cortex does occur. Especially the slow waves in the olf.c. must have another or a second origin than those of the bulb, i.e. respiration. We owe the slight increase in spectral power and the shift of slow waves to lower Fr. after closing the nostril of the contralateral side, to a stronger inhalation with a lower respiration rate.

The mere intake of pheromones causes identical changes in both olf.b. and cortex (exp. 3). The final preference for one of the scents has been preceded by quite different activities in the two projection areas in each of the three animals. These data demonstrate that fieldpotentials function in information processing.

ACKNOWLEDGEMENTS

We thank Dr. F. Lopez da Silva (TNO Utrecht) for valuable advice and use of his computer program. Our students H. Poldermans and V. van Swede were so kind as to provide us with additional data.

REFERENCES

1. Black-Cleworth P and Verberne G: Telemetered and neuronal spike activity in olfactory bulb and amygdala in free moving rabbits. In Biotelemetry (Eds Kimmich, Vos). Leiden, Meander, 1972, pp 317-325.
2. Meyer AA: A simple wide-band F.M. transmitter for telemetering biopotentials. Med.Biol.Eng.12: pp 875-876, 1974.

SESSION VI

TRANSDUCERS FOR BIOTELEMETRY

Chairmen

W.H. KO, M.D.

S. SALMONS, Ph.D.

CHAIRMEN'S OVERVIEW

Transducers are devices for changing energy from one form to another, for example, mechanical energy into thermal energy, light energy into electrical energy. This is a very broad definition and would include hydroelectric power stations, so we try to limit it by saying that we mean energy at signal levels. So, in this context, we are talking about changing information-carrying energy from one form to another. For most purposes we want to use electronic techniques to operate on the signals, so largely we work with transducers which will transduce some other form of energy into electrical analogues. This definition should include electrodes[1] which transduce current carried by ions in tissue to current carried by electrons in metal conductors.

It is clear from some of the earlier sessions that transducers constitute one of the weak links in current telemetry systems. Therefore, it is appropriate that we take stock of the current status of this field.

Transducers are an important link in that transducer design and development is much more difficult than circuit or telemetering systems development. It is interesting to see that various types of transducers have been presented. This session shows that telemetry is in actual use to gain physiological or biological knowledge.

We covered many types of uses including stimulation. There are many common properties between telemetry and telestimulation. That is, the hardware and purpose are very similar. In the near future, we can expect to see many new applications of electrical stimulation in biological and medical research. Research projects using electrical stimulation for sensory prostheses are ongoing. These include visual, auditory and motor-action types for control of the hands or to facilitate walking, phrenic-nerve stimulation to aid respiration, and bladder stimulation to regain control. Researchers are working on electrical stimulation of the central nervous system to control epileptic attacks and other behavioral disorders. Lastly, since there is such interest in transmission across body boundaries, the reverse path should also be a very interesting field to follow.

---

1 Miller HA, Harrison DC: Biomedical Electrode Technology: Theory and Practice. Academic Press, New York and London, 1974

MINIATURE PRESSURE TRANSDUCER FOR BIOMEDICAL APPLICATIONS

Wen H. Ko, Jaroslav Hynecek, Scott Boettcher, John Homa

Engineering Design Center, Case Western Reserve University
Cleveland, Ohio

INTRODUCTION

Miniature pressure transducers and signal processing circuits intended for medical applications are under development. Primary design requirements for this application are absolute pressure, low hysteresis, long-term stability, small size and biocompatibility. Through the use of active temperature compensation and carefully selected packaging techniques, most of these requirements have been met.

DESCRIPTION OF THE SENSOR

The pressure transducer was built from a silicon crystal wafer for several reasons. Among these are: the biocompatibility of silicon (1); the effect of pressure or strain on the material's electrical properties is well understood; and silicon technology is a highly developed art which allows one to incorporate a variety of different devices, sensing as well as amplifying, right on the wafer substrate. The mechanical structure is shown in Figure 1.

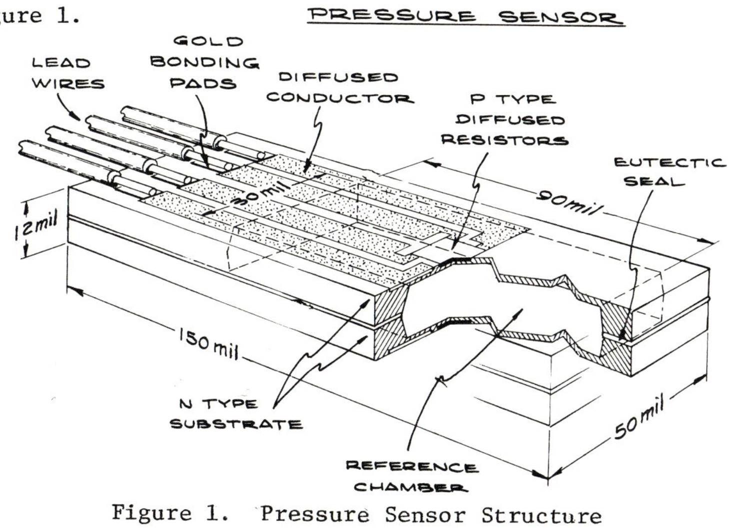

Figure 1. Pressure Sensor Structure

The silicon chip, of dimensions 50 x 150 mils and 5 mils thick, has a preferentially etched cavity from the bottom of the chip in order to form a thin diaphragm on the top of the chip. The sensing bridge elements are fabricated on the top and located in the center of the diaphragm. A similar chip is used as the base to form a device with sealed chamber. The geometrical shape of the device must be conformed to the crystal anisotropy in order to avoid unwanted stresses resulting from possible thermal expansion coefficient mismatch.

If the sensor is placed in an ambient with changing pressure, the diaphragm deforms as a result of the differential pressure. The pressure differential is obtained between the ambient pressure and the pressure in the sealed reference chamber. It is also possible to evacuate the reference chamber prior to sealing and obtain an absolute pressure sensor. The deformation of the diaphragm is detected by the piezoresistance bridge built into it, which changes its resistance in relation to the mechanical stresses in the diaphragm.

FABRICATION PROCEDURE

The starting material used for fabrication is n-type <100> silicon with 1-4Ω/cm resistivity, the wafers are approximately 5 mils thick. The material is polished on both sides.

The first step, prior to standard circuit processing, is the generation of registration marks on the front and back of the wafer. All subsequent processing steps are then aligned to these alignment marks. The subsequent processing steps have two goals: (1) to fabricate low resistivity regions serving as underpassing leads to bring the bridge signal out to bonding pads for wire attachment and (2) to fabricate the piezoresistor bridge in the center of the diaphragm. The standard processing sequence is used for fabrication, the resistors have values of about 5kΩ. The surface concentration, C_s, is about $10^{18}cm^{-3}$, chosen from a compromise between temperature coefficient and pressure sensitivity (2). After the diffusion processing steps are completed, the metallization is performed by vacuum deposition of chrome-gold. Both sides of the wafers are metallized. After metallization, the wafers are gold plated in order to provide enough metal thickness for alloy sealing of the reference chamber and for the wire attachment. The silicon is then chemically preferentially etched, in a heated flask equipped with a reflux condenser in order to maintain a constant concentration of the solution during the etching procedure. The final operation is the vacuum sealing of the reference chamber using a gold alloy preform.

TESTING

The testing procedure includes both pressure and temperature cycling. The transducer is first pressure cycled several times to measure its sensitivity and pressure hysteresis, then the transducer is temperature cycled to determine the temperature coefficient and hysteresis. The temperature data are obtained from a thermocouple, using an ice bath reference junction. A typical temperature cycle curve is given in Figure 2.

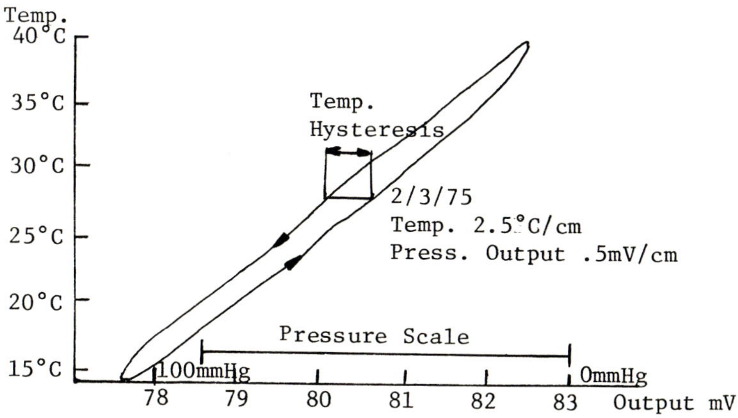

Figure 2

Low drift amplifiers were built, due to the small output voltage of the transducer. With the addition of the electronic compensating amplifiers, stability can be maintained to within 1mmHg between room temperature and 40°C. Pressure hysteresis, as shown in Figure 3, is typically less than .5% full scale. The performance is shown in Figure 4.

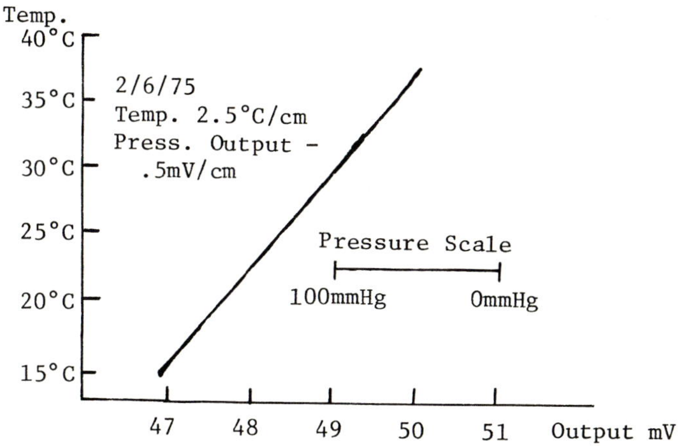

Figure 3

CHARACTERISTICS FOR UNPACKAGED SENSORS

DD WAFERS

PARAMETER	MEAN	S. D.	NO.	(BEST 20 %) MEAN	S. D.
SENSITIVITY	5.4 μV/VMM	4.39	98	7.5 μV/VMM	9.1
TEMPERATURE COEFFICIENT	2 MM/°C	1.6	98	.48 MM/°C	.32
OFFSET	4.16 MV/V	2.88	98	3.04	2.61
PRESSURE HYSTERESIS	.32 MM	.68	98	0	0
TEMPERATURE HYSTERESIS	2.1 MM	4.3	98	.19	.40
NONLINEARITY	0 - 100 M	.25%	8		

Figure 4

REFERENCES

1. Ko WH, "Tissue Reaction to Implanted Semiconductor Crystal" MEL Lab Memo No. 824, CWRU, 1968.

2. Samann, Wise KD, Angell JB, "An IC Piezoresistive Pressure Sensor for Biomedical Instrumentation" Department of Electrical Engineering, Stanford University, California.

3. Lee DB, "Anisotropic Etching of Silicon" J. Appl. Phys., 40 4569, 1969.

4. Tugte ON, Chapman RW, Long D, "Silicon Diffused Element Piezoresistive Diaphragms" J. Appl. Phys., 33, 3322, 1962.

5. Mason WP, Thurston RN, "Use of Piezoresistive Materials in the Measurement of Displacement, Force and Torque" J. Accous. Soc. Am., 29, 1096, 1957.

6. Pfann WG, Thurston RN, "Semiconducting Stress Transducers Utilizing the Transverse and Shear Piezoresistance Effects" J. Appl. Phys., 32, 2008, 1961.

THE CADMIUM-TELLURIDE SEMICONDUCTOR DETECTOR AS A TRANSDUCER FOR BIOTELEMETRY OF RADIONUCLIDE TRACERS

J. Bojsen, S. Vadstrup
The Finsen Institute
2100 Copenhagen, Denmark

P. Siffert, R. Regal, A. Cornet
Centre de Recherches Nucleaires
67037 Strasbourg, France

Biotelemetry of radionuclide tracers has facilitated the study of several physiological functions, e.g. it has been of great advantage in studies of circadian rhythms. The GM detector (1,2), the $CaSO_4$:Dy dosimeter (3) and the Si(Li)(4) detector have previously been used as transducers in biotelemetry. During the last decade several new γ-sensitive semiconductors working at room temperature have been developed, e.g. cadmium-telluride (CdTe), gallium-arsenide (GaAs) and mercuric-iodide (HgI_2) of which the CdTe-detector seems at present to comply best with the demands of an in vivo biomedical transducer.

THE TRANSDUCER

The transducer probe consists of a CdTe-crystal (2-3 mm^3) and a preamplifier. The detector crystal is directly coupled to a FET transistor, which is biased through a 5 KΩ loading resistor. The second stage consists of a voltage amplifier, which is followed by a monostable multi-vibrator having a triggering level adjustable to the energy scale of interest (fig.1). The detector probe is shown in fig. 2.

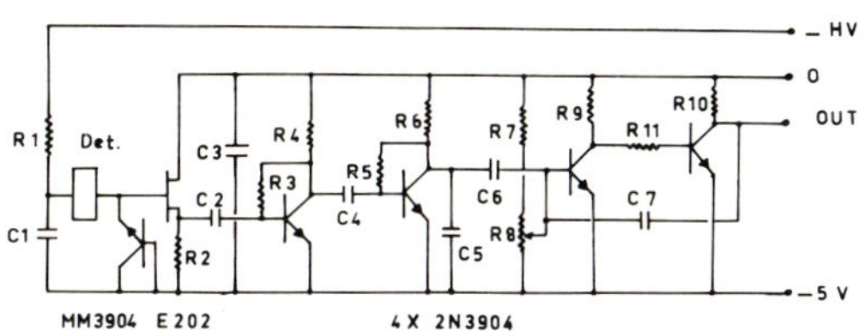

Fig. 1.

The electronic diagram.

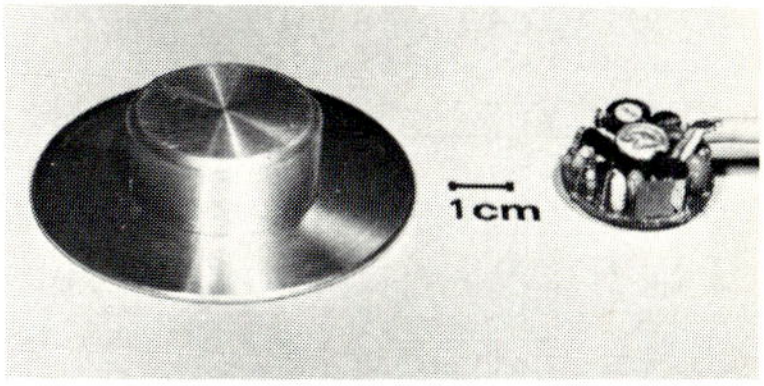

Fig. 2.

The CdTe-detector probe.

R1=100 KΩ; R2=5 KΩ; R3=1 MΩ; R4=R6=R7=R9=R10=R11= 10 KΩ; R5=470 KΩ; R8=470 KΩ pot.; C1=C4=C5=C7=1.2 nF; C2=1μF; C3=47μF; C6=390 pF. -HV=-100 v.

In the production of CdTe-crystals different growing techniques, combined with various dopants lead to the following different categories of crystals; each having specific properties:

-N-type, resistivity 10^6-$10^8\Omega$.cm, grown by the Bridgmann technique, with indium compensation of the cadmium vacancies.
-P-type, resistivity 10^6-$10^8\Omega$.cm, grown by the travelling heater method, with chlorine compensation. Compared to the above technique its advantage is a lower dislocation density but severe polarization problems, until special care is taken.
-P-type, resistivity 10^4-$10^6\Omega$.cm, grown also by the travelling heater method, but without external chemical compensation. No polarization is present. This material gives presently the best resolution in energy, but since the resistivity is lower the active volume is smaller for an equivalent applied voltage. The latter material is used in the detector probe. After cutting and standard surface lapping, followed by cleaning, samples of 1.5 mm in thickness and 5 mm in diameter are obtained. Aluminium contacts 3 mm in diameter were deposited on the opposite faces of the samples and finally the detectors were encapsulated. Under the reverse voltage generally used, the real sensitive thickness is about 600 μ.

IN VITRO EXPERIMENTS

Several requirements have to be fulfilled by a transducer in biotelemetry of radionuclide tracers.

Temperature: The detector must be reliable up to 42°C. The energy spectra of a 241-Am source at 22°C and 41.5°C have been measured with a constant supply voltage. If the threshold level exceeds 30 KeV the integrals of the spectra at the two temperatures are equal, which means the efficiency of the CdTe detector probe is independent of the environmental temperature in the range of interest. The leakage current of the detector is $<$ 10 nA at 42°C.

Supply voltage: The supply voltage of the detector is 100 v d.c. and is generated from a 1.35 volt mercury cell (100 μA). The preamplifier is supplied by 5.4 volt (1.3 mA). As the threshold voltage is a proportion of the supply voltage a compensation of the change in the voltage due to discharge will take place.

Geometry: Point source studies with 99m-Tc and 241-Am in perspex show that the detector is insensitive to transverse displacements up to 5 mm with a point source to detector distance of 24 mm. Studies with 99m-Tc have shown that an increase in volume of a source from 10 μl to a cube of 2 ml with an unchanged center to probe distance of 24 mm increases the count rate about 3%.

Sensitivity: Without any increase in pulse width from the monostable multivibrator the highest count rate of the probe is about 2000 c/s with no deadtime correction, this count rate can immediately be changed to background. At energies higher than 30 KeV at 37°C the background is 1-2 cpm. The sensitivity of the probe to different radionuclides uniformly distributed in water has been determined: 42-K = 3, 113-In = 4, 131-I = 5, 99m-Tc = 12 cpm/(nCi · ml). According to fig. 4 the sensitivity to 133-Xe is > 12 cpm/(nCi · ml).

IN VIVO EXPERIMENTS

Besides regular long term biotelemetry studies we have performed in vivo studies to test the capacity of the probe compared to the conventional NaI-scintillation detector. Local clearance of 133-Xe from adipose tissues and from muscles was measured (Fig.3). In radiocardiographic studies the performance of the CdTe-probe was compared with that of a computerized γ-camera with digitalized regions of interest.

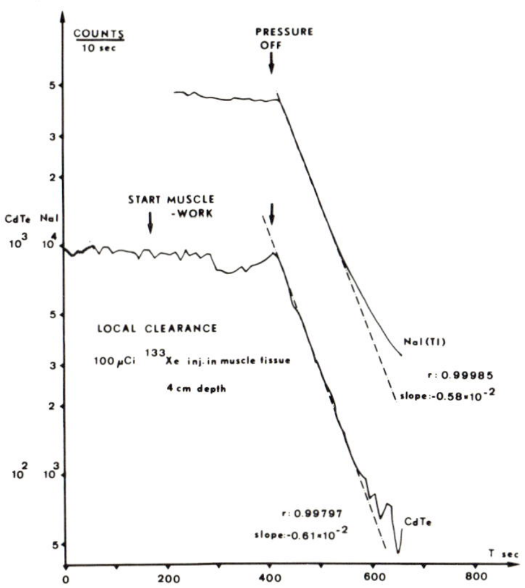

Fig.3. Local clearance of 133-Xe from the anterior tibial muscle. The 133-Xe was injected with a pressure cuff above the knee. The CdTe-probe was compared with a lead collimated NaI(Tl)-detector with a fixed distance of about 30 cm.

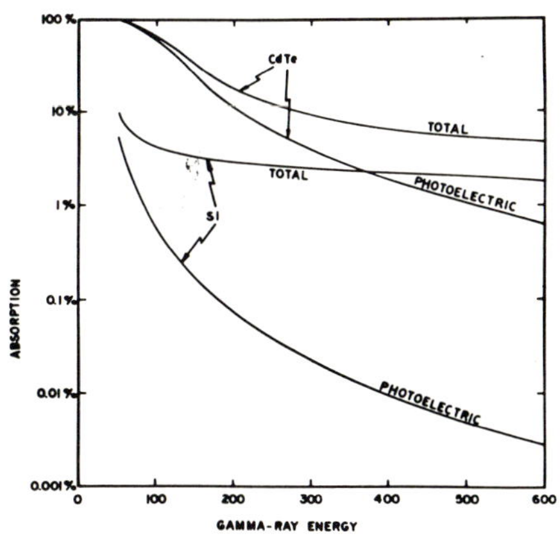

Fig.4. The percentage absorption in 1 mm thick slices of cadmium-telluride and silicon, as a function of γ-ray energy in KeV.

DISCUSSION AND CONCLUSION

The sensitivity of the silicon detector to higher energy photons decreases very rapidly with energy(Fig.4). Consequently, higher Z materials working at body temperature are required. At present, the material fulfilling the requirements best is CdTe. Indeed, the other candidate HgI_2(Z=53-80) has a very high sensitivity, but can in practice be used only below 100 KeV energy, the very low hole mobility (4 cm^2/V.s) restricts the thickness to about 100 μ. The fundamental properties of CdTe can be found elsewhere (5,6,7). Only the scintillators, e.g.NaI(Tl) are more efficient, but they are voluminous and heavy because of the photomultiplier tube, which also requires a high voltage and power consumption. Gas counters have a low density absorption medium, and a high noise level.

The in vitro and in vivo experiments show, that the CdTe-detector is promising as a new surface detector in nuclear medicine and applicable in biotelemetry as well, both in external and in implanted equipment. The good temperature stability is extremely important. Of essential interest in biotelemetry is a small size and a low power consumption. This is fulfilled by the present detector although a further miniaturization of the preamplifier through HYBRID-technology would be useful.

REFERENCES

1. Bojsen J, Wallevik K: A radiotelemetrical measuring device, implantable on animals, for long term measurements of radionuclide-tracers. Internat J Appl Radiat Isotop 23:505,1972.
2. Bojsen J, Vadstrup S: A portable external two-channel radiotelemetrical GM-detector unit, for measurements of radionuclide-tracers in vivo. Internat J Appl Radiat Isotop 25:161,1974.
3. Bojsen J, Møller U, Christensen P, et al: Telemetry of radionuclide tracers by implantable thermoluminescent dosimeters on rats. Proc 2nd Int Symp Biotelemetry. Davos 1974. p.43.
4. Kobayashi T, Sugita T, Sasaki T, et al: A windowless Si radiation detector and biomedical telemetry system for studying the function of organs of free-ranging animals. IEEE Trans Nucl Sci NS-21:735,1974.
5. Siffert P, Ponpon JP, Cornet A: Detecteurs de rayonnements nucléaires à semiconducteurs. l'onde électrique 55:281,1975.
6. Cornet A, Regal R, Ponpon JP, et al: New generation of individual safeguard systems. 2nd ISPRA Nuclear Electronics Symposium. Stresa (Italy) 1975. p.169.
7. Siffert P, Berger R, Scharager C, et al: Polarization in cadmium telluride nuclear radiation detectors. Trans Nucl Sci NS-23:159,1976.

IN VIVO BONE STRAIN TELEMETRY SYSTEMS

Donald R. Young, James M. Glick, and Wayne H. Howard

Environmental Physiology Branch, Ames Research Center, NASA
Moffett Field, California 94035

Studies in our laboratory have been directed toward an evaluation of the mechanisms which give rise to the skeletal demineralization and disuse osteoporosis as observed in hypodynamic and hypogravic environments.

Our experimental approach (1) has been to restrain adult monkeys *(Macaca nemestrina)* in a semirecumbent position for variable periods. Bone mineral loss in the midshaft of the tibia of these monkeys is about 3-4% as determined by in vivo measurements. The balance between the bone formation and bone resorption rates is a principal relationship which must be affected for a modification of mineral content. This paper presents some of our physical science approaches to the evaluation of factors which modify the bone formation rate.

METHOD

Background

Prior studies by others have shown that skeletal stressing achieved by intermittent or steady compression can produce local and regional bone hypertrophy by stimulating the endosteal and periosteal bone formation rate (2). Therefore, the reduction of skeletal loading in our experimental studies could have a significant effect on bone homeostasis. In order to systematically evaluate the influence of skeletal loads in our experimental program, it was important, at first, to measure the normal loading history in our animals, then to examine the departures from normal loading during restraint.

Prior state-of-the-art instrumentation for measurement of loads utilized wire or foil strain gauges bonded directly to the bone surface (3,4) and hard-wired to external data acquisition devices. Those were considered inappropriate for long-term studies. Earlier studies (5) had shown the superior sensitivity of semiconductor gauges. We followed that lead in developing an instrumentation package. Our system consists of an implantable strain sensing element, power supply, transmitting unit, and radio frequency switch.

System Components

The strain sensing elements consist of silicon semiconductor strain gauges bonded to twin beryllium-copper beams with epoxylite 6203 (fig. 1), and sealed into a stainless steel housing. The copper beams are configured to deflect in tension or compression, depending upon the axis through which loads are applied.

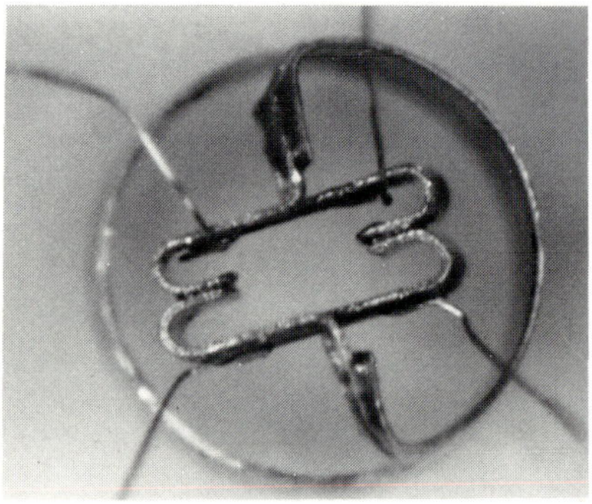

Figure 1. Force collecting ring surrounding twin beryllium-copper beams with attached silicon strain gauges.

Figure 2 shows a completed unit which will be implanted into a 3.2-mm hole drilled into the tibia. The outside wall of the case is threaded to provide channels for bone ingrowth subsequent to insertion. The stiffness of the stainless steel housing is similar to bone. That compliance determines the response of the transducer. The strain gauges are in a half-bridge configuration for accurate temperature compensation.

The half-bridge configuration also reduces the required number of exiting wires to three. The cable from the transducer to the electronics package is comprised of these wires. Each wire is made of 25 strands of number 50 gauge stainless steel wire wound in a helix to provide stress relief. All cables are potted in silastic. The electronics package consists of a power supply, radio frequency switch, and transmitter. Typically, the electronics package is placed in the medial thigh about 6-8 in. from the tibial implant.

Figure 2. Bone strain transducer shown with pencil tip for size comparison.

The basic telemetry scheme utilizes Pulse Interval Modulation (PIRM) which is relatively insensitive to temperature variations and nominal variations in the power supply. Information from each silicon gauge is sampled as follows: A multivibrator generates a set of discrete pulses and simultaneously energizes the strain gauges. The magnitude of voltage change associated with strain determines the time between successive pulses so that the intervals are directly proportional to strain. Those pulses modulate the FM carrier wave in the transmitter. Signals are received by a tuner, demodulated, and displayed as a DC voltage proportional to strain.

Performance Characteristics

Performance specifications are summarized in table 1. The transducer operates over a range of ±4000 microstrains and can withstand 6000 microstrains without damage. The system sensitivity is such that 40 microstrains produces 50 mV at the demodulator output. The response is linear, and the frequency response is flat from DC to above 90 Hz. The noise level was measured to be less than 1% of full scale deflection.

Table 1. System performance

STRAIN GAUGE EXCITATION	XMISSION MODE	BRIDGE CURRENT, μA *500Ω	ELECTRONICS CURRENT, μA	XMISSION RANGE, ft
**PULSATILE	FM	80	310	10 - 15
PULSATILE	PULSATILE	80	35	3 - 10
CONTINUOUS	FM	2700	310	10 - 17

* STRAIN GAUGE OUTPUT ~ 25 mV/V

** WITH 200 mAH, 1.4V BATTERY SYSTEM:

$$\text{LIFE} = \frac{200\ \text{mAH}}{390\ \mu\text{A}} = 512\ \text{hrs WITH GOOD RESOLUTION}$$

IN VIVO RESULTS

The strain gauges are sterilized with ethylene oxide gas. During surgery, general anesthesia is maintained with Surital, i.v. The tibia is approached through an antero-medial incision. A 3.2-mm hole is drilled at midshaft of the posterior aspect of the tibia. The strain gauge is inserted into the hole and sutured with 2-0 nylon suture. An incision is made in the medial thigh and a pocket formed between the sartorius and gracilis muscles. A subcutaneous tunnel is made between the tibial and femoral incisions. The electronics package is brought through the tunnel and inserted into the pocket with all components being affixed by way of stress relief tabs and suture to the surrounding tissue.

Several weeks postsurgery, preliminary calibrations are undertaken. For 10-kg monkeys, the calibrations are performed by applying axial loads of up to 60 kg with frequencies up to 10 Hz through the tibia (fig. 3). In that configuration, loading in the posterior tibia is predominantly compressive. During vigorous physical activity, tibial deformations as high as 1000 microstrains are measured (fig. 4) which correspond to a stress level of 200 kg or about 20 × g in our animals. More typically, 200-400 microstrains are observed. As many as 3-5 events are measured per sec. The onset or rise time is about 30-40 msec or 0.01 strain/sec.

The strain we have observed in walking monkeys is similar to that reported in sheep (6), and therefore, these gauges would be useful for detailed kinesiology studies. In vivo, no measurements as high as 4000 microstrains were made as would be expected during impact loading. The range of these gauges is insufficient for fracture studies. Our companion experiments on excised bone, with deflection rates of 0.007 strain/sec, demonstrate that bones fail in bending at 26,000-38,000 microstrains. Redesign of our gauges for accuracy at those magnitudes would probably result in less sensitivity at the smaller deflections.

Our systems have been evaluated for bio-compatibility in studies in excess of one year. There is no systematic tendency for implant rejection, although in time, bone remodelling and other factors may alter the system sensitivity by as much as 15%, making periodic calibrations necessary.

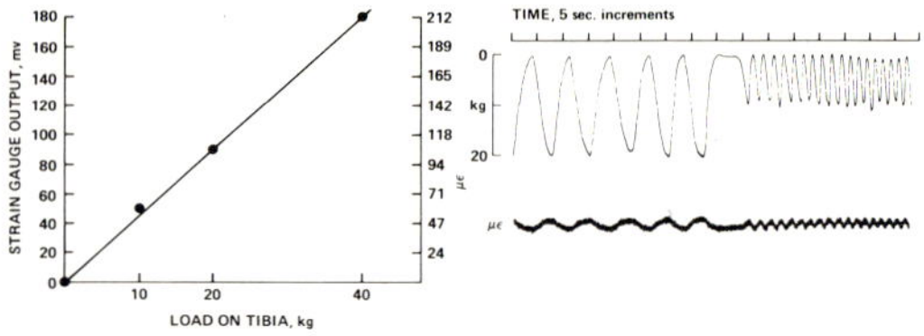

Figure 3. In vivo calibration.

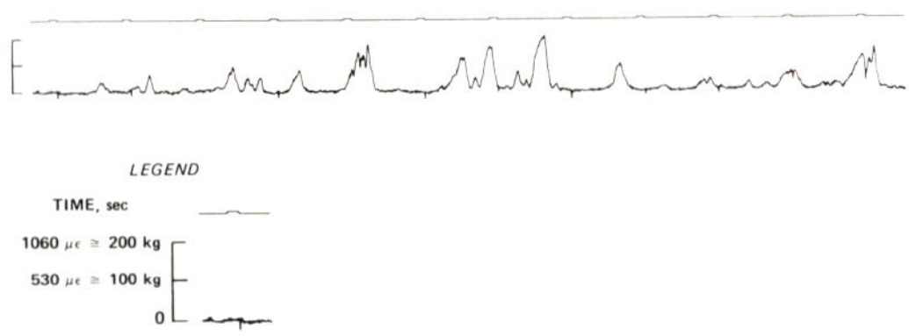

Figure 4. In vivo record of strain in active primate.

REFERENCES

1. Young DR, Adachi RR, and Howard WH: Bone mineral losses during hypokinesis in primates *(Macaca nemestrina)*. Abstr. 46th Ann. Scientific Meeting, Aerospace Med Assoc, San Francisco, California, April 28-May 1, 1975

2. Chamay A, Tschantz P: Mechanical influences in bone remodeling. Experimental research on Wolff's law. J Biomechanics 5: 173, 1972

3. Cochran GVB: Implantation of strain gauges on bone in vivo. J Biomechanics 5: 119, 1972

4. Cochran GVB: Method for direct recording of electromechanical data from skeletal bone in living animals. J Biomechanics 7: 563, 1974

5. Lanyon LE, Smith RR: Measurement of bone strain in the walking animal. Res Vet Sci 10: 93, 1969

6. Lanyon LE: Analysis of surface bone strain in the calcaneus of sheep during normal locomotion. J Biomechanics 6: 41, 1973

AN IMPLANTABLE ELECTRODE FOR TELEMETRIC REGISTRATION OF ELECTROURETEROGRAMS

P.Rosenkilde, E.Thorup, S.Hannibal

Division of Urology and Technical Department,
Gentofte Hospital, University of Copenhagen,
2900 Hellerup, Denmark

INTRODUCTION

The ureter, the fibromuscular tube which conveys urine from the kidney to the bladder, exhibits rhythmical peristaltic contractions which travel at speeds and at frequencies that depend on the volume of urine formed by the kidney. This mechanical function is closely linked to the electrical activity which precedes the peristaltic contractions. Telemetric registration of the electrical activity in the form of action potentials, electroureterograms (EUG), has previously been performed with platinum electrodes with a diameter of 0.07 mm, (1). As several experiments have shown the platinum wires to break at arbitrary sites, they are unsuitable for implantation and telemetric monitoring over long periods.

In the search for a better material in which the different demands to the electrode, i.e. pick-up characteristics (2,3), physical flexibility, small dimensions, high tensile strength, and ease of electrode attachment are combined, we have studied Nikrothal-LX and found it very versatile.

MECHANICAL ASPECTS

The Nikrothal-LX wire has a diameter of 0.05 mm and is made of an alloy of 75 % Ni, 20 % Cr and 5 % "other materials", mainly Mg, C, Al and Co. The wire is coated with a silicone enamel which provides good compatibility with body tissue. The tensile strength of Nikrothal-LX is 110-140 Kg/mm^2 which compared to the 18-24 Kg/mm^2 of platinum, corresponds to an increase in breaking strength of 5-6 times.

In the experiments each electrode is approximately 26 cm long. At both ends the insulation is removed by means of watergrindingpaper (grain 400) for a length of 2 cm. One end is shaped into a coil of 5 or 20 windings with a diameter of 1 mm, yielding an area of 2.5 mm^2/10.0 mm^2. A 23 cm silastic baby-feeding tube

(8F) is slipped over the electrode, and the noncoiled end is either soldered into an implantable telemetry transmitter or connected to an extracorporeal amplifier, depending upon the character of the experiment. The active electrode is sutured with 6-0 atraumatic silk through the 5 winding coil, extraluminal into lamina muscularis of the ureter. The inactive electrode (20 windings coil) is sutured the same way into the ipsilateral fascia of musulus psoas major, which is a noncontractile tissue.

ELECTRICAL ASPECTS

The experiments have revealed the electroureterograms to have amplitudes within the range 100-400 uV peak-to-peak, a duration of a single EUG complex of approximately 1 mSec and a repetion rate less than 10 EUGs per second.

Thus, in order to compare the pick-up characteristics of the Nikrothal-LX electrode to the platinum electrode, we have utilized these data and based our measurements on the ability of the electrodes correctly to reproduce a square-wave voltage of the same amplitude, duration, and repetition rate as the EUG. The square-wave voltage is picked up from a 0.9 % saline solution through which a constant square-wave current is transmitted.(4)

The set-up consists of a 20 cm plexiglass tube with a diameter of 5 cm. Both ends of the tube are sealed with aluminum plates which act as electrodes for the constant square-wave current adjusted for a pulse duration of 1 mSec and a repetition rate of 10 per Sec. The tube is filled with 0.9 % saline of a temperature of 37° C. In the middle of the tube there are holes for the electrodes to be tested, bipolar measurements, and holes for a pair of Ag-AgCl electrodes used for control of the electric field in the tube. When square-wave testing is performed, the electrodes are connected to a Tektronix differential amplifier (3A3, 0.1 Hz - 500 KHz). In case of noise measurements the electrodes are connected to the EMG amplifier (DISA 14C10, 20 Hz - 1 KHz) used for the actual EUG measurements.

Figure 1 shows the voltage pulse picked up by the electrodes: platinum, Nikrothal-LX and Ag-AgCl. As can be seen from the figure, all three pairs of electrodes reproduce the square-wave pulse with no distinguishable differences.

In order to compare the noise of the electrodes, these were connected to the amplifier and no current

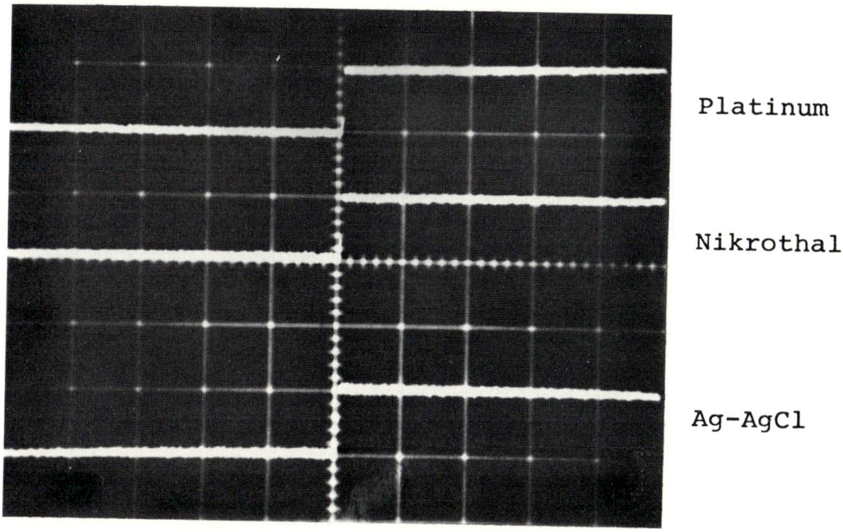

Figure 1. Reproduction of square-wave pulse. Hor.:0.2 mSec/div. Vert.: 200uV/div. (see text)

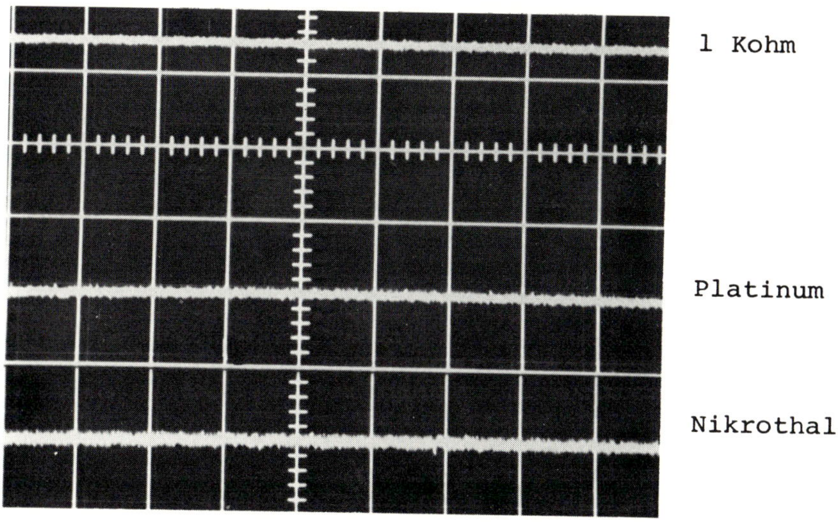

Figure 2. Inherent noise of electrodes and amplifier. Hor.:1 Sec/div. Vert.: 50 uV/div.

was sent through the saline solution. Figure 2 shows the output from the amplifier. For comparison is also shown the resultant noise, when the amplifier input is connected to a 1 Kohm resistor. It is seen that the total noise level for both sets of electrodes is of the order 10 uVolts peak-to-peak, and thus can be disregarded in EUG measurements.

EXPERIMENTAL RESULTS

The Nikrothal-LX electrodes have been used to register monopolar EUGs from the ureter before and after kidney autotransplantation. Up to this point, registration has been performed on 22 adult dogs of Vorsteh-breed with the electrodes implanted from 1 to 11 weeks. The results, regarding the electrodes ability to monitor EUGs, are tabulated in Table 1.

Table 1. Success of EUG-recording with Nikrothal-LX.

Status of recorded EUG	Before tranplantation %	After transplantation %	Mean %
Excellent	50	45	48
Good	32	36	34
Fair	10	14	12
Failure	8	5	6

None of the experiments suffered from broken electrodes, and histological examinations showed very small reactions at the sites of transplantation.

REFERENCES

1: Hrynczuk J.R., Scand.J.Urol. and Nephrol.,6:suppl. 15:105:1972.

2: Geddes L.A.,"Electrodes and measurement of bioelectric events", J.Wiley Comp.,1972.

3: Strong P.,"Biophysical measurements",chap.16, Tektronix, nov.1970.

4: Weiman J & Mahler J.,"Med.Electr.Biol.Engng.", 2: 299:1964.

ADDENDUM

Nikrothal-LX is manufactured by A/B Kanthal, Hallstahammar, Sweden.

SHORT RANGE TELEMETRY OF TISSUE OXYGEN TENSIONS

J.W. Hartley, V.G. Murphy, M.J. Devaney, R.E. Barr and A.W. Hahn

Dalton Research Center, University of Missouri
Columbia, Missouri (USA)

Monitoring of tissue or blood oxygen partial pressure (PO_2) is important in a number of areas of clinical and research medicine (1). Telemetry provides a convenient, if not necessary, means for continuous measurement of tissue and blood PO_2 for long term (2-5 days) studies. Modern integrated circuit technology has made available MOSFET input operational amplifiers where high input impedance, low bias current, and low power drain make them well suited to the monitoring of the nanoampere O_2 electrode currents in a miniaturized PO_2 telemetry system. This paper describes our initial efforts to develop such a system.

METHODS

The transmitter (Fig. 1A) consisted of a current-to-voltage converter with a buffer amplifier driving a voltage controlled oscillator (VCO) to produce a subcarrier for the single transistor frequency modulated (FM) transmitter. The current-to-voltage converter was a single RCA CA3130 FET-input integrated circuit that was followed by a second CA3130 connected as an inverting amplifier to provide both gain and offset control. The signal voltage was input to the VCO section of a RCA CD4046 to generate a frequency of 200 Hz for zero current. The linear range of the VCO extended to approximately 9 KHz, which corresponded to an input current of 18 nanoamperes. The signal was used to swing the frequency of the single transistor oscillator. The fundamental of the transmitter oscillator was approximately 20 MHz but harmonics could be received by broadcast band FM receivers.

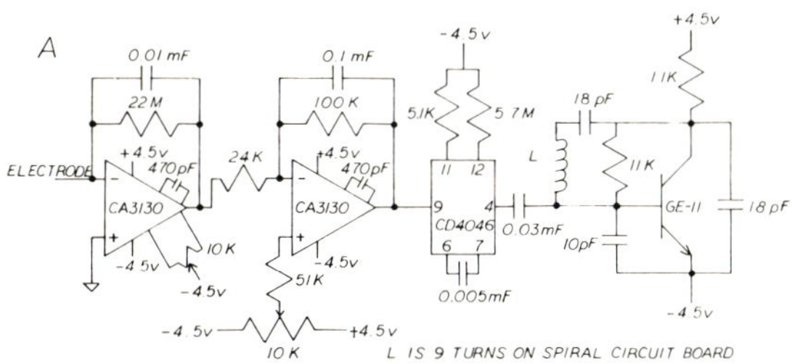

Fig. 1. A. Circuit diagram of the transmitter.

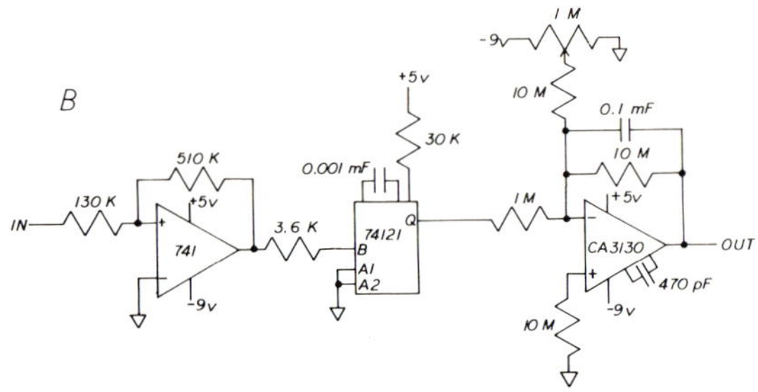

Fig. 1. B. Circuit diagram of the receiver subsystem.

The receiver subsystem (Fig. 1B) consisted of a standard FM receiver equipped with AFC and AGC to demodulate the FM subcarrier. The receiver audio output was passed through a 4-pole active band-pass filter, and into a comparator with a 2 volt hysteresis. The comparator output was used to trigger a 74121 monostable multivibrator with a 10 μsec. pulse width. The multivibrator output was integrated using a RCA 3130 operational amplifier with a one second integrating time constant in the feedback network. The integrator output, proportional to the audio frequency, was recorded on the strip-chart recorder.

For the purposes of testing the system, bare tipped, 25 micrometer diameter platinum (insulated in glass) electrodes were fabricated into an implant package. The implant package consisted of four electrodes, tie down collars, lead wires and a transcutaneous connector device. Details of this package, the surgical implantation procedure and their mode of operation were given earlier (2). After approximately one week post-surgery, after implanting the electrodes, changes in inspired PO_2 produced changes in electrode currents indicating changes in the surrounding tissue oxygenation. The transmitter and batteries were placed in the pockets of a vest worn by the rabbit. To obtain good illustrative data, the rabbit was anesthetized with halothane, although data from unanesthetized restrained and unrestrained states were also collected.

RESULTS

A calibration of curve of the telemetry system, in which output voltage is plotted as a function of input current, is shown in Fig. 2. The response was linear over a range from approximately 5 to 15 nanoamperes input current. At lower currents the response was nonlinear and less sensitive.

Electrode current is dependent on oxygen, polarization voltage, motion artifact, and "aging" and "poisoning" phenomena. The present investigation was not concerned with absolute oxygen tension values but only with relative changes. Hence, the polarization voltage was set for a given run and the system was operated until the electrode current had stabilized before a forcing gas was applied. For short runs, the aging characteristic was unimportant and motion artifact, if it existed, was a small perturbation to the **data.**

A sample of data obtained is shown in Fig. 3. It can be seen that the output voltage of the telemetry system, curve A, closely matches the shape of the input current, curve B. For the sample shown, the lower valued portions of curve A correspond to the relatively insensitive portion of the telemetry system's response curve and, therefore, is not as responsive to changes in the input current at these levels.

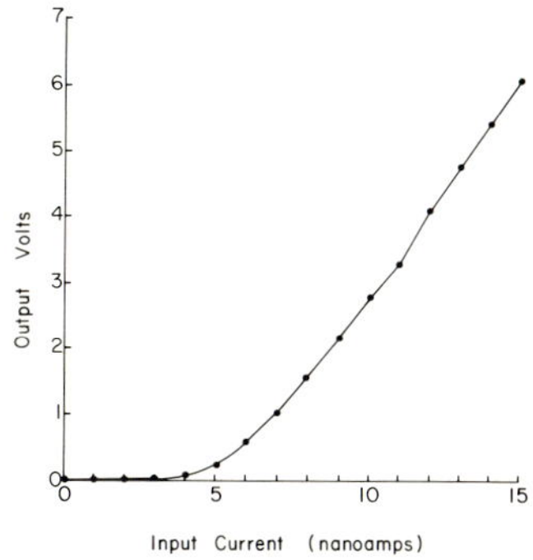

Fig. 2. The current to voltage response curve of the telemetry system.

DISCUSSION

Continuous monitoring of tissue or blood oxygen tension for a period of days has many problems associated with it. One of these problems is that it would be impractical to attempt to collect such data by "hard-wire" methods. Even short term

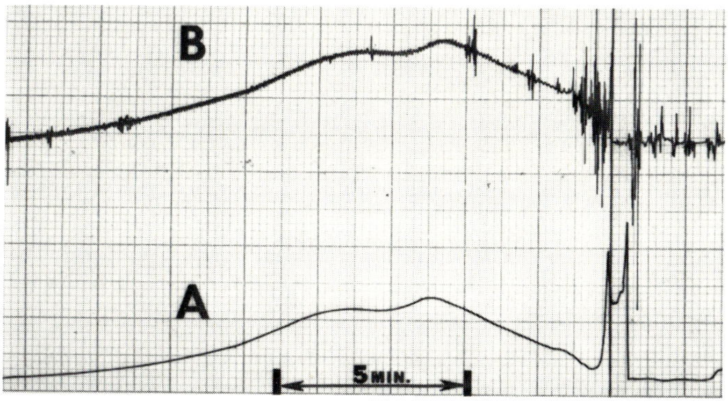

Fig. 3. Sample data taken from electrodes implanted in rabbit striated muscle. Curve A: The input current to the transmitter. Curve B: The output voltage of the telemetry system.

monitoring requires restrained or anesthetized animal conditions if hard wire methods are used. The use of a telemetry system is a practical alternative in both instances.

One of the reasons for developing a telemetry system for use with the electrode package described was that previous experiments using restrained unanesthetized rabbits yielded data that could have been influenced by the restraining stress. Using the telemetry system permitted animal mobility, eliminating the possible restraining influence. However, in this experiment, no significant difference was found between data obtained for the restrained and unrestrained animal and, in both cases, there was no response to changes in respiratory oxygen or carbon dioxide concentrations. Therefore, in order to adequately demonstrate the applicability of the telemetry system, an anesthetized animal was required.

There were a number of features about the telemetry system that need improvement. The main problems encountered were instability of the transmitter frequency due to varying stray capacitance antenna and receiver drift due to the pulsed FM nature of the transmitter. It would also be beneficial to reduce the size of the overall package.

In conclusion, while our present system needs additional work on further miniaturization and other technical features, we have demonstrated the feasibility of monitoring oxygen tension from biological subjects. Results of this study show that it is possible to transmit data from chronically implanted oxygen electrodes. It permits unrestrained, unanesthetized preparations to be studied, with local oxygen tensions being monitored under the influence of many types of forcing agents.

ACKNOWLEDGEMENTS

Thanks are in order for H. Whittlinger of RCA, Summerville, New Jersey, for his assistance in circuit design, and for Bonnie L. MacAlmon and Michael F. Nichols for their invaluable technical assistance.

This work was done in part with the support of the U.S. Army Surgical Directorate under contract DADA17-71-C-1104, NASA grant NGR26-004-099 and NSF grant NGH 38957.

REFERENCES

1. Bruley DF and Bicher HI: Oxygen Transport to Tissue. New York, Plenum Press, 1973.

2. Murphy VG, Barr RE and Hahn AW: Development of chronically implantable electrodes for monitoring oxygen tension in skeletal muscle. In Biomedical Sciences Instrumentation - Vol. II (ed KC Rock). Pittsburgh, Instrument Society of America, 1975, pp 31-36.

A STRAIN GAUGE JAW FORCE TRANSDUCER*

Ved R. Singh

Institute of Biomedical Engineering
University of Toronto,
Toronto, Canada

INTRODUCTION

A jaw force transducer, using semiconductor strain gauges, has been developed. The transducer consists of stainless steel plates in the form of a fork to fit into the jaws. The strain gauges are bonded to both limbs of the fork, the output of which (gauges) is fed to an electrical detecting system. The jaw forces and movements cause changes in the resistance of the gauges which, in turn, give an electrical output. This output can be telemetered by the use of the conventional techniques of voltage controlled oscillator (VCO), modulation, and digital counting, if desired.

The author's work on the development of different transducers, using semiconductor strain gauges, has achieved realization of old applications in a novel way, and investigation of a host of novel uses and applications of these devices in various branches of medicine, engineering and science. Some of the biomedical transducers and their applications were reported in previous conferences and published in various journals (1-6). These transducers were mainly used for the measurement of physical parameters in medicine like blood and other physiological pressures, blood flow, respiration rate, drip rate, pulse rate and minute physiological displacement. This paper describes the development of a transducer, using strain gauges, for the measurement of jaw forces in living humans and experimental animals.

*This investigation was carried out at the National Physical Laboratory of India in New Delhi.

CONSTRUCTIONAL DETAILS

Fig. 1 shows the block diagram of the jaw force transducer with its electronic equipment. The transducer consisted of stainless steel plates in the form of a fork to fit externally or internally, as the case was, to the jaws. Two semiconductor strain gauges, each having a resistance of 120 ohms and both connected in series, were cemented to both limbs of the fork of the transducer. The output of the gauges was applied to the electrical detecting system which consisted of a conventional Wheatstone bridge, a high-gain amplifier, and a recorder or an output meter.

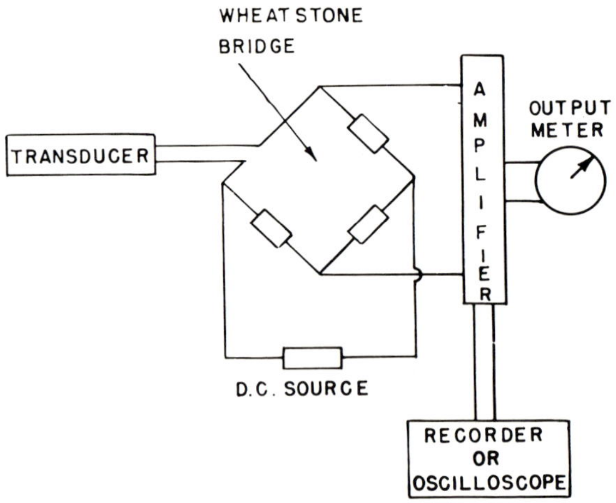

Figure 1: Block diagram of measuring apparatus for strain gauge jaw force transducer

OPERATION

The free limbs of the fork which were designed to suit the place of application were placed to the inside of the jaws and the teeth, for taking measurements. The jaw forces and movements caused, through the minute displacements of the limbs of the fork, changes in the resistance of the gauges which, in turn, gave an electrical output. This output was then calibrated in terms of jaw forces and movements. The electrical output can be telemetered by the use of conventional techniques of voltage controlled oscillator (VCO), modulation, and digital counting, if desired.

APPLICATIONS

The jaw force transducer is useful in monitoring the motion of the entire jaw as a function of time during chewing and speech. By standardizing the technique, the medical doctors will be able to distinguish the words by machine to get additional clinical information about paralyzed persons in controlling devices such as motorized wheel chairs and artificial limbs. Jaw forces are, generally, measured by keeping the teeth in contact with hard objects. This transducer is very useful to carry out research work in humans on mastication, i.e. measurement of jaw reaction to foreign objects in the food.

CONCLUSION

A transducer using a semiconductor strain gauge has been designed and developed for the measurement of jaw movements and forces.

REFERENCES

1. Singh VR, Parshad R: Transducers for biotelemetry. Proc 2nd Intn'l Conf on Biotelemetry, DAVOS (Switzerland) May 20-25, 1974
2. Singh VR: A urinary flow transducer. Proc Int'l Conf on Bio-Medical Transducers (Biocapt 75), Paris (France) Nov. 3-7, 1975
3. Singh VR, Parshad R: An account of Biomedical instrumentation done in NPL. Proc Symposium on Transducer Technology, Bangalore (India) Sept, 1974
4. Singh VR, Parshad R: Strain gauge transducers for measuring blood pressure. Research and Industry (New Delhi) 18: 48, 1973

5. Singh, VR, Parshad R: Strain gauge instrumentation for measurement of small displacement of body parts: applications to body pulsations with respiration. Digest 10th Intn'l Conf on Med & Biolog Engrg, Dresden (Germany) June 17-23, 1973 p153

6. Singh VR: Biomedical transducers. The Clinician (India) 38: 80-88, 1974

DYNAMIC MEASUREMENTS OF VISCO-ELASTICITY FOR ACTIVE HUMAN MUSCLE

G.Matsumoto, T.Furukawa and Y.Tsuchida
Research Institute of Applied Electricity
M.Kato
Dept. of Physiology, School of Medicine
Hokkaido University, Sapporo, Japan

INTRODUCTION

Subjectively-sensed information such as palpation has been widely used to diagnose patients. Recently, objective means which take the place of subjective diagnosis are receiving attention, for example, to quantify the edema and to obtain information on stati and dynamic characteristics of muscular diseases.

This paper describes a stiffness measuring device which is developed to measure quantitatively the visco-elasticity for active human muscle. Musclus brachioradialis and M. extensor carpi radialis longus are mainly aimed from the viewpoint of their clinical importance. Several experimental results are given with discussions.

PRINCIPLE AND CONSTRUCTION OF DEVICE

The device consists of two measuring parts, one is the stiffness and the other is the displacement measuring device.

Stiffness measuring device: This device consists of a magnet, driving and pickup coils, spring, and piston motion shaft with pellotte (1) as shown in a later figure. The principle is that a self-exciting vibration of the mechanical system will continue when the coils and amplifier form a positive feedback loop. A circuit diagram is shown in Fig.1. Suppose a simple mechanical vibrating system as seen in Fig.2 (a), in which the upper part of the dotted line represents a model of the measuring device and the lower part represents a model of the visco-elastic media. Let notations of the mass and the stiffness of this system be M_T and M_s, S_T ans S_S. Then, the equivalent electric circuit is given in Fig.2 (b). From the analogy between the mechanical and electrical systems, the resonant frequency of this system is given as

$$f = \frac{1}{2\pi}\sqrt{\frac{S_T + S_s}{M_T + M_s}} \qquad (1)$$

Considering the model of the measuring system as seen in Fig.3, resonant frequency is given by Eq.(2), where M_O is the additional mass.

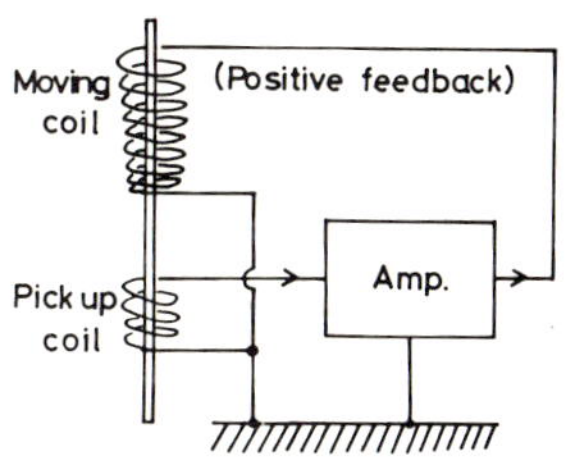

Fig.1. Circuit diagram of stiffness measuring device.

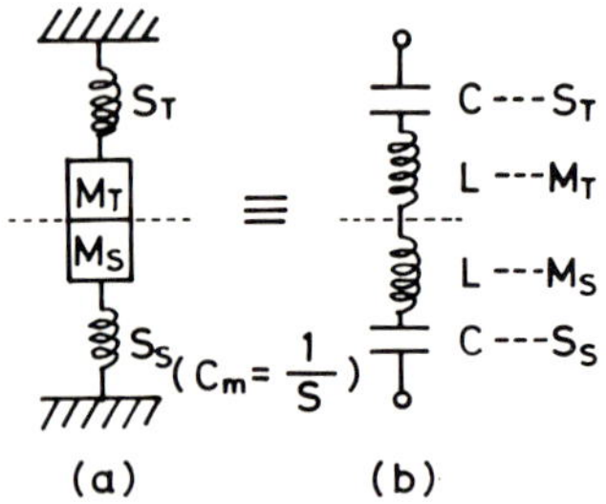

Fig.2. A model system and its equivalent circuit.

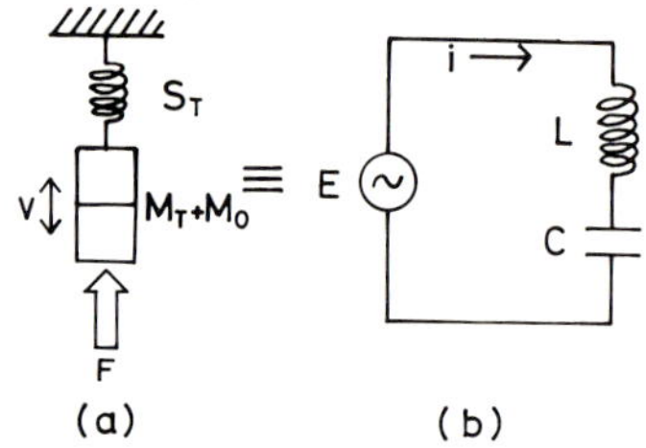

Fig.3. A model of the measuring system with equivalent circuit.

$$f = \frac{1}{2\pi}\sqrt{\frac{S_T}{M_T + M_o}} \tag{2}$$

This equation can be rewritten as

$$\frac{1}{\omega^2} = \frac{1}{S_T}(M_T + M_o) \tag{3}$$

where $\omega = 2\pi f$. Eq.(3) means that the relation between $1/\omega^2$ and $(M_T + M_o)$ is linear and M_T is given by the extrapolated intersection with x axis and also S_T given by the tangent of this linear curve as seen in Fig.4. According to the experimental results for our prototype device which is shown in Fig.5, it was $M_T = 16$ gr. and $S_T = 1.5 \times 10^6$ dyne/cm respectively. Assuming that $M_T >> M_S$, S_S will be decided through the measurement of the resonant frequency.

Displacement measuring device: In order to measure a center offset followed with the static pressure to the visco-elastic media and to measure the vibration amplitude, a displacement measuring device was developed. The displacement detector as seen in Fig.6 (a) consists of a light source, two lenses (focal length 2.5 cm), and a photo-transistor. The LED emits a carrier light with 10 kHz which is collimated, passed through an aperture, and focused on the

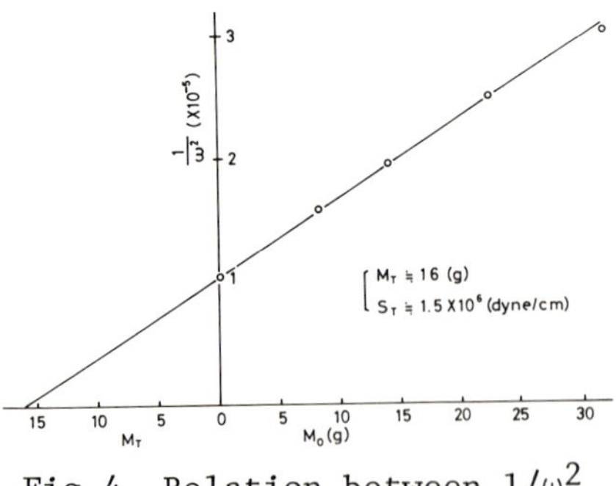

Fig.4. Relation between $1/\omega^2$ and $(M_M + M_o)$.

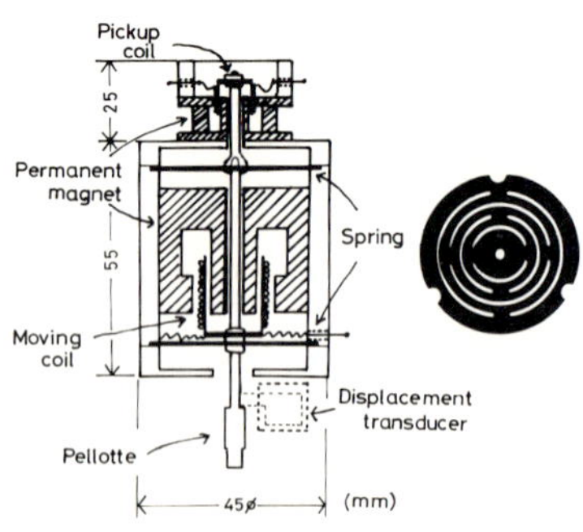

Fig.5. Construction of device.

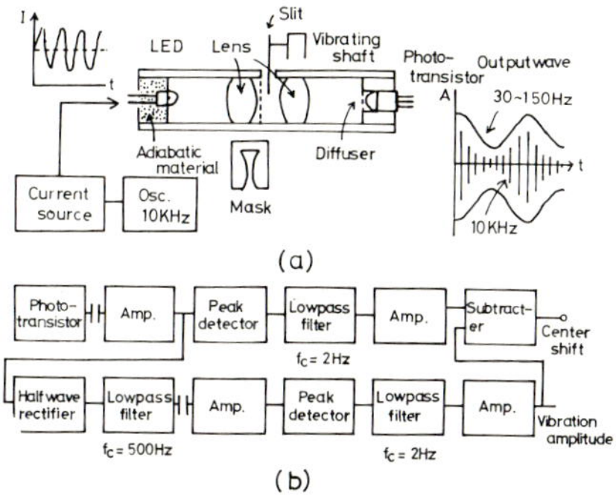

Fig.6. Displacement detector and the output wave form (a), and whole circuit diagram (b).

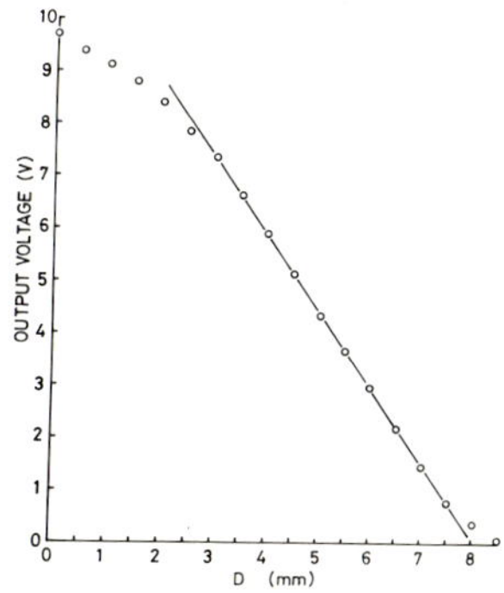

Fig.7. Relation between displacement and output voltage.

photo-transistor. The carrier light is modulated by the aperture opening which corresponds to the displacement, and detected by the photo-transistor. The detected carrier output is demodulated, amplified, and rectified by a peak detector to obtain the vibration amplitude, while the signal is directly rectified without demodulation to obtain the center offset. In the prototype system, a peak detector is used in the center offset channel; the vibration amplitude should be subtracted from the rectified output. As a result, linear region in the displacement is about 6 mm and the lower limit of the detection is about 0.2 mm as seen in Fig.7.

EXPERIMENTAL RESULTS

At first, we tried to check this device using a fish paste as a model. The center offset, with the static pressure of 75 gr. applied to this sample, is only 0.1 mm ;and the variation of the resonant frequency as time elapses is negligible as seen in Fig.8.

Musclus brachioradialis: Measurements were made on the contracted and relaxed states of the muscle. To produce these two states the subjects were asked to grip a handgrip of 10 kgw and then asked to relax the muscle. The measurement was first made on the relaxed muscle, then after a pause of several minutes the contracted muscle was tested. Results were successful, and averages S_s of these two states were given 0.18×10^6 and 0.94×10^6 dyne/cm respectively.

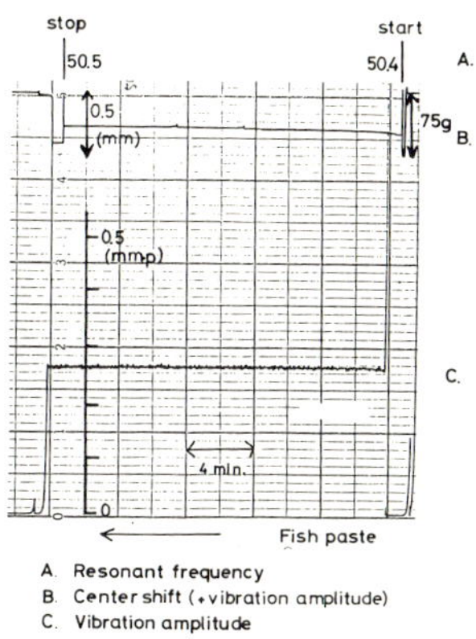

Fig.8. Experimental results of a fish paste.

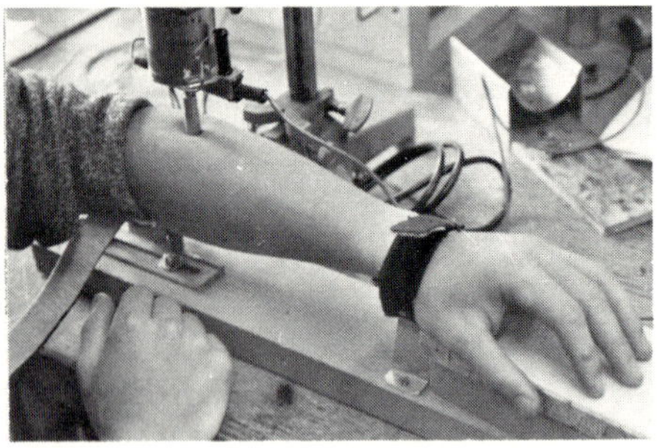

Fig.9. Holding method of the forearm.

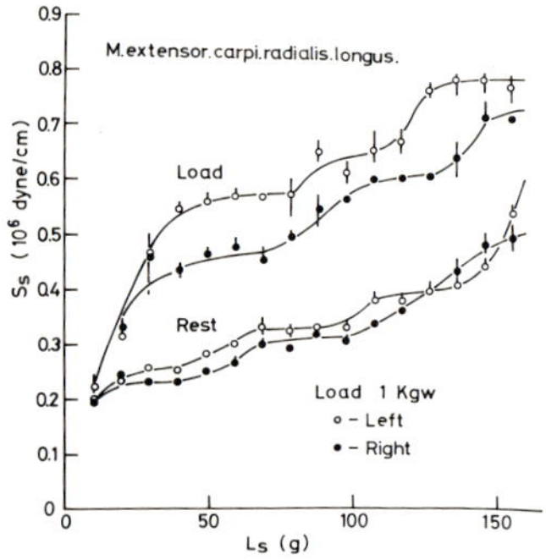

Fig.10. Comparison of the resting and 1 kgw loading states under the change of static load.

M. extensor carpi radialis longus: The reason we studied the extensor carpi radialis muscle is that it is easy to identify the muscle percutaneously and also easy to apply the load against the muscle. The resting state means that the subject was asked to hold his fingers extended against the load. Holding method of the forearm is shown in Fig.9. Following characteristics were investigated: (1) Time variance of the loading state, (2) variation of S_s for gradual increase of loading, (3) recovery process of S_s after critical loading state, and (4) relation between static load and S_s in the resting and loading states. One of the results is shown in Fig.10; this is the comparison between the resting and 1 kgw loading states under the change of static load.

CONCLUSION

Quantitative measurements of visco-elasticity for two parts of the active human muscle were achieved utilizing the stiffness measuring device. Through our experiments which showed a clear difference between the resting and loading states, one should be able to apply this device to patient monitoring, clinical usage, and physiology. Further investigations should be conducted to measure the mass of human muscle, and more experiments should be done on the dominant effect of the viscosity.

We are greatly indebted to Prof.S.Tsukahara for his co-operation and advice. This work is also indebted to the Grant No.010901, Specified Category 1975, Grant-in-Aid for Scientific Research from the Ministry of Education of Japan.

REFERENCE

(1) T.Kobayashi, S.Sakamoto, M.Hogaki et al: Studies on the abdominal and Cervical Stiffness during Labour (As determined by a Compliance Measuring Device). 6th. Int. Conf. on M.E. and B.E. (1965)

IMPLANTABLE ACCELEROMETER FOR TELEMETRY

E. Konigsberg; L. Cothran*, E. W. Hawthorne

Konigsberg Instruments, Inc., Pasadena, California;
and *Howard University School of Medicine, Washington, D. C.

INTRODUCTION

A miniature acceleration sensor with characteristics suitable for the measurement of the contractile acceleration of heart muscle has been developed, and has been utilized in animal physiological experiments. The size and shape of this sensor are similar to miniature pressure transducers previously developed. The unit is used in the same way as the implantable pressure sensor, i.e., surgically attached near the apex of the myocardium. The output of the sensor is analogous to that of the pressure gages, and is suitable for telemetering, either by wireless or transcutaneous hardwire connection to the outside world.

METHODS

The miniature accelerometer, a cross sectional view of which is shown in Figure 1, consists of a suspended seismic mass attached

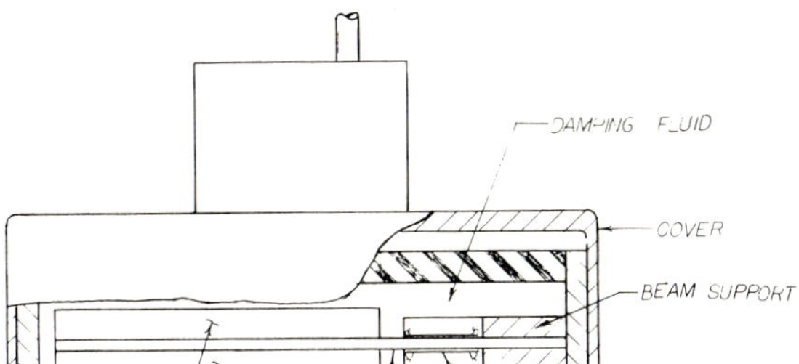

Figure 1
Cross-Section View of Accelerometer

to twin cantilever beams on which semiconductor strain gages are bonded. The beam and mass are enclosed in a titanium housing approximately 6.5 mm in diameter and 2 mm in height. Electrical connection is made to the strain gages by a cable which emerges axially along the center line of the titanium package. The configuration is similar to that of previously reported implantable pressure transducers (1). Surgical techniques established for the implantation of pressure gages may therefore be used for the new component.

The chamber containing the seismic mass and beam is filled with silicone oil and sealed. A damped resonant frequency of approximately 500 Hz is thereby obtained with a damping ratio of 0.55.

These units were fabricated from raw materials using processes similar to those used for the previously developed pressure transducers. Their sensitivity was calibrated using the earth's 1 g field by rotating the transducer through 180 degrees so that input acceleration between -1.0 g and +1.0 g are obtained.

Measurement of the accelerometer natural frequency and damping factor were accomplished by observation of the electrical response to a mechanical shock. The accelerometer was struck sharply and its response recorded on a memory oscilloscope. The natural frequency and damping factor may be calculated from the theoretical relation between the ringing frequency and decay envelope thus obtained (2).

Implantation of the accelerometer in an experimental animal was carried out according to a procedure previously disclosed (3). In addition to measurements of apical acceleration, precordial acceleration using another instrument of the same type was simultaneously measured. The fluid dynamic cardiac parameters, pressure and dP/dT were measured in order to investigate the relations and correspondences between acceleration measurements and the more traditional measures.

RESULTS

The static tests of sensitivity demonstrated the accelerometer output to be approximately 7 mV per g with a cross axis sensitivity of less than 5%. On the basis of the resonant frequency and damping ratio measurements, the frequency response of these units is predicted to be flat within ±1 dB from dc to 225 Hz.

In the animal experiments, accelerometers were sutured to the free wall of the left ventricle and their outputs recorded along with various cardiac pressure measures. An example of the results thus obtained is shown in Figure 2, which demonstrates the apparent sensitivity of these accelerometers to the physiologic changes induced by the infusion of dobutamine.

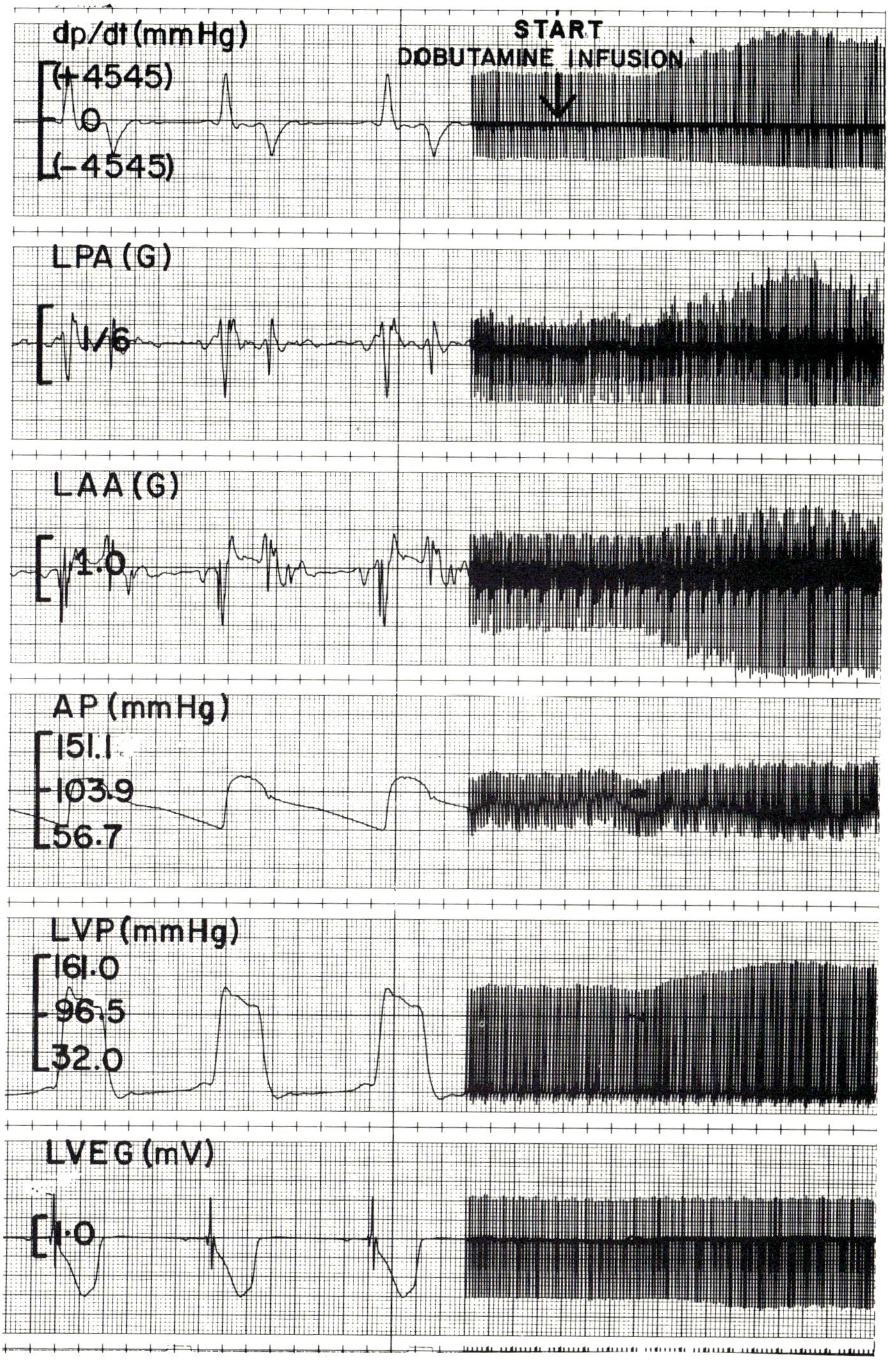

Figure 2
Multiparameter Response to Dobutamine Infusion

REFERENCES

1. Konigsberg E: A pressure transducer for chronic intravascular implantation. Proceedings of the Fourth National Biomedical Sciences Instrumentation Symposium, Anaheim, CA 1966.
2. Stein PK: Measurement Engineering, Vol. 1, 3rd ed., 1964. Stein Engineering Services, Phoenix, Arizona.
3. Tearney RJ, Cothran LN, Hawthorne EW: Left ventricular apical acceleration. 60th Annual Meeting, Federation of American Societies for Experimental Biology, Anaheim, CA, 1976.

USE OF IMPLANTABLE STIMULATORS TO INVESTIGATE FUNCTIONAL ADAPTATION IN SKELETAL MUSCLE

S. Salmons*, F.A. Sréter**, and C.J. Morris†

* *Department of Anatomy, University of Birmingham Medical School, Birmingham, England.*

** *Department of Muscle Research, Boston Biomedical Research Institute, Boston, Massachusetts 02114, U.S.A.*

† *Department of Experimental Pathology, University of Birmingham Medical School, Birmingham, England.*

Muscles are extremely efficient machines for converting chemical energy into mechanical work. The cells of which they are composed become specialised for this purpose as a result of developmental processes in which the motor nerve is known to play an important part. The nature of this neural influence on muscle differentiation has been the subject of vigorous controversy, but recent evidence shows that it derives at least partly from the long-term effects of the activity imposed on the muscle fibres by their motoneurones.

There is a division of labour among these motoneurones which is determined more by their intrinsic properties than by the manner in which they are interconnected. Small motoneurones have a low firing threshold and therefore tend to discharge almost continuously at a low impulse frequency. These motoneurones, and the small groups of muscle fibres which they innervate, are involved in postural activity. In keeping with this role the muscle fibres have properties which are appropriate for sustaining continuous use for long periods without fatigue. They have a slow contractile speed, which enables them to maintain a given tension with less frequent activation than a faster-contracting muscle would require. In this way their energy requirements are reduced to a point where they can be met via the highly efficient oxidative metabolism pathway. The fibres therefore contain numerous mitochondria to which oxygen diffuses from a rich capillary network, a process assisted by the small diameter of the fibres and the additional myoglobin which they contain.

Large motoneurones, on the other hand, have a high firing threshold and discharge less often, but at a high impulse frequency. Such motoneurones are recruited for making rapid postural adjustments and movements. The corresponding muscle fibres may therefore be subjected to bursts of intense activity interspersed with relatively long recovery periods. In these fibres economy is subjugated to speed of action. Energy is derived from the breakdown of stored glycogen, a pathway not subject to the constraints imposed by a continuous requirement for oxygen. Such fibres are large in diameter, have a sparse distribution of mitochondria, and are only seldom associated with capillaries. If they should be subjected to continuous activity they cannot replenish their reserves and so fatigue rapidly.

Thus in response to fluctuations in the level of excitation in the spinal cord, motoneurones are enlisted in order of increasing size. The position of any particular motoneurone in the hierarchy of recruitment determines the functional demands which will be made on the corresponding muscle fibres, demands which are reflected in their physiological and biochemical properties (1).

During the initial stages of differentiation, muscle fibres do not differ in their physiological or biochemical properties; these differences appear at a later stage, which in many mammals is not complete until a few weeks after birth. It is around this time that the adult patterns of muscle activity are becoming established. We believe (a) that the characteristic properties of slow-contracting muscle fibres develop in adaptive response to the continuous low-frequency activity which is imposed on them by their motoneurones, and (b) that the presence of this pattern of activity is a necessary condition for the retention of these properties (2,3,4). This hypothesis implies that the capacity for responding adaptively to postural patterns of activity may be a feature of adult, as well as of neonatal, skeletal muscle. We have demonstrated that this is so by using an implantable electronic stimulator to impose such activity on a mature muscle composed largely of fibres of the fast-contracting type.

The stimulator is based on a simple complementary astable circuit (3) accommodated in one half of a stainless steel capsule. The other half of the capsule contains a battery of three mercury cells, and the device begins to operate when the two halves are screwed together. The electrode leads are of helical stainless steel pacemaker wire sleeved in silicone rubber tubing and fitted with welded eyelets which are secured by screws to terminal nozzles on the capsule. Before implantation the device is coated with Dow Corning Silastic 382 silicone rubber into which a web of Dacron mesh is incorporated for suturing purposes; it is then sterilized by immersion in 1% benzalkonium chloride for not less than 24 hours. The stimulator, of length 25 mm and weight 9 g, is implanted in the abdominal cavity of the animal under strictly aseptic conditions. The leads are taken subcutaneously to the left hind limb where they are fixed near to the common peroneal nerve. Accurate placement of the electrodes at this stage is essential if the nerve is to remain in an adequate stimulating current field for the full range of movement of the limb. Stimulation is immediately evident as an oscillation of the hind paw, resulting from excitation of muscles in the front compartment of the leg. After the operation the animals are returned to conventional housing. They are completely unrestrained and appear quite unperturbed by the stimulation, as evidenced by their excellent condition and normal gain in body weight throughout an experiment. After periods of stimulation up to 5 months the physiological characteristics of the muscles are measured in a terminal procedure, after which the muscles are quickly excised, weighed and frozen for subsequent histochemical and biochemical examination.

We have found that muscles subjected to stimulation in this way undergo striking changes. These consist of an increase in the density of the capillary bed, a switch from a glycolytic to an oxidative enzyme profile, a decrease in the capacity of the sarcoplasmic reticulum for sequestering calcium ions, a decrease in the specific activity and alkali stability of calcium-activated myosin ATPase, a change in the complement of light chains associated with the globular head of the myosin molecule, and changes in the structure and amino acid composition of the myosin heavy chains. The physiological characteristics reflect these changes; after five months the stimulated fast muscle fibres have acquired the contractile characteristics of slow fibres and indeed surpass fibres of this type in their resistance to fatigue (3,5,6,7,9). These changes are fully reversible; if stimulation is stopped after six weeks the original characteristics of the muscle are eventually reinstated. It is not a little surprising that a tissue as highly differentiated as muscle can change so profoundly. Changes in the contractile proteins are particularly significant, since they provide clear evidence of a re-direction of protein synthesis within the cells.

Practical implications

These results have important implications for the clinical use of muscle stimulation. These may be illustrated by reference to the following example. There are some forms of paralysis in which the motoneurones have become disconnected, as a result of trauma or disease, from the higher centres which are responsible for the initiation of voluntary movement. In such cases the integrity of the final common pathways to the muscles is preserved. The possibility therefore arises of restoring function to the paralyzed muscles by applying appropriate patterns of electrical stimuli to their motor nerves. However when such stimuli are applied direct to a motor nerve it is the axons of largest diameter which are recruited at the lowest threshold, and these activate the largest, most readily-fatigued muscle fibres. As has been pointed out by others (8) this represents an inversion of the natural order of recruitment and places the greatest functional demands on the muscle fibres least equipped to deal with them. This would seem at first sight to constitute a major obstacle to the use of functional electrical stimulation for all but the most phasic of activities. What our studies indicate, however, is that the adaptive capabilities of skeletal muscles will enable them ultimately to accommodate these totally unaccustomed patterns of efferent stimulation. The apparent deficiences which such muscles may exhibit in the short term need not, therefore, be regarded as a serious objection to the implementation of this technique. This example should serve to emphasize that when procedures involving chronic stimulation are being contemplated it is essential to take proper account of the long-term consequences of changes in muscle activity.

ACKNOWLEDGEMENTS

The work was supported by grants to S.S. from the Medical Research Council and to F.A.S. from the National Institutes of Health (HD-06203), the National Science Foundation (GB-41383) and the Muscular Dystrophy Association of America, Inc. We thank Mr. R.A. Prysor-Jones and Mr. D.R. Gale for their excellent technical assistance.

REFERENCES

1. Henneman E and Olson CB: Relations between structure and function in the design of skeletal muscles. J Neurophysiol 28: 581-598. 1965.
2. Salmons, S: The development and use of implantable electronic devices to investigate the influence of activity on the contractile speed of mammalian skeletal muscle. Ph.D. thesis, University of Birmingham, 1968.
3. Salmons S and Vrbová G: The influence of activity on some contractile characteristics of mammalian fast and slow muscles. J. Physiol 201: 535-549, 1969.
4. Salmons S: Skeletal muscle - an adaptive machine? Brit Sci News No. 83: 2-4, 1971.
5. Sréter F.A., Romanul FCA, Salmons S and Gergely J: The effect of a changed pattern of activity on some biochemical characteristics of muscle. In Exploratory Concepts in Muscular Dystrophy II (ed AT Milhorat) International Congress Series No 333: Amsterdam, Excerpta Medica, pp 338-343.
6. Romanul FCA, Sréter FA, Salmons S and Gergely J: The effect of a changed pattern of activity on histochemical characteristics of muscle fibres. Ibid, pp 344-348.
7. Sréter FA, Elzinga M, Mabuchi K, Salmons S and Luff AR: The N^{τ}-methylhistidine content of myosin in stimulated and cross-reinnervated skeletal muscles of the rabbit. FEBS Lett 57: 107-111, 1975.
8. Peckham PH: Electrical excitation of skeletal muscle: alterations in force, fatigue and metabolic properties. Ph.D. thesis, Case Western Reserve University, 1972.
9. Salmons, S: On the feasibility of using diaphragm muscle as a myocardial substitute (Letter). Med & Biol Engng 13: 608-609, 1975.

A PRESSURE TELEMETRY SYSTEM UTILIZING A CAPACITANCE-TYPE TRANSDUCER

Thomas B. Fryer

Ames Research Center, NASA
Moffett Field, Calif. 94305

The measurement of pressures such as CSF, RV, RA, and LA in the body is of great physiological significance to researchers and clinicians, but requires very stable, accurate measurements to be of value. Miniature pressure transducers designed for physiological measurements and suitable for implantation in the body have been manufactured and commercially available in recent years. (5) These transducers are solid state strain gage devices approximately 7 mm in diam. and 1 mm thick, attached to the end of a 1 mm-diam. cable. Although such devices have been widely used for arterial pressure measurements (9) where the pressure levels are large (100-200 mm Hg), zero drift has limited their use for low pressures such as encountered in venous or intracranial measurements (5,7,11). Dynamic pressure changes can be measured quite accurately with these devices, but drifts of the order of 1 mm or more per day make them unsuitable where static changes of a few mm Hg are of significance. Accurate static pressure measurements have been made in the past using fluid-filled catheters attached to a manometer; but this is cumbersome, movement of the subject is limited, catheters are a source of infection, continuous recording is not feasible, and the system does not allow for telemetry from an unrestrained subject. For continuous recording and telemetry applications, the use of pressure transducers is essential, but in order to accurately measure static changes of a few mm Hg, better pressure transducers than those currently available are essential.

The system described here uses a capacitance-type pressure transducer to achieve better static accuracy. The size of the transducer is approximately the same as those presently available in commercial units (7 x 1 mm), but it is hoped that the elimination of the epoxy glues and other plastic materials that are used to attach the strain gages to the diaphragm will improve the long term stability. The construction of the transducer is relatively simple (fig. 1), but does require precision machining because of the small dimensions and tight tolerances. The movement of the diaphragm with pressure changes the capacitance between the two insulated metal surfaces. The magnitude of the capacitance and its change per mm Hg is very minute (table 1). The obvious

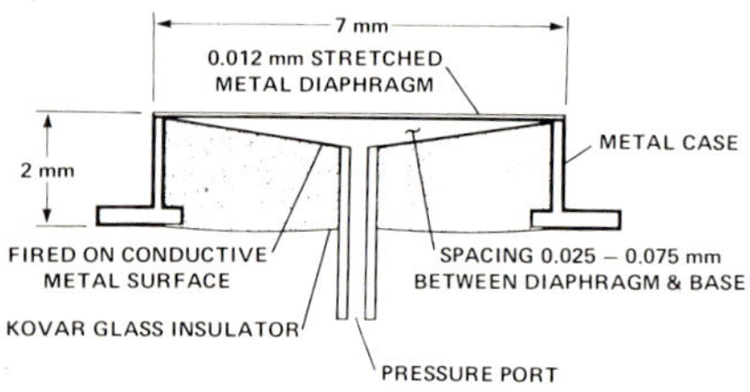

Figure 1. Capacitance pressure transducer.

Table 1

Pressure, mm Hg	Capacitance, pF
0	7.8
100	8.5
200	9.2
300	10
400	11
500	12
600	13.5
700	15.5

difficulty, of course, is the accurate measurement of this small change in capacitance. There are reports in the literature on the use of inductive (8,2) and capacitance (1) transducers, but to effectively measure the small changes designers have usually attempted to use plastic diaphragms (10,12) and mechanical leverage arrangements (12) to increase the electrical change. Hysteresis in plastic materials and their susceptibility to moisture leakage (10) create other unstable factors that do not solve the long term stability problem. By using a carefully selected circuit, the system described in this report is able to function accurately with changes in capacitance of the order of 1 pF full scale.

One of the circuit designs selected for use with the capacitance transducers is shown in fig. 2. Transistor Q_2 with its associated components forms a Hartley-type oscillator. The inductance of the collector tank circuit is wound to obtain an operating frequency of approximately 10 MHz with a nominal 10 pF transducer capacitance. The exact operating frequency is not a critical design factor. As the transducer capacitance changes with pressure, the frequency changes according to $f = \frac{1}{2\pi\sqrt{LC}}$. Although the LC tank is the principal frequency-determining element, other factors, such as supply voltage, transistor bias, and output loading will have second order effects. To obtain maximum accuracy with the small changes in capacitance available from the transducer requires that these effects be reduced to a minimum. Transistor Q_1 and diode D_1 provide a regulated supply voltage. Transistor Q_3 operated as an emitter-follower isolates the output loads from the tank circuit. Most importantly the entire circuit, except for

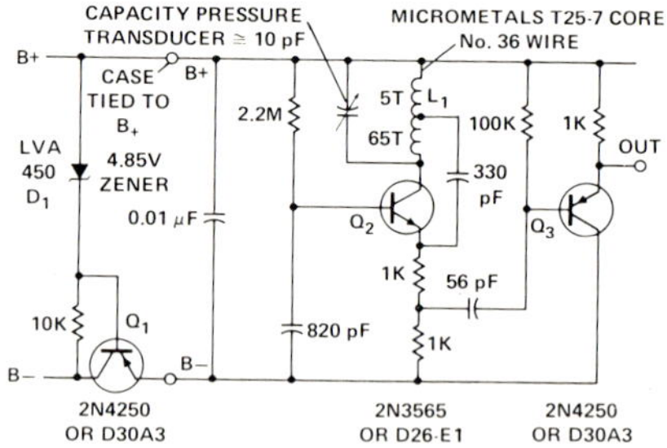

Figure 2. Variable frequency oscillator for use with capacitance-type pressure transducer system.

the regulator Q_1 and D_1, is located in a hermetically sealed metal case to eliminate extraneous distributed capacitance effects.

Figure 3 shows a typical calibration curve. The stability is shown with two calibrations made 9 months apart. This cell is vented to atmosphere and has high sensitivity for low pressure measurements. For total implant use, a hermetically sealed absolute cell (vacuum reference) or one with partial pressure would be used. By adjusting the diaphragm thickness and spacing during manufacture, it is possible to build cells for different pressure ranges and sensitivities. Both the capacitance change versus pressure and the frequency versus capacitance change are non-linear relationships (approximately the square law), but for the small changes normally encountered, a reasonably linear output can be obtained (fig. 3).

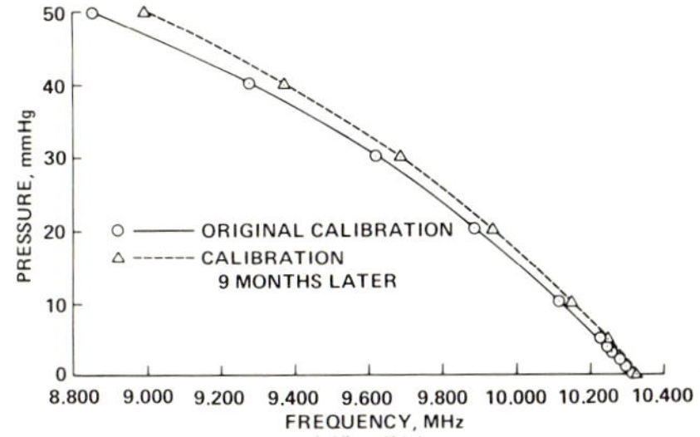

Figure 3. Data from two calibrations of complete capacitance pressure transducer system: calibrations made 9 months apart.

Figure 4. Transducer mounted for intracranial pressure measurements.

The circuit of fig. 2 and the capacitance transducer have been packaged into a complete system for intracranial pressure measurements (fig. 4). The holder is designed to mount into the skull and to use the coplanar technique (6) so that the dura is not penetrated. Although the frequency readout is useful in some applications, there are other occasions when an analog voltage is preferable. A diode bridge circuit, as previously described in ref. 4, can be used to obtain a voltage instead of a frequency output from a capacitance transducer. By locating two diodes at the transducer, the cable capacitance effects are eliminated and accurate results, similar to those of the frequency system, can be obtained.

Although the number of systems completed is insufficient to provide statistical verification of accuracy, early results are promising. The assembly of the transducer into a completely implantable telemetry system without transcutaneous wires has not been described; however, such a system is being designed. The system will be inductively powered (3); the low power requirement, approximately 10 mW, makes this quite simple. Telemetry of the oscillator frequency variation is also quite easy with available telemetry systems. The significant problem in the measurement of low level physiological pressures is not the telemetry system design, but the lack of transducers of suitable size and stability. Tests on the capacitance pressure cell described herein and data from previous non-physiological applications show promise of overcoming the transducer problems.

REFERENCES

1. Atkinson JR, Shurtleff DB, and Foltz EL: Radio telemetry for the measurement of intracranial pressure. J Neurosurg. 27: 428-432, 1967.
2. Currie JC, Riddle HC, and Watson BW: The use of telemetry to study the physiological and clinical variations of intracranial pressure in man. In Biotelemetry, HP Kimmich and JA Vos, eds., pp. 326-331, 1972.
3. Fryer TB: Power sources for implanted telemetry systems. Biotelemetry 1:1, 31-40, 1974
4. Harrison DR, Dimeff J: A diode-quad bridge circuit for use with capacitance transducers. Rev Sci Inst. 44:10, 1468-1472, 1973
5. Konigsberg E: Capabilities of implantable transducers. In Res. Animals in Med., LT Harmison Ed. DHEW (NIH) 72-333, 1145-1150, 1973
6. Majors R, Schettinie A, Mahig J, et al.: Intracranial pressure measured with the co-planar pressure transducer. Med and Bio Eng, 10: 724-733, 1972
7. McCutcheon EP, Evans JM, Stanifar RR: Evaluation of miniature pressure transducer. In Chronically Implanted Cardiovascular Instrumentation. EP McCutcheon, ed., 21-35, 1973
8. Olsen ER, Collins CC, Loughborough WE: Intracranial pressure measurement with a miniature passive implanted pressure transensor. Am J Surg, 113: 727-729, 1967
9. Sandler H, Fryer TB, Datnow B: Single-channel pressure telemetry unit. J Appl Physiol, 26: 235-238, 1969
10. Timm GW, List JS, Bradley WE, et al.: Frequency stability of an implantable pressure telemetry capsule. IEEE Trans Biomed Eng, BME-22:3, 250-251, 1975
11. Topham WS: Review of animal instrumentation--transducer capability. In Res Ani in Med., LT Harmison, ed. DHEW (NIH)72-333, 1229-1241, 1973
12. Walker AE, Viernstein L, Chubbuck JG: Intracranial pressure monitoring. Amer. Assoc. of Neurological Surgeons, Bal Harbour, Fla., 1975

A SEMICONDUCTOR MEMORY SYSTEM FOR DATA STORAGE IN BIOTELEMETRY OF RADIONUCLIDE TRACERS

J. Bojsen

The Finsen Laboratory, The Finsen Institute
DK-2100 Copenhagen, Denmark.

A semiconductor memory system based on the latest CMOS technology is a realistic alternative to other recording media. The acquisition of data in biotelemetry of radionuclide tracers in humans has been performed by radiotelemetry (ref.1) or storage telemetry methods (ref.2).

This paper describes a CMOS semiconductor memory system for data storage in biotelemetry of radionuclide tracers. The calculation of the count rate takes place during the measurements before storage in the memory. The accumulation period is chosen before the measurements have started, and consequently the storage capacity can be considerably reduced.

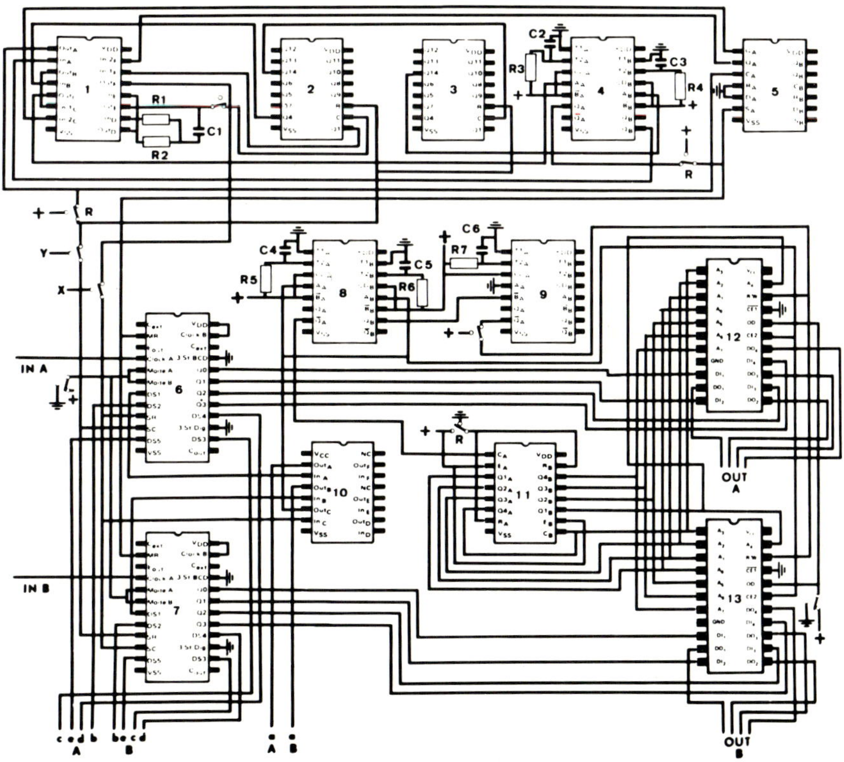

Fig.1. Two channel semiconductor memory system. All switches are shown in "write" position. 1.MC14572 Hex

Gate. 2,3. MC14020 14 Bit Binary Counter; 4,8,9.MC14528 Dual Retriggerable/Resetable Monostable Multivibrator. 5. MC14013 Dual Type D Flip-Flop. 6,7. MC14534 Real Time 5-Decade Counter. 10. MC14049 Hex Inverter/Buffer. 11. MC14520 Dual Binary Up Counter. 12,13. INTEL 5101 CMOS STATIC RAM 1,024 bit. R1=470 KΩ; R2=47 KΩ;R3, R4,R6=10 KΩ; R5,R7=5 KΩ; C1=163 pF; C2,C3,C6=1nF; C4=33 pF; C5=470 pF. VCC, VDD and + = + 5,4 volt.GND, VSS and ⏚ = 0v. All unused inputs on +5.4 volt;R is reset.

METHODS

A wiring diagram of the two channel semiconductor memory system is shown in fig. 1. In order to achieve a high degree of circuit integration, the combination of the 1,024-bit semiconductor memory CMOS static RAM (Intel 5101) organized as 256 words x 4 bits (12,13), and the real time 5-decade counter with internal scanner and multiplexer (Motorola MC 14534) as the signal accumulation unit (6,7), appeared to be the best choice at present. Detailed information about the "5101"-memory and the MC-circuits can be found elsewhere (ref.3,4).

The time reference is based on a RC-generator f = 32 kHz established by two inverters (1). By means of two serialized 14-bit counters (2,3) different accumulation periods can be selected ($T \leq 1h$). The NAND- and NOR-gates (1) and the Dual Type D Flip-Flop (5) control the "write" cycle. Proper standard impulses are generated by the monostable multivibrators (4).

In the 5-decade counters (6,7) "mode A" and "B" are connected to + 5 V during the measuring period, which means that the counter will operate as a 5-decade counter but only 4-decades are sequentially coming out in binary code, as the "units" are neglected.

The static random access memories (12,13) are operated with the $\overline{CE1}$ terminal at ground potential and the OD terminal on + 5 V. The CE2 is only "1" when the memory is active which happens every time one decade is read or written. The READ/WRITE control is kept "low" when a decade is written in the memory and kept constant "high" when a decade is read out. The write cycle is 0.5 msec. The three monostable multivibrators (8,9) connected to the terminals READ/WRITE (12,13) and the address register counter (11), are only used for proper delay of the control logic connected to the semiconductor memories.

At the end of the measurements the memory system is connected to the read out control logic shown in fig.2. The timer and control logic (1,2,3,4) is basically analogous to that in the memory system. The printer we use has BCD parallelized inputs, which are fed from 4x2 latches (5,6,7,8).

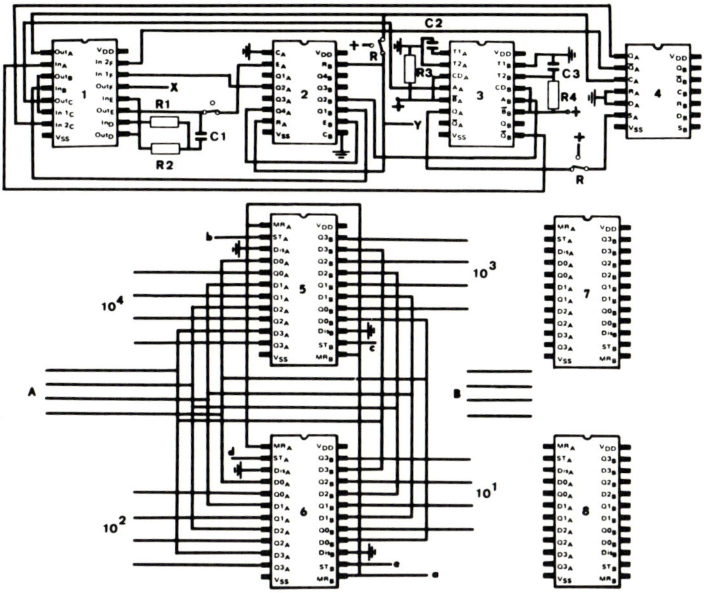

Fig. 2. Two channel "read" control unit. All switches are shown in "read" position. 1. MC14572 Hex Gate. 2. MC14520 Dual Binary Up Counter. 3. MC14528 Dual Retriggerable/Resetable Monostable Multivibrator. 4. MC 14013 Dual Type D Flip-Flop. 5,6,7,8. MC14508 Dual 4-Bit Latch. R1=470 KΩ; R2=47 KΩ; R3,R4=100 KΩ; C1=1μF; C2=10nF; C3=1nF.

In the read cycle the "mode A" and "B" (Fig.1 (6,7)) are connected to OV. The internal scanner outputs are then applied for proper sequential transfer of the BCD decades from the memories to the eight latches in the "read out" control logic (Fig.2). The scanner impulse from the "units" (DS1)(6,7)(Fig.1), which are active in this mode, controls the printer.

RESULTS

Due to the statistical nature of the radioactive decay the "units" can be neglected, this means that if 1 hour accumulation intervals are selected and 4 decades are stored in each memory at the end of each accumulation period, the present storage capacity will contain in maximum 64 hours measurements. The maximum count rate will in this example be 99,990 counts/hour. The memory of each channel in the two-channel semiconductor system is 256 x 4 bits, but it can easily be extended.

The supply voltage is 5.4 volts (Mallory 2× TR-152R,

350mAh) and the current consumption averages about 2 mA. A single "5101"-memory has in standby a current consumption of only 15 μA. The weight of the system is 150 g and the volume is 270 cm^3.

DISCUSSION

GM-detectors applied in the radio- and storage telemetry equipment (ref. 1,2) can be used directly in connection with this memory system, but the small cadmium telluride γ-detector (2-3 mm^3) which seems to be applicable as a new biotelemetry transducer will make things move more quickly towards a further miniaturization of portable biotelemetry equipment for detection of radionuclide tracers. It is considered to integrate the CdTe--detector and its preamplifier in HYBRID circuity in a single To-5 house (8 mmØ). If the present memory system is made in "chips" and connected by thick film technology the consequence will be a reduction in size to about 5 cm^2.

CONCLUSION

In long term storage telemetry of radionuclide tracers, as in studies of circadian variations in tracer disappearance, this CMOS static RAM memory system is a strong competitor of the analog electromechanical tape recorder. The weight, size and power consumption are much less than for the tape recorder system.

REFERENCES

1. Bojsen J, Vadstrup S: A portable external two-channel radiotelemetrical GM-detector unit, for measurements of radionuclide tracers in vivo. Int J Appl Radiat Isotopes 25:161,1974.

2. Bojsen J, Vadstrup S: A four channel storage telemetrical measuring system for monitoring radionuclide tracers in vivo. Proc. 2nd ISPRA Electronics Symp. Stresa May. 1975.

3. Oliphant J: Designing non-volatile memory system with Intels 5101 RAM, Application Note Intel AP-12 1975.

4. Semiconductor Data Library. Vol.5, Series A Motorola Semiconductor Products Inc., 1975.

ACKNOWLEDGEMENTS

The author wishes to thank Mr. K. Jørgensen for excellent technical assistance, Mrs. Tove Nørager for typewriting, and Mr. P. Riel for photographic assistance.

DISCUSSION

SALMONS: Dr. Young, what led you to develop a small implant for your studies rather than attaching a pair of strain gages directly to the surface of the bone?

YOUNG: The wire and foil gages described by Cochran and Lanyon tend to be useful for approximately two to three weeks. Then remodeling of the bone occurs beneath the gages. In fact, the bone may be destroyed in the process of bonding the gages. Since our investigations extend over several months, we need something for longer terms. Secondly, the monkey is a wild animal. Any device having wires does not last for more than minutes or days when applied to these animals.

SESSION VII

BIOTELEMETRY OF

CARDIOVASCULAR PARAMETERS

Chairmen

E.P. McCUTCHEON, M.D.

R.D. RADER, Ph.D.

EDITOR'S OVERVIEW

In this session we return to the subject of biotelemetry from animals, but with emphasis on research of the cardiovascular system. In most cases the subjects were dogs and the telemetry systems consisted of implanted ECG and Doppler ultrasound flow-measurement devices. In the final paper, Perachio et al discuss an unusual radio-controlled system for drawing multiple blood samples from rhesus monkeys.

As in previous sessions, the authors make quite evident the benefits that have accrued from recent hardware miniaturization and other advances in biotelemetry systems. As noted previously, the benefits include the ability to obtain research data that are less subject to distortions from the stresses of capture and restraint of the animal subjects. In several of the papers the benefits noted are improvements in the accuracy and utility of ultrasonic blood-flow measurements.

A PULSED DOPPLER BLOOD FLOWMETER FOR CHRONIC PHYSIOLOGICAL STUDIES

H.V. Allen, J.W. Knutti, M.F. Anderson, J.D. Meindl

Stanford University
Stanford, Calif.
94305

INTRODUCTION

A totally implantable pulsed Doppler ultrasonic blood flowmeter (PDUF) has been developed using custom integrated circuits. This system performs all the functions of a normal pulsed Doppler flowmeter by generating the timing operations on one custom silicon chip and processing the Doppler signals on a second IC.

The implantable system has been used to monitor changes in vessel diameter, velocity profile, and volume flow in the portal vein of healthy dogs. Significant variations in both the profile and volume flow have been noted following both feeding and changes in posture.

BASIC INSTRUMENTATION

The pulsed Doppler flowmeter functions as a miniature sonar system determining both the range and the velocity of objects in the path of the sound beam. The system operates by emitting a short burst (~ 1μ sec) of high frequency sound (6 MHz) every 50 μsec from a piezoelectric transducer into the vessel (see Figures 1A and 1B). Sound is reflected back to the transducer from red blood cells with a frequency shift proportional to the velocity of the cells according to the Doppler equation. The returned signal is heterodyned in the receiver against a continuous signal having the same frequency as the transmit burst. The receiver output is a wide bandwidth (0 to 0.5 MHz) signal which contains all the information on the Doppler shift as a function of range.

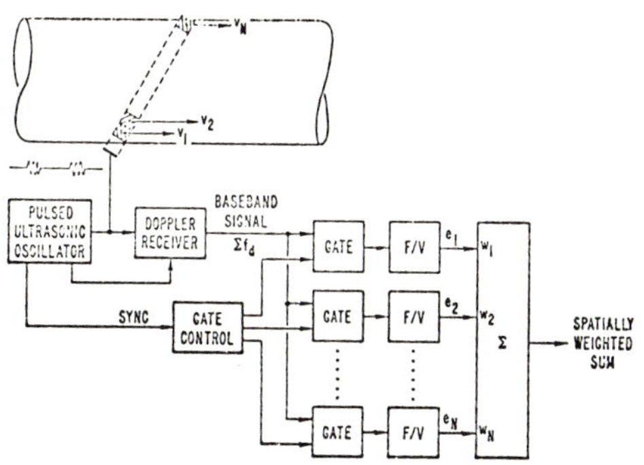

Fig. 1A. Block Diagram of the Pulsed Doppler Ultrasonic Blood Flowmeter.

This information is sampled at fixed time delay after each burst to extract the Doppler information at a given range, the time delay being proportional to projected distance across the vessel. The frequency (and thus the velocity) is estimated in a zero crossing counter (ZCC) that give a voltage output proportional to the RMS input frequency (2).

The ZCC outputs are multiplexed together sequentially to give a flow profile as shown in Figure 1B. In addition the outputs are fed to a summing network which weights the individual outputs according to their relative distance from the center of the vessel (3). This spatially weighted sum when multipled by the vessel area produces the instantaneous volume flow (see Figure 1B).

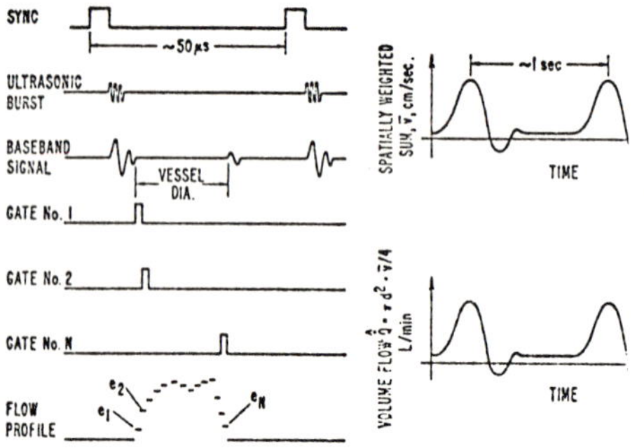

Fig. 1B. Timing Diagram for the Flowmeter Showing Data that is Provided by the Flowmeter.

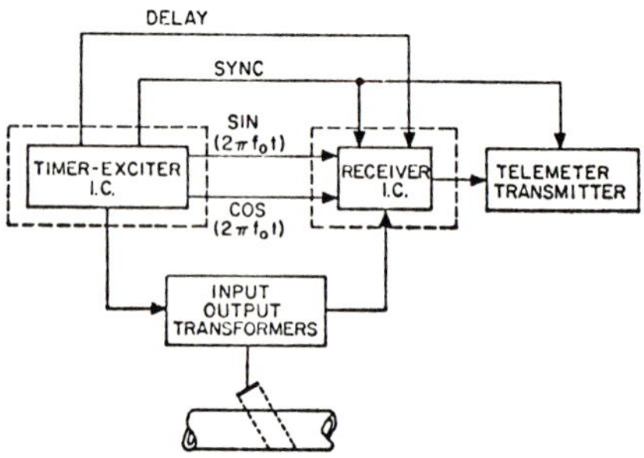

Fig. 2A. Diagram of the Implantable Portion of the Integrated Circuit Flowmeter.

THE IMPLANTABLE SYSTEM

The totally implantable telemetry system is similar to the system described above except that the baseband signal is telemetered. A block diagram of the implantable portion is shown in Figure 2A. The two integrated circuits are diagrammed in more detail in Figure 2B.

The timer-exciter is realized with a novel timing scheme, employing a free running repetition rate oscillator (see Figure 2B). The ultrasonic oscillator is synchronized to the repetition oscillator by the sync one-shot. The ultrasonic burst is then gated to the input-output transformer network (Figure 2A) after a short delay to allow the ultrasonic oscillator to reach steady state following resynchronization. To assure coherence from one transmit burst to the next, both the repetition rate oscillator and the ultrasonic oscillator have been specially designed for high stability.

Directional flow sensing is accomplished by using a quadrature detection system (5). The two quadrature local oscillator signals are generated on the timer-exciter chip using RC-CR phase splitters. A total of 73 transistors, 9 capacitors, and 45 resistors are contained on a 3 mm square chip of silicon.

The receiver integrated circuit (Figure 2B) amplifies the microvolt returns from the transducer and heterodynes this signal

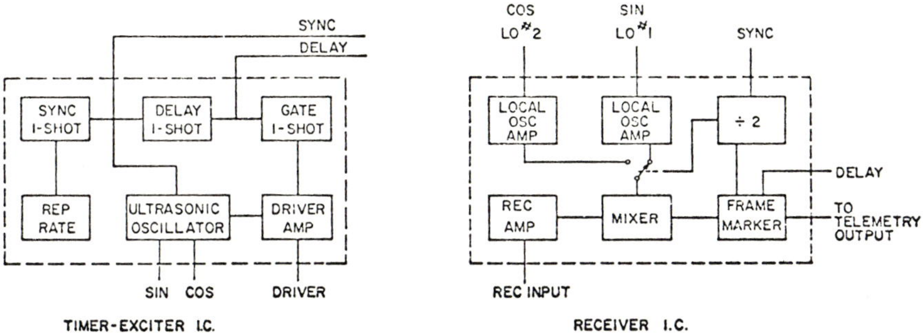

Fig. 2B. Block Diagrams of the Two Integrated Circuits.

down to baseband, removing the ultrasonic carrier from the output through the use of a fully balanced multiplying mixer (4). Directional information is retained by alternating the quadrature local oscillator signals used in the mixer. At the beginning of each resynchronization, a ÷2 circuit is toggled selecting the correct local oscillator signals to be fed to the mixer. The output of the mixer is then a multiplexed composite baseband signal containing all the flow information including the direction of flow. This signal is marked on alternate frames in order to allow identification of the two quadrature baseband signals externally and thus to determine the absolute direction of flow.

Telemetry of the wide bandwidth, large dynamic range baseband signal can be accomplished using either an inductive data link or a wideband R.F. link. The inductive data link employs two coils of wire; one inside the package and the other directly above the package on the surface of the animal form a loosely coupled transformer pair. Gain variations across the link due to changes in coupling are compensated for by detecting the amplitude of the sync signal in the external electronics.

A wideband R.F. telemetry link is currently being developed involving more complex circuitry which will eliminate the need for the external coil attached to the animal.

Currently the implantable package measures 1.8 x 6.1 x 6.1 cm^3; however, the pulsed flowmeter circuitry occupies less than 25% of the total volume. Batteries occupy one third the volume while the transcutaneous inductive data link and non-integrated circuitry used in support of the flowmeter accounts for the remainder of the space.

EXPERIMENTAL RESULTS

The non-directional prototype of this system (6) has been used to measure flow in the portal vein of healthy dogs for periods of up to 5 months. This vessel is an ideal site for the pulsed Doppler flowmeter. The internal diameter on the vein is typically in the 7 to 10 mm range. This vessel has significant flow variations as a function of posture, feeding, and conscious state.

A flow transducer placed on the portal vein in the plane of the superior mesenteric and splenic veins clearly shows the relative contributions to portal vein blood flow from these two veins as is shown in Figures 3A and 3B. The data is from a flowmeter in its third week of operation.

The photographs show the baseband telemetry signal in the top trace and the velocity profile in the bottom trace. The internal diameter of the vessel can be determined from these figures by Doppler imaging and is 4 mm. This data is vital since the low pressure portal vein is easily constricted by the flow cuff. In the first case (Figure 3A) the flow prior to feeding is 770 ml/min; the flow profile is relatively flat. Immediately following feeding, (Figure 3B), the flow has increased on the side close to the transducer and is slightly decreased on the opposite side; the total volume flow shows a small increase of 4% to 800 ml/min.

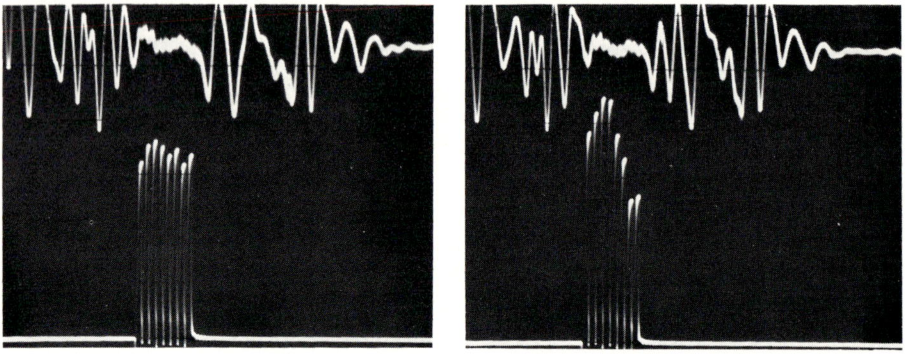
(a) (b)

Fig. 3. Flow Profiles in the Portal Vein Before (a) and After (b) Feeding.

ACKNOWLEDGMENTS

This research is supported under PHS Research Grant No. 1 P01 GM17940-06 from the Department of Health, Education, and Welfare.

REFERENCES

1. Brody WR: Theoretical Analysis of the Ultrasonic Blood Flowmeter. SEL Technical Report No. 4958-1 Stanford University, 1971.
2. Rice SO: Mathematical Analysis of Random Noise. Bell System Technical Journal 23: 282-332, 1944.
3. Peronneau P, et al: Velocimetre Sanguin Par Effet Doppler a Emission Ultrasonore Pulsee: Description de L'Appareil. Resultats L'Onde Electrique 50: 369, 1970.
4. Gill RW: An Implantable Pulsed Doppler Ultrasonic Blood Flowmeter Using Custom Integrated Circuits. SEL Technical Report No. 4958-3 Stanford University, 1975.
5. Kalmus HP: Direction Sensitive Doppler Device. Proceedings of the IRE: 698-700, June 1955.
6. Gill RW: Low Power Integrated Circuits for an Implantable Pulsed Doppler Ultrasonic Blood Flowmeter. JSSCC 10: 464-471.

A TRUE VOLUMETRIC FLOW MEASUREMENT TECHNIQUE

Roland D. Rader

Department of Physiology
University of Southern California
School of Medicine
Los Angeles, California

The major features of pulsatile blood flow in mammals are: (1) the flow may be either laminar or turbulent, (2) the axial flow velocity profile changes shape throughout the flow cycle and is often asymmetrical, (3) the vessel boundary and area may change because of a number of factors, and (4) the flow velocity may have both radial and axial components at partially blocked or curved sections and at bifurcation.[1,2]

Devices designed to measure pulsatile flow must therefore deal effectively with these characteristics. The ideal requirements of such devices are: (1) the volumetric flow sensitivity factor should be equal for either laminar or turbulent flow and should be independent of the geometric shape of the vessel and the spatial velocity of flow, and (2) the zero flow base line should be independent of temperature, time, and blood and vessel properties.

Such a device must determine the instantaneous product of the cross-sectional area of flow and the average axial velocity of flow in that area. The simplest equation of performance for such an electronic device is:

$$\text{Eout} = KA\,\overline{V}z = K\,\dot{Q}z \tag{1}$$

where Eout is an output voltage; K is a constant; A is the instantaneous area; and $\overline{V}z$ is the mean velocity in the area A. The equation which is typically realized has the following form:

$$\text{Eout} = Kx\,\dot{Q}x + Ky\,\dot{Q}y + Kz\,\dot{Q}z + b \tag{2}$$

where $\dot{Q}z$ is the axial-directed flow, $\dot{Q}x$ and $\dot{Q}y$ are orthogonal radial components of flow, and where b is the value of the output voltage when $\dot{Q}x = \dot{Q}y = \dot{Q}z = 0$.

THEORY OF FLOW MEASUREMENT BY DELTA PHASE METHODS

Figure 1 illustrates a vessel with an arbitrary boundary in which flow with x, y, and z components exists. The time for a burst of ultrasonic energy originating at any point y on crystal X_1 to arrive at crystal X_2 is a function of both x and y since the velocity of ultrasonic energy in the vessel is a function of x and y. This time is a function of the x and z flow velocity components as follows:

$$Td = \frac{1}{\sin\alpha}\left[\int_{-R}^{x_1}\frac{dx}{C} + \int_{x_1}^{x_2}\frac{dx}{C + Vx\sin\alpha + Vz\cos\alpha} + \int_{x_2}^{R}\frac{dx}{C}\right] \tag{3}$$

where x_1 and x_2 are the intersections of the diagonal line between crystals X_1 and X_2 with the boundary of the vessel, R is the position of the crystal in the x-y plane, C is the velocity of ultrasonic energy, and α is the angle between the flow velocity and the ultrasonic vectors.

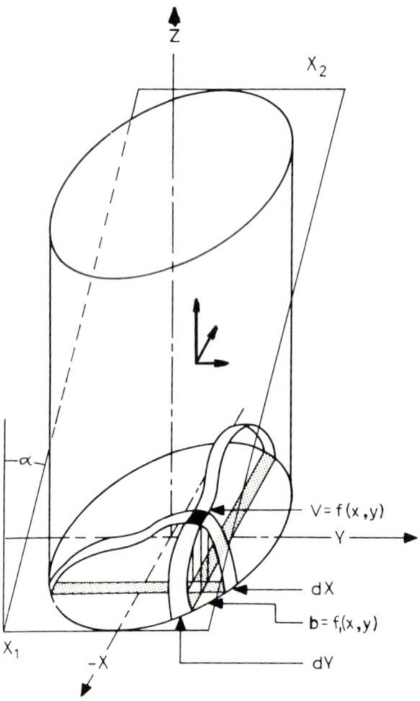

Figure 1

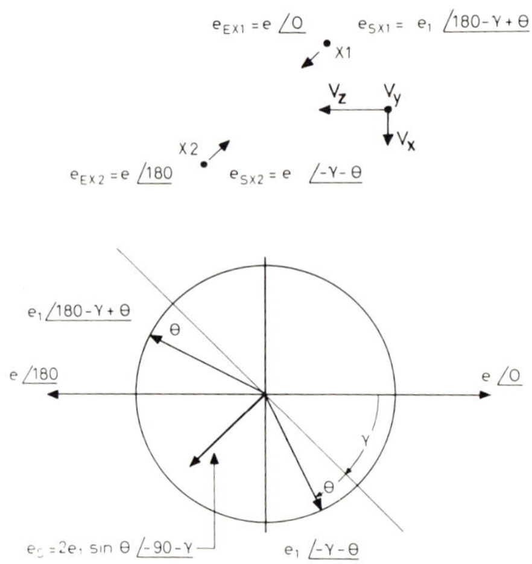

Figure 2

Simplification of Equation 3 and conversion to a phase shift defined by $\theta = 2\pi f\, Td$ yields:

$$\theta d = \frac{4\pi f R}{C \sin\alpha} - \frac{2\pi f}{C^2 \sin\alpha} \int_{x_1}^{x_2} (Vx \sin\alpha + Vz \cos\alpha)\, dx \qquad (4)$$

Now consider that crystal X_1 is excited by a brief burst of 5 MHz sinusoidal voltage $e\underline{/0^\circ}$, and crystal X_2 by a similar burst $e\underline{/180^\circ}$ (Fig. 2). The voltage received at crystal X_2 associated with the excitation voltage on X_1 will be $e_1\underline{/-\gamma - \theta}$, and the voltage received at crystal X_1 will be $e_1\underline{/180 - \gamma + \theta}$. Addition of the two signal voltage vectors gives:

$$e_s = 2e_1 \sin\theta \; \underline{/-90 - \gamma} \qquad (5)$$

where γ is the first term and θ is the second term in Equation 4, and e_1 is the magnitude of each incremental voltage on each small segment of the crystals. Regardless of the value of γ or θ, the composite vector will be perpendicular to the line $-\gamma$ (Fig. 2). Since θ is a small angle, sine θ is approximately equal to θ, and the magnitude of the composite voltage vector becomes:

$$e_c = 2\, e_1 \int_{-R}^{R} \theta \, dy$$

$$= \frac{4\pi f\, e_1}{C^2 \sin\alpha} \int_{-R}^{R} \left[\int_{x_1}^{x_2} (V_x \sin\alpha + V_z \cos\alpha)\, dx \right] dy \tag{6}$$

But the flow through the diagonal plane through the vessel in the x-and-y direction is defined by:

$$\dot{Q}_x = \iint_A V_x \, dA_x = \int_{-R}^{R} \int_{x_1}^{x_2} \frac{V_x \cos\alpha}{\sin\alpha} \, dx \, dy$$

$$\dot{Q}_z = \iint_A V_z \, dA_z = \int_{-R}^{R} \int_{x_1}^{x_2} V_z \, dx \, dy \tag{7}$$

and thus Equation 6 becomes:

$$e_c = \frac{4\pi f\, e_1}{C^2} \left[\dot{Q}_x \tan\alpha + \dot{Q}_z \cot\alpha \right] \tag{8}$$

This is in the form of Equation 2, except constant b has been omitted and Q_y is not detected. Well developed flow in long, straight tubes has no radial components, and under these conditions, Equation 8 will reduce to the form of Equation 1.

TEST RESULTS

Figure 3 presents the test results from a probe used to measure well developed flow in long tubes. For a circular lumen, the sensitivity factor is independent of Reynolds numbers up to 10,000. On the basis of a probable change from a parabolic profile in a laminar flow region to a flat profile when flow is turbulent, it appears that the calibration factor is indeed independent of the flow profile. The independent relationship of the calibration factor to the average flow velocity in two different luminal configurations further verifies that the measurement is independent of the profile and the vessel geometry. The slope of the calibration factor as a function of average flow velocity for the full and the half-area lumen is 0.02 and -0.01, respectively. The respective intercepts are 39.74 and 43.72. Rejection of reflected energy by using asymmetrical probes results in an improvement in zero flow base-line stability. Insertion of simulated vessels in 6 mm asymmetrical probes, for example, causes a base-line shift of approximately 20% of fullscale. Disturbed flow is not accurately measured, since flow normal to the crystal plane results in zero sensitivity, whereas flow parallel to this plane produces maximum sensitivity.

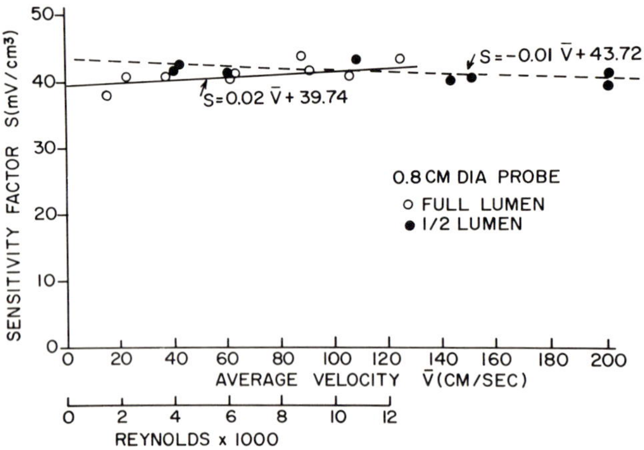

Figure 3

CONCLUSIONS

Well developed flow is unambiguously measured by the delta phase method. Measurements by this as well as other standard techniques are ambiguous in regions where radial flow components exist, such as close to junctions or partially blocked vessels. Of the standard electronic methods of measuring flow, the delta phase technique has the unique ability of inherently measuring flow changes independent of vessel area, vessel configuration, and flow velocity profile. For well developed axial flow, the sensitivity factor is also independent of the expected values of the average velocity and Reynolds number.

ACKNOWLEDGMENT

This research was supported in part by the National Science Foundation Grant ENG75-19049.

REFERENCES

1. Nerem RM, Rumberger JA, Jr; Gross DR, et al: Hot-film anemometer velocity measurements of arterial blood flow in horses. Circ Res 34: 193, 1974

2. Peronneau P, Gisbertz K-H, Steckmeier B, et al: In vitro and in vivo experimental studies of pulsatile flow patterns in curved and stenotic vessels. In Proceedings from a Specialists Meeting on Fluid Dynamic Aspects of Arterial Disease (ed., RM Nerem). Columbus, Ohio State University, 1974

A MINIATURE IMPLANTABLE ECHO ULTRASONOMETER

G. K. Kojima, E. P. McCutcheon, E. L. Carlson, and R. D. Lee

Ames Research Center, NASA, Moffett Field, California 94035

INTRODUCTION

The dimensions of internal body structures are significant in many investigative studies, and many approaches to making these measurements have been described. Of the various alternatives explored for use in animals, ultrasonic techniques have proven particularly useful (1). Static and dynamic dimensions can be obtained by measuring the time taken for an ultrasonic pulse to travel either between a pair of transducers placed on opposite sides of the organ or vessel (transmission method, 2), or from a single transducer out to a reflecting surface and back to the transducer (echo method, 1, 3).

The resolution of the transmission method is acceptable (>0.75 mm at 3 MHz), but the two transducers must remain in precise axial alignment. Maintaining alignment is particularly difficult with large changes in shape of an organ, especially when the alterations are not of a simple translational form (3). Successful chronic implantation of these transducers has been described in studies of the heart (4, 5) and abdominal organs (6). A totally implantable battery-powered device was designed to overcome the problems created by the presence of wires penetrating the skin (7).

The great expansion of the clinical use of transcutaneous reflected (echo) ultrasound has stimulated interest in studies with animal models, and a chronically implanted, directly-connected echo transducer was recently described (8). The echo technique offers the advantages of measurements from multiple intervening surfaces with only slightly less resolution than the transmission method, and surgical trauma is reduced since only one transducer is required. To avoid the problem of penetrating the skin with leads and to provide long-term dimensional data, a miniature implantable ultrasonic echosonometer has been developed. This new device is safe for long periods of real-time imaging; it provides acceptable resolution of fine structures, it images structures inaccessible by external probes, and it contains no batteries.

The unit is powered by an external, power oscillator/transmitter coupled to a subcutaneous 7.3-cm diameter power receiver loop via transcutaneous electromagnetic induction (8, 9), see fig. 1. The 2.25 MHz, planar, ultrasonic transducer is placed on the organ surface so that it is oriented in the direction desired for observation. The 2.25 MHz echo signals are coupled from the body by transcutaneous EM induction in a manner similar to the way in which the power is coupled. The signal can then be processed for display in a standard echo device.

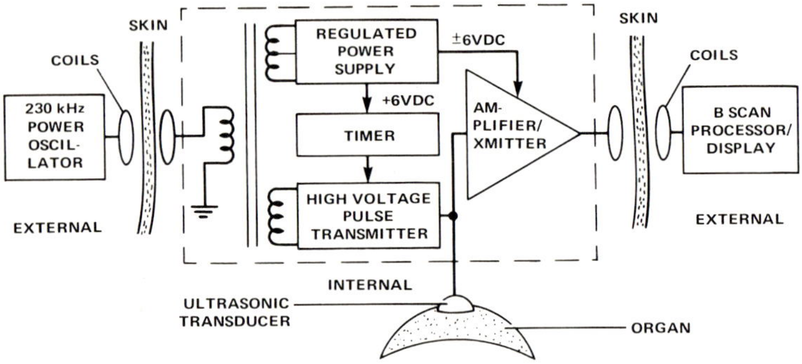

Figure 1. Basic sections of pulsed echo ultrasonometer.

OPERATION

The implantable echosonometer consists of six basic sections (fig. 1):

1. a power receiving section
2. a dual dc power supply section
3. a timing section
4. a high voltage pulse transmitter section
5. a piezoelectric ultrasonic transducer
6. an amplifier/transmitter section.

The power receiving section includes an 8-turn, 7.3-cm diameter coil and a miniature toroidal transformer. The coil is positioned beneath the skin of the animal and receives 230 kHz power transcutaneously from an external oscillator coil placed in proximity over the subcutaneous coil. The toroidal transformer transforms the received power to appropriate voltage levels. The 12.7-mm diameter toroidal core of the transformer was selected for its small size and high permeability.

The dc power supply section converts the 230 kHz output from the center-tapped, low voltage secondary of the power transformer to a regulated ±6 Vdc, providing power to the timer and to the amplifier/transmitter.

The timing section controls the time between successive pulse transmissions. The main component of the timer is a quad 2-input NAND gate CMOS flatpak connected as an astable multivibrator. The output consists of a series of pulses with a repetition rate of approximately 350 Hz. These pulses trigger the high voltage pulse transmitter.

The high voltage pulse transmitter is a high-speed switch triggered by the timer. The negative output pulse, which rings the transducer, is generated by charging a capacitor to approximately 200 Vdc; the capacitor is rapidly discharged when the positive side of the capacitor is grounded by switching silicon controlled rectifiers.

The lead titanate-zirconate transducer, at the output of the high voltage pulse transmitter, receives the -200 V peak impulse from the transmitter and rings. This ringing generates a 2.25 MHz ultrasonic impulse which is transmitted outward from the transducer into the organ to which it is attached. A concave face on the transducer provides moderate focusing. As the ultrasonic pulse propagates through the organ, any acoustic impedance mismatch due to varying tissue densities will reflect a portion of the pulse energy. The returning echos generate time-delayed electrical pulses. Spatial dimensional information is contained in the time delay between returning echos, so that a one-dimensional image of the organ may be constructed. The 350 Hz repetition rate is adequate for real-time dimensional imaging of moving organs.

The low amplitude 2.25 MHz electrical echo pulses are amplified by a high gain amplifier. The last stage of the amplifier has a 15 turn, 7.3-cm diameter loop located subcutaneously. The loop serves to couple the signal out of the body for display processing.

CONSTRUCTION

The units are constructed in a single 7.5 cm × 2.3 cm × 2.3 cm copper-clad PC board enclosure with a Silastic outside coat (fig. 2). The enclosure has three shielded compartments. One compartment holds the toroidal transformer. The second compartment houses the dual dc supply, timer, and the high voltage pulse transmitter; these three circuits are constructed in a cordwood design. The third chamber shields the amplifier/transmitter from the other circuits. The amplifier/transmitter is constructed on a ground-plane PC board for improved electrical stability.

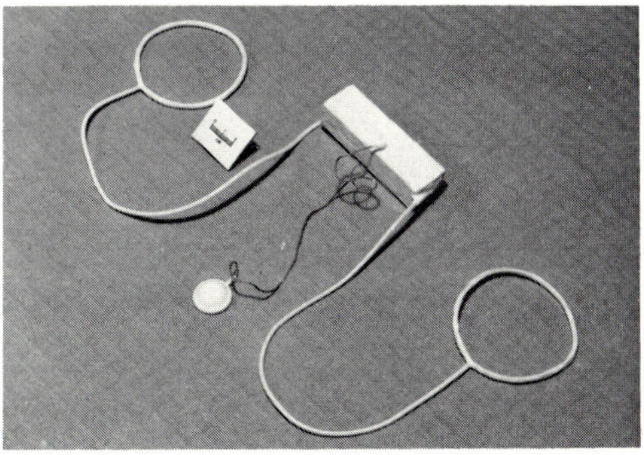

Figure 2. Implantable ultrasonometer.

The transducer is a truncated form of the standard type used clinically. The cylindrical sensing element is enclosed in a Silastic mounting containing mesh reinforcement. Sutures can be placed through this material at any point desired and its moderate flexibility provides improved conformity to the organ surface.

Acute and chronic evaluation indicates that the complete unit may be placed in either the chest or abdomen. During surgery, beam-orientation may be verified with a second, directly-connected transducer positioned opposite the placement site. After fixation of the main body of the unit, the two flexible coils are brought through separate exit sites and stabilized subcutaneously.

RESULTS

Initial experience indicates the unit is well tolerated and the image quality is comparable to that obtained transcutaneously. Power and signal coil placement were not critical; moderate relative motion could be tolerated without signal distortion.

REFERENCES

1. Stegall, HF, Kardon, MD, Stone, HL, and Bishop, VS: A portable, simple sonomicrometer. J. Appl. Physio. 23:289, 1967

2. Cobbold, RSC: Transducers for Biomedical Measurements: Principles and Applications. New York, Wiley, 1974, pp. 149-154

3. Baker, DW: Sonocardiometer. In Methods in Medical Research (ed RF Rushmer). Chicago, Year Book, 11:1966, pp. 31-34

4. Guntheroth, WG: Implantation of dimension transducers. In Chronically Implanted Instrumentation (ed EP McCutcheon). New York, Academic Press, 1973, pp. 313-316

5. Kardon, MB, Horwitz, LD, and Bishop, VS: The ultrasonic pulse transit time technique. In Chronically Implanted Instrumentation (ed EP McCutcheon). New York, Academic Press, 1973, pp. 117-121

6. Vatner, SF, and Patrick, T: Redistribution of regional blood flow during "fight or flight" response in conscious unrestrained primates. Circulation 52 (Suppl 2):223, 1975

7. Lee, RD, and Sandler, H: Miniature implantable sonomicrometer system. J. Appl. Physiol. 28:110, 1970

8. Myerowitz, PD, Griffith, J, Roberts, AF, et al: Long-term canine model for echocardiography. Am. J. Card. 34:72, 1974

9. Schuder, JC, and Stephenson, HE, Jr.: Energy transport to a coil which circumscribes a ferrite core and is implanted within the body. IRE Trans. BME, BME 12:3, 4:154, 1965

A REMOTELY CONTROLLED BLOOD SAMPLING SYSTEM

A. A. Perachio, D. H. Rice, E. Lide[†], *and J. Sikes*

Yerkes Regional Primate Research Center, Emory University and Lockheed-Georgia Company[†], *Atlanta, Georgia*

The investigation of correlations between circulating hormone levels and stress has been hampered by the limitations in blood sampling techniques which involve either venipuncture or some degree of restriction of movement associated with the mechanics of an indwelling catheter. In association with experiments concerned with neuroendocrine mechanisms related to stress and aggression, we developed a remotely controlled blood withdrawal system that can be carried by a completely unrestrained animal living in a social environment. The device is attached to a chronically implanted venous catheter and through a two-way telemetry system, allows the investigator to both sample venous blood and to inject substances intravenously.

SYSTEM DESCRIPTION

The system consists of a console unit and an animal unit which is worn in a backpack. The console transmits a 8-bit PCM/PDM instruction word at a rate of approximately 60/sec. The instruction word is composed of: (1) the address of each of four units, (2) selection codes for two motors controlling the backpack hydraulics and (3) the code for controlling the motor direction. The backpack unit also generates an 8-bit word; this word contains status information on the contents of the catheter, the occurrence of each blood sample taken and the location of a 4-port valve controlling fluid flow in the backpack.

The purpose of two-way telemetry between the console and the animal-worn unit is to ensure that blood samples can be taken at particular points in time for the purposes of correlating hormone levels with a behavioral event. In addition to data on the occurrence of a blood sample, telemetric feedback from the unit also serves to indicate that the hydraulic lines have been valved to appropriate channels. The present system allows the selective withdrawal of ten separate blood samples of pre-selected volumes up to 3 ml each. Samples are separated by a flushing operation that clears the catheter of residual blood. The flushing capability of the unit can also be used for intravenous injection.

CONSOLE UNIT

The console contains a PCM/PDM transmitter that operates at 138.5 MHz and a broadband receiver for the signal broadcast from the animal unit at 148.5 MHz. The logic within the console produces the 8-bit instruction word as programmed through panel switches. The 8-bit word received from the animal unit is decoded within the unit and displayed on a series of indicator lights associated with separate unit operations. Each of four positions of the hydraulic valve is represented by a separate light, i.e., flush, vent, blood sample, injection or drug port. When the valve is in one of these positions, an additional light is illuminated indicating that the valve is in detent. When the command is to be issued for the valve to move, a detent override switch is activated which causes another panel light to be illuminated. A final panel light is controlled by information from the backpack that indicates whether the catheter is filled with blood or contains the clear flushing solution.

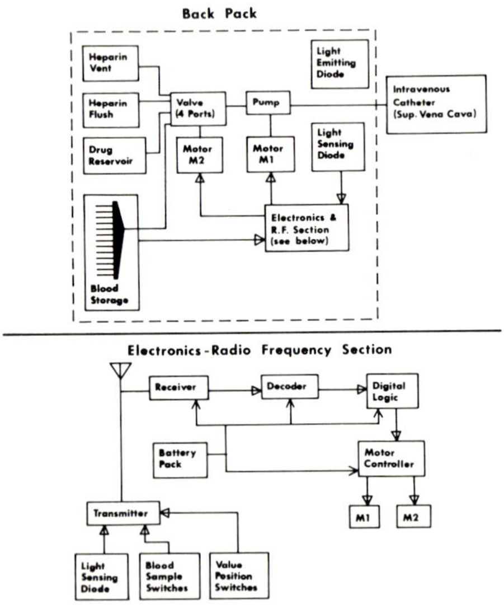

Fig. 1 A block diagram of the backpack unit

BACKPACK UNIT

The receiver, power supply, logic, pump and valve motors and fluid reservoirs are contained within a 2.75" x 4" x 6" Lexan container that is secured within a harness attached to the animal. The catheter, exteriorized percutaneously, is coupled securely to the input port of the Lexan housing. The input line is straddled by a light-sensing diode and a miniature light source (see Fig. 1). The line then passes to a miniature peristaltic pump that is driven by a variable reluctance stepping motor. The flow produced by this pump is directed to one of four major ports by a valve that is driven by a separate variable reluctance motor through a gear-reducer and gear train. The hydraulic line is led through a rotating assembly, driven by the valve motor over the cylindrical frame that contains the fluid reservoirs and ten blood sample containers. Both pump and valve motor are operated in either direction as selected by the instruction word. Three switches are mounted on the valve assembly: two of these produce the binary code for the four valve positions; the remaining switch enables the logic associated with the detent position. Additional switches located within each sample tube are mechanically closed when the chamber has been filled with blood. This switch information, along with the valve and detent switches, operates on the logic, which in turn controls the operation of the motors and also keys appropriate bits in the word broadcast to the console receiver.

TABLE 1
MAKE-UP OF 8-BIT WORDS

BIT NO.	CONSOLE	ANIMAL UNIT
1	Detent override	Undesignated (always "1")
2	Valve forward	Valve location
3	Valve reverse	Valve location
4	Pump forward	Detent
5	Pump reverse	Detent override
6	Animal selection	Catheter status
7	Animal selection	Sample filled
8	Undesignated ("1")	Undesignated (always "1")

OPERATING SEQUENCE

Prior to an individual blood sample, the catheter contains a heparin/saline solution. The valve, in this first step of the sequence, is in detent in the heparin vent position. This is signified to the operator by illumination of the detent light and the vent light on the console. An instruction is broadcast from the console unit, addressing a particular animal unit with a code, involving bits 4, 6 and 7, for the activation of the pump motor to operate in the forward direction (see Table 1). As the pump draws blood into the catheter within the backpack, the light falling on the light-sensing diode changes in intensity, causing a change in the output of the diode which keys the logic (bit 6). The console unit decodes the word, causing the catheter clear light to be extinguished on the operator panel. Transmission is stopped momentarily, causing the pump to halt. The instruction word is altered by re-programming the panel switches so that the valve may be moved to the blood sample port. This instruction word, when decoded by the backpack logic, activates the valve motor causing the valve to be rotated to the detent position of the next valve port. When the valve is in detent position, a word is broadcast by the backpack unit, received and decoded by the console, causing the appropriate panel light to be illuminated. Simultaneously, the valve motor is stopped by the internal logic within the backpack at the time of closure of the *detent switch. Instructions for the pump to be activated in the forward direction are transmitted. Fluid is pumped into a single blood sample storage chamber until the switch within that chamber is closed. This produces a change in the state of the logic which keys the transmitter with a new 8-bit word and simultaneously stops the pump. The blood sample light is switched on by the decoded word. The valve now can be commanded to move to either the drug reservoir or the heparin flush position, pausing at each location as the detent switches are closed. At the end of the sequence, when the valve is in the heparin flush position, the pump is driven in the reverse direction, causing the catheter to be flushed with heparin/saline. When the catheter is cleared, the light-emitting diode output is altered and that information is coded and transmitted back to the console unit and the console transmission is terminated. This entire sequence can be accomplished in less than one minute. The maximum frequency of blood sampling for all ten samples with catheter flushing between samples is less than five minutes.*

In addition to the use of this device for psychoendocrine research, numerous applications can be envisioned. These include a variety of pharmacological and physiological problem areas. The present unit has been designed for and used in non-human primates; improvements in the design of the valve, that are under development, might allow eventual use of this instrument for human research and therapeutic purposes.

ACKNOWLEDGMENTS

This work was supported in part by N.I.H. grants RR-00165 and NSO 9688 and NASA grant NGR 11-001-065.

TELEMETERED CARDIOVASCULAR CHANGES WITH FREE RANGING (FRE) AND TREADMILL (TME) EXERCISE IN THE DOG

E.L. Carlson, E.P. McCutcheon, W.R. Freund,
T. Fryer, and H. Sandler
Stanford University and NASA, Ames Res. Ctr.
Moffett Field, CA 94035

INTRODUCTION:

Within the last decade there has been an increasing use of radio telemetry systems to obtain basic cardiovascular data (1, 2). This mode of data acquisition is especially appropriate for multiple observations in unrestrained or untethered animals. A major concern in our laboratory has been evaluating and utilizing telemetry systems to assess cardiovascular function, emphasizing 24-hour monitoring and exercise applications. The purpose of the present study was to use a new telemetry system to compare cardiovascular changes in dogs exposed to restricted, unidirectional exercise on a treadmill, and unrestricted free-ranging exercise in an open field. Data from free-ranging animals are very limited and no comparisons of the two types of exercise have been made previously in the same animal.

METHODS:

The study group consisted of 4 dogs. After passing a 3-week quarantine protocol, they were anesthetized with pentobarbital anesthesia, 30 mg/kg. A left thoracotomy was made through the fourth intercostal space. A multichannel telemetry system (Konigsberg Inst., Pasadena, CA) was implanted into the left thoracic cavity as previously described (3). Two of the five channels of physiological data were assigned to pressure. The solid state transducers from these two channels were placed into the left ventricle and the descending thoracic aorta. A third channel was used for electromagnetic flow; this transducer was placed on the ascending aorta. The fourth channel was assigned to the electrocardiogram (ECG). The ECG lead was obtained from the ground electrode of the electromagnetic flow transducer and the case of the left ventricular pressure transducer. The fifth channel was temperature, with the thermistor located inside the electronic package. A coil for transmission of physiological signals and recharging the nickel cadmium batteries was sutured to the lateral chest wall. This positioning allowed operation of the unit, including switching on and off, by a coplanar coil placed on the skin surface.

The transducers were calibrated at 2-week intervals, using

a manometer-tipped catheter (Millar Inst.), indocyanine green determination of cardiac output, and an esophageal thermistor-tipped catheter (Yellow Springs Inst.).

During the 3 to 4 week recovery period, the animals were trained to perform the following exercise protocol, patterned after Barger (4). This consisted of 2 minutes of exercise (familiarization exposure), and 20 minutes of rest, followed by 5 minutes of maximal treadmill (TME) or free-ranging exercise (FRE). The data were averaged for 30 second periods of maximal exercise, defined by the occurrence of heart rate levels greater than 170 beats/min. The response of each animal to standing on its hind legs, with the front legs supported, was also evaluated routinely.

RESULTS:

The changes with standing considered pressoreceptor mediated included a transient decrease in aortic pressure and stroke volume, followed by an increase in heart rate and a return of aortic pressure.

The results comparing the 2 types of exercise are shown in Table I. The control heart rates in the field were significantly greater than the controls prior to TME. However, the maximum heart rates were not significantly different. During TME, mean aortic pressure increased 54% compared to a 27% increase during FRE. Statistically, the differences in pressure were of borderline significance (p value between 0.10 and 0.05). The peak left ventricular pressure was significantly lower during FRE compared to TME, 177 $\pm$ 11 to 212 $\pm$ 6 mmHg respectively ($p < 0.05$). Stroke volume during FRE was significantly lower in both the control and exercise states when compared to the TME phase of the study. TME and FRE stroke volumes did not change significantly from the control to the exercise state. Cardiac outputs increased significantly with both types of exercise. Left ventricular end-diastolic pressure was not different in control states, but was significantly higher with FRE.

Review of the continuous polygraph records showed the decreased mean aortic pressure during FRE was produced by a fluctuating decrease in aortic diastolic pressure. Additionally, in 2 of the dogs, a marked ST segment depression present during FRE was absent at comparable heart rates during TME.

DISCUSSION:

The orthostatic response was intact and elicited repeatedly in all dogs. The cardiovascular systems of these animals responded appropriately and rapidly, with similar adjustments in each. In contrast, the cardiovascular responses to both TME and FRE differed even at comparable heart rates, as evidenced by a tendency to a lower mean aortic pressure, and a significantly lower peak left ventricular pressure with FRE. Cardiac output and stroke volume were decreased, and end-diastolic pressure did not change during FRE exposure in comparison to TME. The results with TME agree with those of Vatner, et al (1), where similar changes in heart rate, mean aortic pressure and stroke volume were noted in animals running at speeds of 15 to 20 mph. The responses with FRE are similar to those described previously in sled dogs (2). The study of VanCitters and Franklin (2) also showed a widening of aortic pulse pressure and a lesser increase in mean aortic pressure with exercise. There have been no previous reports of ST changes with exercise in dogs. However, the ST segment changes did not impair the dogs' ability to exercise at maximum rates.

The mechanisms of these differences are not established. Candidate significant factors include behavior alterations by the changing environment, and variations in acceleration levels affecting fluid dynamics and mechanoreceptor reflexes. Future experiments will determine acceleration levels and neural pathways by use of multiaxis accelerometers, cardiac denervation, and pharmacologic receptor blockade.

REFERENCES:

1. Vatner, SF, D Franklin, CB Higgins, T Patrick and E Braunwald: Left ventricular response to severe exertion in untethered dogs. J. Clin. Invest. 51:3052-3060, 1972.

2. VanCitters, RL and DL Franklin: Cardiovascular performance in Alaska sled dogs during exercise. Circ. Res. 26:33-42, 1969.

3. Fryer, TB, H Sandler, W Freund, EP McCutcheon and EL Carlson: A multichannel implantable telemetry system for flow, pressure, and ECG measurements. J. Appl. Physiol. 39:318-326, 1975.

4. Barger, AC, V Richards, J Metcalfe and B Gunter: Regulation of the circulation during exercise. Cardiac output (direct Fick) and metabolic adjustments in the normal dog. Amer. J. Physiol. 184:613-623, 1956.

TABLE I: CARDIOVASCULAR CHANGES WITH EXERCISE

	MEANS ± SEM	HEART RATE bpm	MEAN AORTIC PRESSURE mmHg	PEAK LEFT VENTRICULAR PRESSURE mmHg	STROKE VOLUME ml/beat	CARDIAC OUTPUT L/min	LEFT VENTRICULAR END-DIASTOLIC PRESSURE mmHg
Treadmill Exercise n = 4 dogs, 7 exposures	CONTROL	109±3	114±9	154±7	33±2	3.57±.27	12.1±1.7
	EXERCISE	198±9	176±11	212±6	42±3	8.48±1.03	8.8±0.8
Free Ranging Exercise n = 4 dogs, 5 exposures	CONTROL	124±3*	113±12	153±9	20±2*	2.20±.20*	12.6±1.2
	EXERCISE	226±20	144±16	177±11*	23±2*	4.93±.12*	13.6±2*

*denotes p = < .05 when comparing treadmill exposure to free ranging exposure for both control and exercise states.

EXPERIMENTAL USE OF AN IMPLANTABLE DIRECTIONAL DOPPLER FLOWMETER TO MEASURE THE FLOW IN PULMONARY ARTERY AFTER LUNG HOMOGRAFT

D. Cathignol[1], C. Fourcade[1], P. Vuillard[2]

1. INSERM, Bron, France. 2. Unité de Chirurgie Vasculaire, Bron, France.

Direct measurement of pulmonary artery flow is an important parameter in grafted dogs. Until recently we have been trying to place Doppler probes at the level of the pulmonary artery and the aorta. However the connecting wires from the probe to the exterior flow meter present two inconveniences :

- they are a source of infection particularly in the case of dogs treated with immunodepressors,
- they can result in traumatic destruction of the artery if for any reason these wires are pulled on.

We have thus decided to implant telemetric flowmeters which have been developed in our laboratories (1).

EQUIPMENT

- The flowmeter (figure 1). The part which is implanted is composed of :
  - a single transistor Hartley type oscillator which feeds a transducer emittor.
  - an amplifier which amplifies the signal received by the transducer receiver and which transmits it directly to the exterior by means of an aerial.

The carrier frequency is determined by parasite coupling between the transducer emittor and the receiver.
This system makes it possible to use a classical directional flowmeter as long as the oscillator is replaced by a phase lock oscillator. The technical characteristics are as follows :

overall size	Ø 15 mm X length 45 mm
weight	120 g
power supply	15-20 mA X 1,35 V
life	3000 measurements of one minute in length
range	30 to 50 cm

The short working range of the system is for us a minor inconvenience as the principal aim is to have no connection wires between the probe and apparatus exterior to the animal. The apparatus is switched on by placing a magnet near to the flowmeter which acts on a magnetic switch. The probe is relayed directly to the flowmeter thus eliminating all problems of leakage at this level. We have flowmeters of all possible diameters. The probes are made out of moulded acrylic resin and articulated by means of a silastic ring. All other parts of the flowmeter are coated in medical silastic. Sterilization is obtained by placing the flowmeter and its probe in a saturated atmosphere of trixymethylene for a period of 48 hours.

METHOD

20 adult mongrels weighing between 14 and 20 kg were used. These dogs have received a homograft left diaphragmatic lobe with preservation of the two apical lobes.

1 ; includes 7 dogs which have undergone no specific treatment.
2 : includes 5 dogs which have been administered 2mg/kg Azathioprine and 2 mg/kg Prednisolone.
3 : includes 8 dogs which have been given the same treatment as for group 2 + Mg_{++} Heparinate 0.1 ml X 2.

The surgical technique has been previously described for the implantation of the probe outside the suture zone (2).The probes are placed, one on the pulmonary artery, a second on the aorta, the flow of which serves as a reference. The emittor as well as the aerial are placed beneath the skin.

RESULTS

The flow rate measurements are made daily by bringing the aerial and the magnet near to the dog. Pulmonary ventilation is also measured by means of a transthoracic impedance measurement.

- Flowmeter results. The normal curve (figure 2).Before showing the changes in flow, we shall briefly mention the normal velocity profile (this has been obtained by means of a classical Doppler flowmeter). A first positive wave ABC is followed by a wave of short reflex and low amplitude CDE. At the end of the diastolic stroke, the flow is practically zero.

Different pathological curves. We have been able to distinguish in these 20 transplantations two types of curve noted as type 1 and type 2 (fig.2b, 2c). Type 1 is characterized by an increase in the continuous flow, we interpret this as being due to shunts. Type 2 is characterized by the appearance of a reflux wave, which can be explained by an increase in peripheral resistance.

- The different flow profiles and their anatomophysiological correlations (figure 3). We have been able to classify the charges of flow rate in the pulmonary artery into 4 groups :

- the mean flow remains stable and the instantaneous profile normal (type 312). The two dogs included in this category showed no signs of rejection and died as a result of intercurrent disease
- the mean flow decreases with a progressive appearance of an inverted flow wave (type 293). The five dogs included in this category showed signs of intense rejection and large permeable blood vessels.
- the mean flow increases to a high value (possibly attaining the flow rate of the descending aorta) with the appearance of a continuous flow (type 314). The two dogs in this category died of asphyxia, their PO2 being extremely low.

- the average output decreases rapidly without inverse output wave. The necropsy shows artery thrombosis (type 306).

DISCUSSION AND COMMENTS

The appearance of a reflux wave is due, in our opinion, to an increase of the peripheral resistance. In fact, we have been able to reproduce these results by selective injection in healthy dogs in the APG of a Baryum sulphate solution, the macromolecules of which block the capillary vessels. At this advanced stage of the rejection it would seem to be necessary to refer this hypertension of the pulmonary artery to the perivascular infiltration and to the large reduction of blood vessel caliber which results.

The second type of curve brings to mind the flow in an arterioveinal fistula. This would seem to indicate a decrease in peripheral resistance due to arterial-veinal shunts. Scintigraphic measurements have shown a complete image absence although flow remains high, there being no embolism of the microspheres. This explains why some workers have not been able to show such high flow rates.

The apparatus which we have developed allows us to continuously record the flow rate of the pulmonary artery and that of the aorta. Having used both a classical type Doppler flowmeter and an implantable type, we can affirm that the latter has considerable advantages both with respect to the risks of infection as accidental traumatism of the arteries by the probes.

REFERENCES

1. Cathignol D, Fourcade C, Descotes J: An implantable directional Doppler flowmeter. Biotelemetry, 2nd Int Sym, 1974, p 16-18

2. Vuillard P, Gadot P, Tiano R: Une nouvelle technique d'homogreffe monopulmonaire experimentale. J Chirg 208:6, 1975 p587-590

ACKNOWLEDGMENTS

P. PILLARD,R. TIANO, and CHUZEL for their surgical contribution.

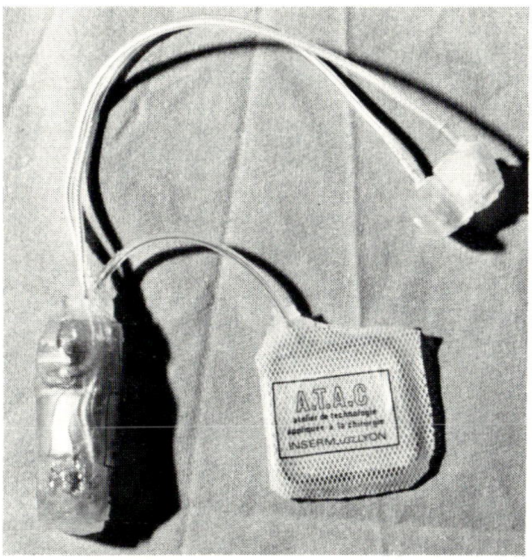

FIGURE 1

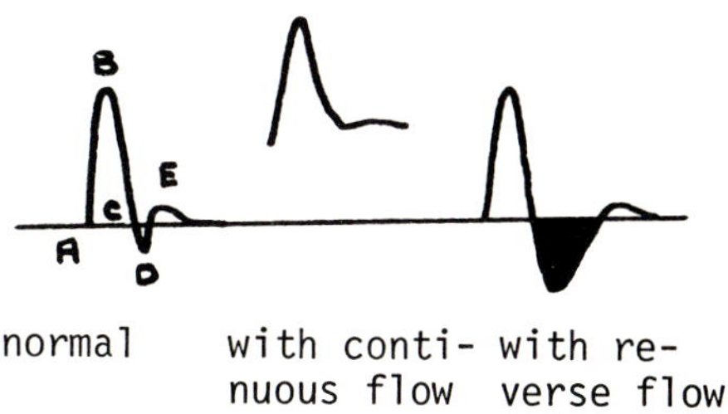

FIGURE 2

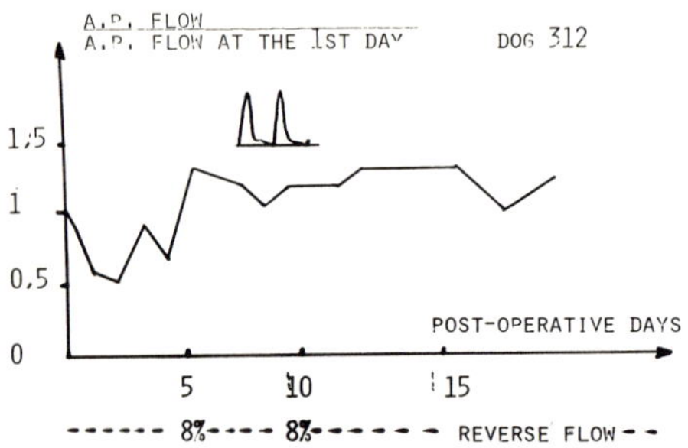

FIGURE 3a

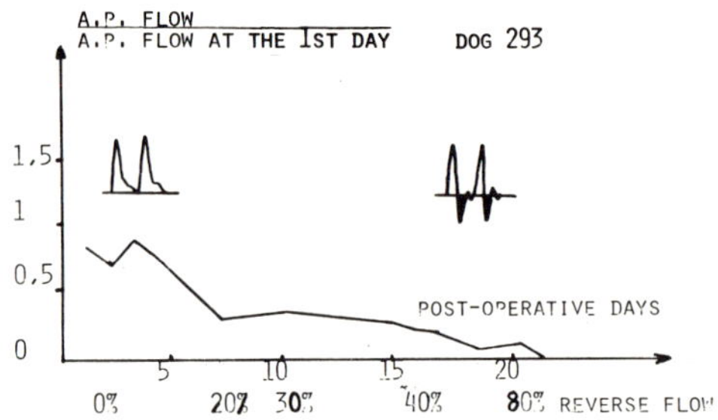

FIGURE 3b

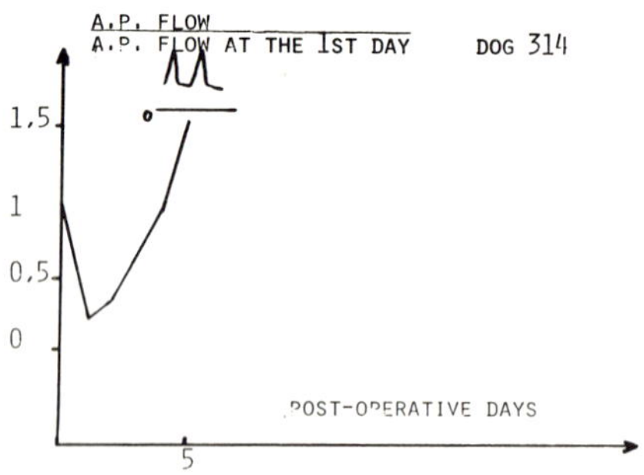

FIGURE 3c

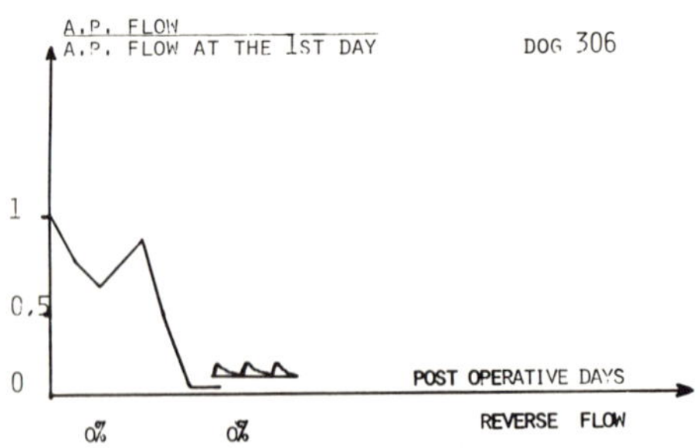

FIGURE 3d

RADIOTELEMETRY OF RENAL AND ILIAC BLOOD FLOW FROM EXERCISING HEART AUTOTRANSPLANT RECIPIENTS

W.R. Freund[1], E. Dong, Jr.[2], and J. D. Meindl[3]

1&3: Stanford Electronics Laboratories, Stanford, California
2: Department of Cardiovascular Surgery, Stanford University School of Medicine, Stanford, California

Previous studies have demonstrated the sustained capacity of the transplanted heart to respond in a directionally appropriate manner to the metabolic demands of exercise (1,2). In comparison to the normal heart, however, the absolute level of cardiac output from the transplanted heart is consistently lower (3). Little information is available to describe the adaptive peripheral circulatory changes which allow normal levels of activity despite reduced cardiac outputs. The purpose of this study was to investigate these adaptations by comparing peripheral flow distribution during rest and exercise in the dog.

METHODS

Totally implanted CW Doppler flowmeters (4) were employed to measure blood flow velocities through the renal and iliac arteries of adult mongrel dogs. A total of 16 flowmeters were implanted in 8 animals (5 controls and 3 heart autotransplant recipients) to compare their resting and exercise flow values. The use of totally implanted telemetry systems allowed frequent measurements from the unanesthetized animals without the inconveniences often associated with percutaneous leads.

Flowmeter operation was controlled by a 500 KHz "command transmitter" which activated the implanted circuitry. The transducer elements (1.5 mm square) were operated at an ultrasonic frequency of 6MHz. The Doppler audio signals were telemetered in the 88-108 MHz band to commercial FM receivers. In the absence of interference from adjacent signal sources, telemetry range exceeded 50 meters. The received Doppler audio signals were processed by conventional zero-crossing techniques.

Following a 3 week recovery period, the animals were treadmill exercised (30° incline) in accordance with the protocol developed by Barger, et al. (5). Studies were repeated over a 2 week period to verify consistency of response. Arteriograms were performed prior to recovery of the probes to estimate lumen diameter and verify that the transducers presented no flow obstruction. Direct bleed-out studies were also performed with 3 animals to verify flowmeter linearity.

RESULTS

At the time of study, the transplant recipients exhibited characteristics often associated with cardiac denervation i.e. peripheral edema, absence of a startle response, and complete absence of normal sinus arrhythmia (6). For both groups of animals maximum response was achieved after 1 min. of exercise. The hemodynamic findings for normal and transplant recipient animals are summarized in Table 1.

Table 1. Flow and Heart Rate Response to Exercise

	Autotransplant	Normals
RH	56.7(7.4)	108(8)
EH	131.7(27.4)	190(21.1)
RR	45(3.5)	62.7(19.7)
ER	41.7(2.1)	53.8(15.7)
RI	20(0)	33.8(9.2)
EI	69(1.4)	74(18.3)

Numbers in parentheses indicate ± S.E.M; RH= Resting heart rate (beats/min); EH= Exercise heart rate (beats/min); RR= Resting mean renal flow velocity (cm/sec); ER= Exercise mean renal flow velocity (cm/sec); RI= Resting mean iliac flow velocity (cm/sec); EI= Exercise mean iliac flow velocity (cm/sec).

In comparison to the normal animals, the absolute levels of heart rate, mean renal and iliac flow velocities were consistently lower for the transplant recipients during rest and exercise. In contrast, however, the <u>relative changes</u> with exercise in heart rate and flow velocities were similar for both groups of animals (see Table 2). For both groups, heart rate approximately doubled, mean iliac flow velocities increased more than two-fold, and mean renal flow velocities decreased by less than 20%.

Table 2. Relative Change in Flow and Heart Rate with Exercise

	Autotransplant	Normals
Heart Rate	+130%(23)	+81%(24)
Mean Renal Velocity	-3%(25)	-13%(6)
Mean Iliac Velocity	+182%(68)	+125%(12)

Numbers in parentheses indicate ± S.E.M.

DISCUSSION

These results demonstrate the capacity of the transplanted heart to fulfill the metabolic demands of exercising muscle without compromising renal perfusion. In contrast to the findings of Willman, et al. (6) resting heart rates dropped following cardiac autotransplantation. This is possibly due to differences in transplantation technique and the relative location of the SA node with respect to the suture line. The possibility of cardiac renervation should also be considered in the interpretation of these results. In conclusion, we found that normal and cardiac autograft recipient dogs did not differ significantly in their renal and iliac flow response to treadmill exercise.

REFERENCES

1. Dong E Jr, et al: Performance of the heart two years after autotransplantation. Surgery 56: 270

2. Stinson EB, et al: Hemodynamic observations one and two years after cardiac transplantation in man. Circulation XLV: June 1972

3. Stinson EB, et al: Hemodynamic observations after orthotopic transplantation of the canine heart. J Thorac Cardiovasc Surg 63, 3: 344, 1972

4. Curtis PJ: Specifications for implantable ultrasonic flowmeters. (Unpublished information)

5. Barger AC, Richards V, Metcalfe J, et al: Regulation of the circulation during exercise. Cardiac output (direct fick) and metabolic adjustments in the normal dog. Am J Physiol 184: 613, 1956

6. Willman VL, et al: Physiologic alterations following transplantation of the heart. Proc 1st Intl Cong Transpl Soc: 661, 1967.

TELEMETRIC MONITORING OF PNEUMATICALLY POWERED TOTAL ARTIFICIAL HEARTS IN CALVES

Don B. Olsen, Lee M. Smith, Jack Kolff, Steve Nielsen, Judd Lunn and W. J. Kolff

Institute for Biomedical Engineering and Division of Artificial Organs, Department of Surgery, Bldg. 518
University of Utah, Salt Lake City, Utah 84112

INTRODUCTION

Continuous monitoring of physiological parameters in chronic studies is a problem shared by all investigators in biomedical research (1). A total mechanical heart (TAH) has been used to replace the natural heart of calves in this laboratory. The marked increase in survival times of calves with a TAH (one month (2) and 3 months (3) in 1974 and 4 months in 1975 (4) necessitates improved methods of monitoring and evaluating cardiac performance (5,6). Open sampling and pressure ports into the cardiovascular system are sources of infection and thromboembolism with occlusion. These events complicate critical evaluation of the prosthetic device and may result in termination of the experiment (3,7). Preliminary evaluation of the miniaturized biotelemetry equipment developed by NASA-Ames, used in calves with TAH, has been reported (8). This report includes the results of calves with TAH instrumented with biotelemetry for up to 2 months.

MATERIALS AND METHODS

The characteristics (3) and fabrication (9) of the pneumatically powered TAH (Jarvik) has been described. It consists of a flexible diaphragm and semi-rigid housing of polyurethane fastened to a solid base. Two Bjork-Shiley valves were placed into each ventricle and a heart driver console pulsed the heart via a 6-foot percutaneous drive line. The telemetry unit was connected to the ventricle, aortic graft and atrial cuff prior to surgery (Fig. 1). Open ports were used in the aorta, pulmonary artery and left atrium to calibrate and zero the telemetry unit. The Konigsberg pressure transducers were protected from the blood by a thin layer of polyurethane, maintaining a smooth, continuous layer at the blood interface (Fig. 2). The electrocardiographic leads were attached to the right atrium.

The TAH with the telemetry unit was implanted via right lateral thoracotomy under cardiopulmonary bypass after the natural heart was excised. The calf was slowly weaned from the bypass and the TAH took over the total pumping responsibility. The heart driver

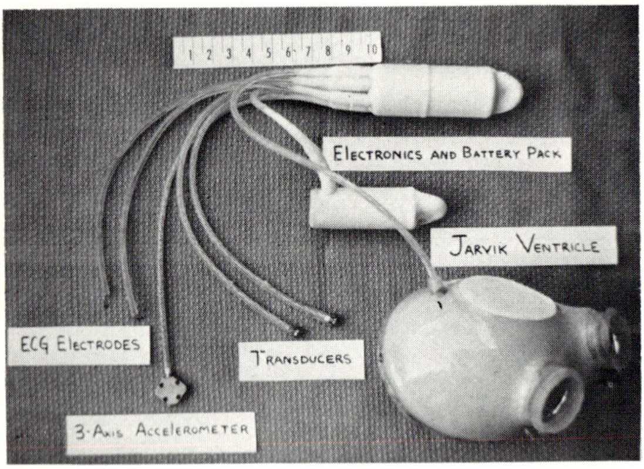

Figure 1.

Figure 2.

has individual right and left side driving pressure selectors and single, adjustable heart rate and percent systole modes. These variables were tuned for optimum cardiac output and low left atrial pressure and the calf was allowed to recover.

Continuous monitoring was done for the first 24 hours during intensive care and careful maintenance of fluid balance. After recovery, pressures were taken with the telemetry and compared with corresponding open port catheters connected to Statham transducers. All data were recorded on a Hewlett-Packard multichannel recorder. The telemetered data was picked up on a receiver and a demodulator. Pressures were recorded at rest (lying down), standing and walking on a treadmill. Cardiac output was determined with the Fick technique. Changes in flow and pressures were compared with data obtained with a non-invasive cardiac monitor on the air driving lines. This electronic monitor displayed peak systolic pressures and ejection times. Previous measurements in the mock circulation suggested that cardiac output was a function of ejection time (10).

RESULTS

The two longest surviving calves lived 31 and 36 days. The ports and catheters remained open in the longest survivor receiving anticoagulants. Both calves performed on the treadmill 3 times each week, beginning the second week. The data are presented in Figure 3 with fixed heart driving parameters. The mean aortic pressure was lower (10 mm Hg) during exercise. The peak aortic and left ventricular pressure (telemetry checked against open port) did not change from rest through exercise. The ejection time and peak air drive pressure did not change. The oxygen consumed by the calf increased 45% from rest to exercise and the peripheral resistance was 10 dynes-sec/cm^{-5} and reduced to 6.3

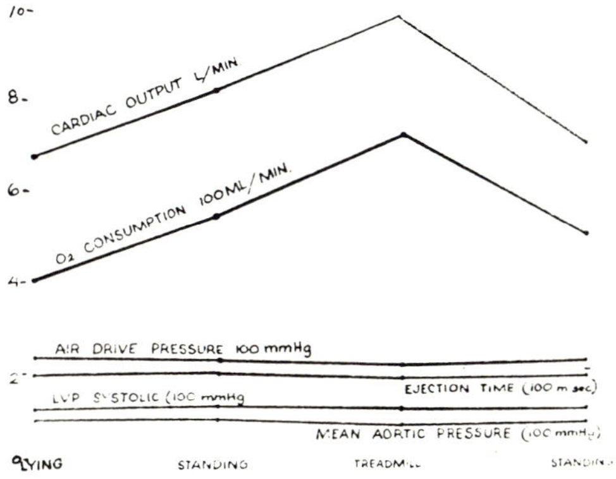

Figure 3.

with exercise.

DISCUSSION AND CONCLUSIONS

Telemetry has become a sophisticated bonafide technique for collecting physiological data (1). Evaluation of the telemetered intraventricular pressures facilitated fine tuning of the driving system for optimum cardiac function, as reported previously (8). The lengthening of the ejection time as a function of increased flow in the mock-circulation with fixed compliance (10) did not hold in vivo with a changing peripheral vascular bed. The stroke volume increased without lengthening of the ejection time. The stability of the peak to peak pressures was consistent throughout the experiments, but the drift in position (zero) created problems. We had companion pressure catheters to zero the telemetry. However, reasonable adjustments could be made by turning the heart driver off for 8-10 seconds at the beginning of diastole. The left atrial pressure and intraventricular pressure equilibrated at a near zero in about 4 seconds. The peak aortic pressure was then equalled to the peak ventricular pressure, thus establishing a zero for all channels. The TAH calf, devoid of multiple catheters, is more easily managed in the laboratory setting (3,8) and the elimination of anticoagulants with a marked reduction of infection is a benefit (2). The motion-artifact free pressure tracings telemetered during 15 minute exercise periods aided in defining the reduced peripheral vascular resistance as an adjunct to enhancing cardiac output, as described in other reports (5,6). It had previously been demonstrated that the artificial heart responded with increased cardiac output to an increase in atrial pressure. Both increased filling pressure (volume) and decreased peripheral vascular resistance are important in the "Starling-like" response of the TAH.

ACKNOWLEDGEMENTS

This work has been supported in part by a grant from the National Institutes of Health, the National Heart and Lung Institute Grant No. PO1-HL13738 and the National Aeronautics and Space Administration Grant No. NCA2-0R21-601. Telemetry equipment was manufactured for NASA-Ames by Konigsberg Instruments Inc., 2000 East Foothill Boulevard, Pasadena, California 91107.

REFERENCES

1. Monitoring. Chapter XII in Research Animals in Medicine, (ed. L.T. Harmison), Maryland, N.H.L.I., National Institutes of Health, pp 1085-1270, 1973.

2. Olsen DB, Unger F, Oster H, et al: Thrombus generation within the artificial heart. J of Thor and Cardiovasc Surg 70:248, 1975.

3. Lawson J, Hershgold E, Kolff J, et al: A comparison of polyurethane and silastic artificial hearts in 10 long survival experiments in calves. Trans ASAIO 21: 368, 1975.

4. Olsen DB, Kolff J, Lawson J, et al: Saving the aortic and pulmonary artery valves with total heart replacement. Trans ASAIO 22 (in press), 1976.

5. Olsen, DB, Nielsen M, Lunn J, et al: Total artificial heart performance in the calf. Abstracts Am Col Cardiol 37: 160, 1976.

6. Lunn J, Liu W, Stanley T, et al: Effects of treadmill exercise on cardiovascular and respiratory dynamics before and after artificial heart implantation. Trans ASAIO 22 (in press), 1976.

7. Kolff WJ and Lawson J: Status of the artificial heart and cardiac assist devices in the United States. Trans ASAIO 21: 620, 1975.

8. Smith LM, Olsen DB, Sandler H, et al: Total artificial heart monitoring with biotelemetry. Abstracts of 28th ACEMB, p 98, 1975.

9. Kessler TR, Arnett G, Foote J, et al: Urethane construction techniques for the artificial heart. Abstracts ASAIO 22 (in press), 1976.

10. Foote JL: Noninvasive measurement of inflow and outflow pressure in the artificial heart. 26th ACEMB, p 22, 1973.

SESSION VIII

BIOMEDICAL TELEMETRIC EQUIPMENT

Chairmen

T.B. FRYER, Ph.D.

G. MATSUMOTO, Ph.D.

EDITOR'S OVERVIEW

This final session consists of a large number of papers which concentrate on the circuit-design or hardware aspects of recent advances in biotelemetry. Several of the papers provide additional information on equipment used in investigations reported in earlier sessions. The papers consider such diverse subjects as antenna design, the use of surface-acoustic-wave (SAW) oscillators, new and unusual telemetry systems, telephone monitoring of cardiac pacemaker performance, studies of error sources in EEG, ECG, and ultrasonic blood-flow systems, and improved data-handling processes.

The new and unusual telemetry systems include transmission in the IR range of the electromagnetic spectrum, transmission via the back scatter from incident radar signals, transmission via ultrasound, and use of time sequencing to allow several telemeters to operate on the same carrier frequency. The ultrasound unit utilizes novel forms of pulse-width modulation and pulse-rate modulation in which data are represented by the dwell times of 0 and 1 of flip-flop circuits and by the rate of 0, 1 switching of the flip-flop.

A WIDEBAND EMG TELEMETRY SYSTEM

R. M. Westbrook, T. B. Fryer, and S. A. Rositano

Ames Research Center, NASA
Moffett Field, California 94035

INTRODUCTION

Electronic gait monitoring of children with cerebral palsy has proved extremely useful in determining the type of orthopedic surgical correction that is indicated. The technique involves placing electrodes on six opposing muscle groups on each leg, and mounting pressure-triggered step indicator switches on both shoe soles. These myoelectric (EMG) and pressure sensors are monitored as the child ambulates; the information obtained, in a desired EMG bandwidth of 20 to 2000 Hz, is used to correlate the activity of the leg muscles with each phase of walking.

Telemetering the data has the obvious advantage of eliminating cabling between subject and recording equipment, thus allowing greater mobility. Ordinarily, transmission of several data channels would be done using a multiplexed data transmitter with a single RF link. But because of the EMG bandwidth requirement of 2000 Hz, and the 12 channels needed, such a multiplexed system would be difficult to implement in a low-powered device having a reasonable transmitting range. By comparison, a low-powered, single-channel transmitter for each sensor can be readily designed to meet the range and bandwidth requirements. Also, the single transmitter concept allows data channels to be easily added or subtracted as dictated by study requirements. Furthermore, using individual transmitters for each sensor, instead of one master transmitter assembly, eliminates the need for interconnecting cabling on the subject. This provides maximum mobility during gait analysis, more comfort for the subject, and reduces preparation time.

This paper describes a system consisting of individual, crystal-controlled, single-channel transmitters small enough to be placed at the sensor sites, yet having a linear frequency response from 20 to 2000 Hz and an operating range of 15 m.

DESCRIPTION

A wideband EMG telemetry system has been developed which consists of miniature, individual, crystal-controlled RF transmitters located at each electrode site. The transmitters are assigned operating frequencies within the 174 to 216 MHz band. The receivers are crystal-controlled also so that there is no ambiguity in channel identification. The effective transmitting range is 15 m. Subcarriers are used to provide good amplitude linearity with a data bandwidth of 20 to 2000 Hz. The amplifiers have an input impedance of 10 MΩ and are designed to be used with two sensing electrodes per channel. Figure 1 is a block diagram of one transmitting channel.

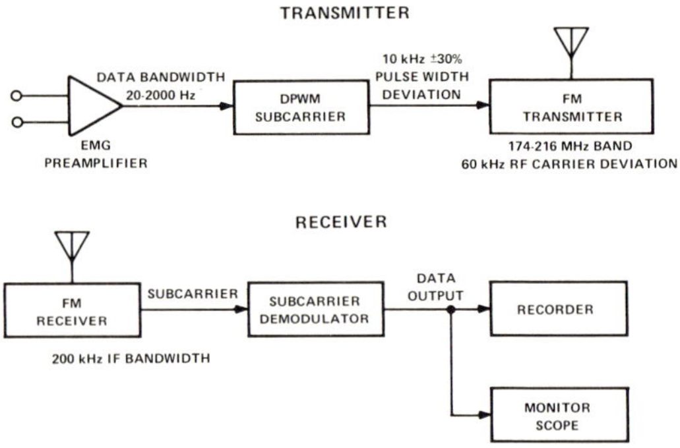

Figure 1. Transmitting and receiving system.

The circuit for the EMG transmitter is quite simple. Because the entire unit including batteries had to be small enough for attachment at the electrode site, the circuit was designed to minimize the number of components and the current drain. Figure 2 shows the circuit diagram of the complete unit with preamplifier, subcarrier oscillator, RF transmitter, and battery supply.

The preamplifier uses two operational amplifiers, one as a voltage follower and the other as a gain stage. Gain is set at 56 so that a full scale input signal of ±3 MV will provide ±168 MV at the modulator input, with a bandpass of 20 to 2000 Hz.

The subcarrier oscillator employs a single operational amplifier as a differential pulse-width modulator (DPWM)(1). Pulse widths are varied ±30% with a fixed frequency of 10 kHz. All three operational amplifiers are on one integrated circuit chip. Total current drain is less than 200 μA with a ±1.5 supply voltage.

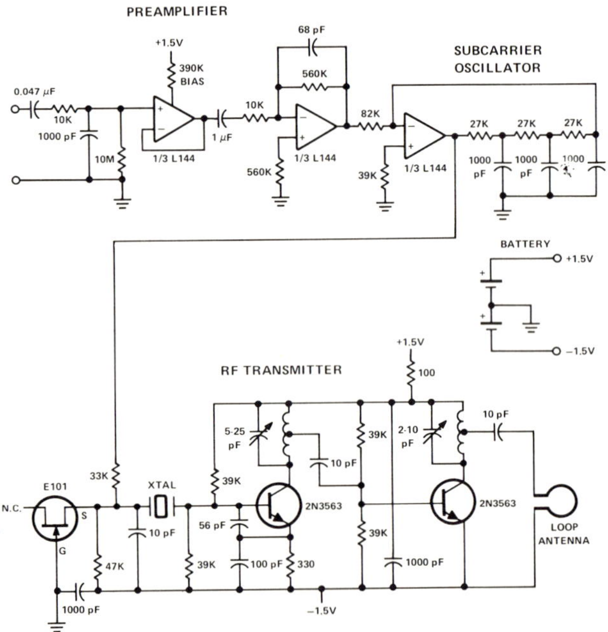

Figure 2. DPWM–FM transmitter.

The RF transmitter consists of a class A oscillator-tripler stage followed by an output tripler stage; current drain of the complete RF section is less than 3 mA. The oscillator is tuned to the fundamental frequency mode of the crystal in the emitter circuit and is sharply tuned to the third harmonic in the collector circuit. The third harmonic output is applied to the tripler output stage for a total frequency multiplication of 9. Frequency modulation is accomplished by applying the rectangular waveform from the subcarrier oscillator to the junction field-effect transistor. The resulting capacitance change deviates the fundamental frequency about 7 kHz. When multiplied to the 174 to 216 MHz operating range, this results in a RF carrier deviation of nearly 60 kHz. Crystals are selected so that the RF carriers of the transmitters are spaced 750 kHz apart. Because a subcarrier is used, the deviation of the RF carrier need only insure a good signal-to-noise ratio in the transmission link and remain within the selected bandpass. The transmitter is packaged in the form of a disc having a diameter of 3.75 cm and a thickness of 0.9 cm. Two silver oxide cells are used as a battery supply. Each cell weighs 1.2 g, has a diameter of 0.76 cm, and is 0.53 cm thick. A cell of this size provides 22 hr of continuous operation.

A circuit diagram of the subcarrier demodulator is shown in figure 3. At its input, the demodulator has a low pass filter which is set for a 3 dB point at 40 kHz with a 36 dB per octave rolloff. This filter allows sufficient rise time for threshold detection of the 10 kHz subcarrier, yet greatly reduces any spurious responses of a higher frequency that might occur during modulation and transmission.

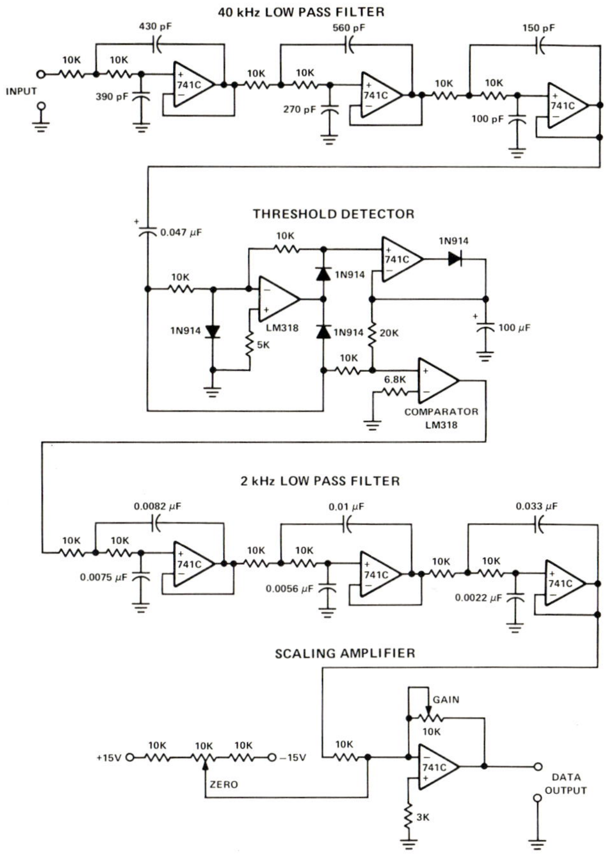

Figure 3. Subcarrier demodulator.

The threshold detector automatically measures the peak-to-peak amplitude of the subcarrier and generates a voltage that is centered precisely at the midpoint of the rectangular waveform. Using this information, the comparator reshapes the subcarrier to a rectangular waveform with fast rise and fall times and a constant amplitude. Low pass filtering of the modulated rectangular waveform reproduces the original EMG. For a subcarrier frequency of 10 kHz, a low pass filter with 36 dB per octave rolloff at 2000 Hz eliminates the sampling frequency while retaining accurate amplitude information within the 2000-Hz bandwidth. An output amplifier is used to scale the data so that all transmitters can be adjusted for equal system gain.

CONCLUSION

The use of individual transmitters for each data channel has a number of advantages. By eliminating the need for cabling between the several EMG electrode sites and a master transmitter assembly, they provide maximum mobility during gait analysis; and the elimination of cables between sites reduces subject preparation time while improving subject comfort. Data channels can be easily added or subtracted to meet the needs of the particular study.

REFERENCES

1. Westbrook RM, Fryer TB: A narrowband crystal controlled biomedical telemetry system. Proceedings of International Telemetering Conf., Los Angeles, Calif., Oct. 1972, Vol. VIII, pp. 214-220.

Tuned Loop Receiving Antennas For Indoor Telemetry

R.W. Vreeland, B.B. Rutkin and C.L. Yeager
University of California San Francisco Medical Center, UB-10
San Francisco, Calif. 94143

Location of seizure foci prior to surgical treatment of epilepsy requires continuous telemetry from indwelling brain electrodes for a period of several days to record spontaneous seizures. During this period the patient is ambulatory and may roam as far as a solarium located at the end of a long corridor.

Our nine channel 100 MHz pulsed telemetry system (1 and 2) was first used with a single receiving dipole located in the patient's room. This was satisfactory when the patient was in his room, but did not adequately cover the corridor. When we constructed a special room for telemetry studies (with an adjacent room for the telemetry receiver and EEG recording equipment), we considered the possibility of installing several half wave receiving dipoles in the corridor. Because of the size and appearance of commercially available 100 MHz dipoles, they were deemed not acceptable. (Concealing the dipoles above the false ceiling was not feasible because of the metal lath which supports the ceiling.)

The problem was solved by installing tuned loop antennas: three on the corridor ceiling and a fourth on the ceiling of the

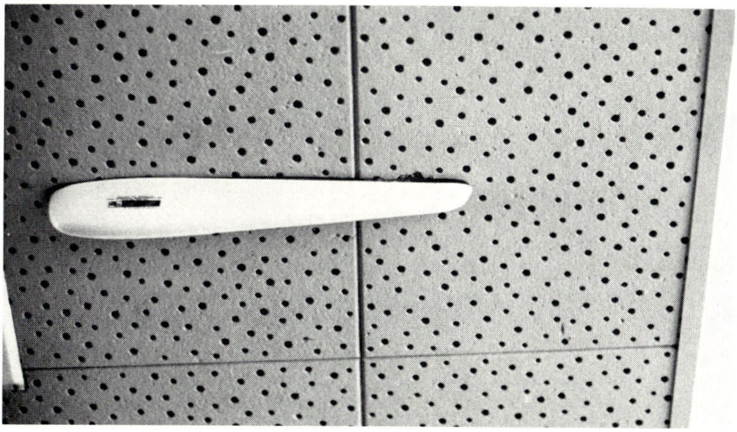

Fig. 1. One of three antennas attached to the corridor ceiling. The antenna is only 12 inches long.

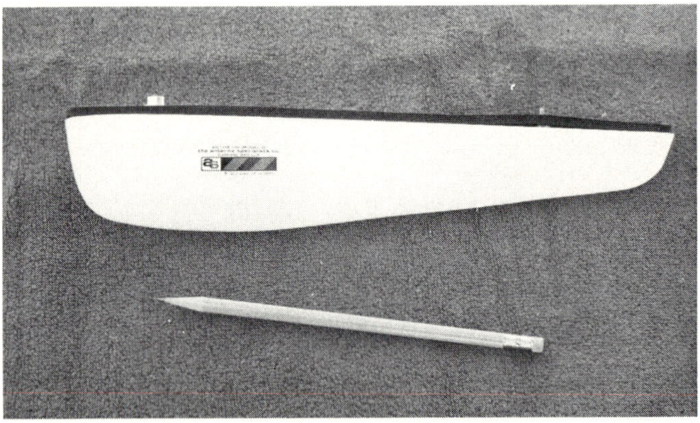

Fig. 2. An unmounted aircraft marker beacon antenna with the black neoprene mounting gasket in position. Note the BNC cable connector protruding through the gasket at the left end.

patient's room (fig. 1). Each antenna has a separate RG-58A/U foam coaxial transmission line, which terminates in a Mini-Circuits Laboratory ZSC-4-1 four-way combiner located in the recording room. Phasing of the antennas was found to be unnecessary.

We use modified Antenna Specialists Co. AV-569 aircraft antennas. As supplied by the manufacturer, they are fixed-tuned to 75 MHz for the reception of marker beacon signals. They are packaged in streamlined plastic "boats" (fig. 2) which are designed to be attached to the aircraft with two 10-32 machine screws. To avoid electrical contact with the metal lath in the false ceiling we have mounted the antennas with ¼-20 nylon studs passing through the ceiling. The lower ends of the studs have been machined to fit the 10-32 mounting holes in the antennas. This makes a very neat installation. The antennas are only 12 inches long and extend downward only 2-3/8 inches below the false ceiling (fig.1).

An antenna which has been removed from the styrofoam filled plastic "boat" is shown in figure 3. The unmodified loop is tuned by a 1-7 pf variable piston capacitor which is connected in parallel with a 15 pf fixed capacitor. For 90 MHz-105 MHz operation, the 15 pf capacitor is replaced by a 6.8 pf unit. Modification does not require removal of the antenna from the plastic boat. It is only necessary to separate enough of the styrofoam from the boat to permit lifting the small end of the antenna far enough to expose the capacitor. This must be done carefully in order to avoid damaging the piston capacitor. It is not necessary to move the 50 ohm tap.

Like all loops (and dipoles) the antenna is directional. The optimum location and orientation were determined experimentally for each antenna before permanent installation.

A modified loop was compared with a 100 MHz half-wave dipole

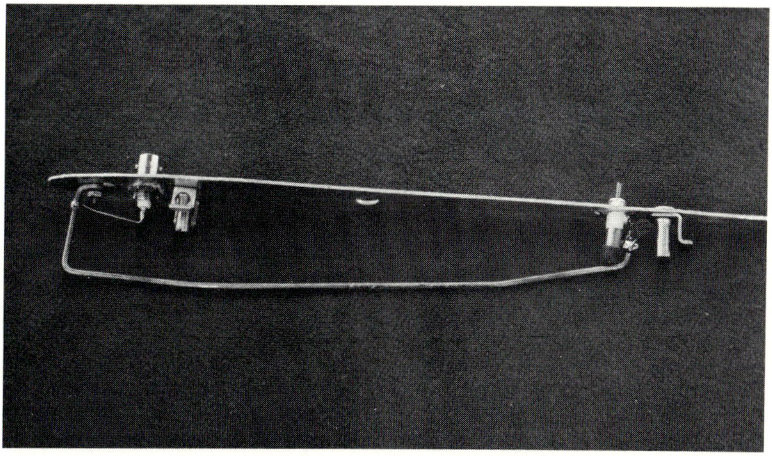

Fig. 3. The loop after removal from its housing. Note the BNC connector and the 50 ohm output tap at the extreme left. At the right is the piston tuning capacitor with the 6.8 pf fixed capacitor soldered to its terminals. The bright metal cylinders at the extreme right and just to the right of the BNC connector are the captive 10-32 mounting nuts.

and found to be 2 dB to 12 dB less sensitive depending upon the orientation of the loop. Both the normal background noise and noise from a sparking motor commutator were 15 dB less for the loop than for the dipole. This 3 dB to 13 dB improvement in signal to noise ratio resulting from the use of a loop antenna is especially important for pulsed telemetry systems. Pulsed systems are affected by severe impulse noise of the type generated by automobile ignition systems.

These measurements were made with the loop tuned to the telemeter frequency. An FM broadcast station 8 MHz from this frequency was attenuated 42 dB when the loop was substituted for the dipole. Another FM station only 2 MHz from the telemeter frequency was attenuated 28 dB. The distance from our hospital to a tower from which four FM broadcast stations transmit is only 1000 meters. For this reason the selectivity of the tuned loop is highly desirable.

The completed four-antenna installation was tested by carrying the telemeter along the corridor and monitoring the received signal strength. There were several very sharp signal dropouts due to cancellation of the signals from two or more antennas or to reflections. These nulls were only several inches wide, and the probability of a patient standing in one at a critical time was quite low. At all other locations the signal was 35 dB to 80 dB above the noise level.

The problem of signal dropouts might possibly be solved by automatic antenna switching to the strongest signal, but this additional complexity is not warranted in our application.

Our receiver is an Astro Communications Laboratory model SR-209. It has a peak holding automatic gain control system which sets the AGC at the peak pulse level. This is particularly important for low duty pulsed telemetry of the type we use. The receiver also has an AGC actuated relay which is used to disable the demodulator so that noise will not be recorded during signal dropouts.

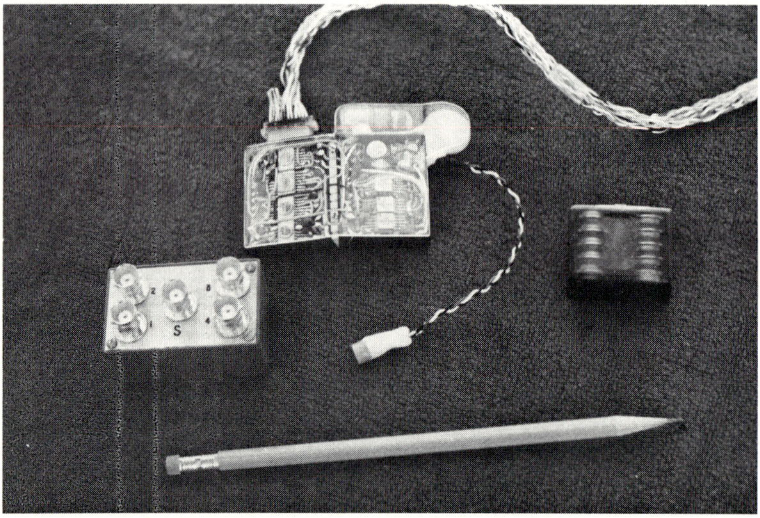

Fig. 4. From left to right are the four-way antenna combiner, the nine-channel telemeter and the rechargeable battery.

The antennas are connected to the receiver via the four-way combiner shown in fig. 4. For depth electrode recording, the telemeter and battery are attached to the patient's head bandages under a stocking cap. The battery shown in fig. 4 is useful for continuous recording periods of twelve to eighteen hours. When 200 to 300 hour life is required, we use two disposable Burgess H146X batteries. For scalp electrode recording, a separate FET pre-amplifier unit (not shown) is used.

References

1. Vreeland RW, Yeager CL: A Nine Channel "COS/MOS" EEG Telemeter for Long-term Seizure Monitoring. Proc. 26th ACEMB, 1973, Minneapolis, Minn.: 402
2. Vreeland RW, Yeager CL: Low Level "COS/MOS" Multiplexing for Simplified EEG Telemetry. Biotelemetry II. 2nd Int. Symp., Davos 1974, pp. 55-57 (Karger, Basel 1974)

SHORT-RANGE PASSIVE TELEMETRY BY MODULATED BACKSCATTER OF INCIDENT CW RF CARRIER BEAM

Alfred R. Koelle, Steven W. Depp, Jeremy A. Landt, Ronald E. Bobbett

Los Alamos Scientific Laboratory, University of California, Los Alamos, New Mexico

Modulated Backscatter Telemetry was developed as a means for reporting the temperature of cattle, along with an identification code for each animal. Objectives included low cost and passive (no battery) operation of the implanted transponders. This novel telemetry method should be useful in other applications. The paper describes the principle of operation and solutions to some of the problems encountered during the development.

PRINCIPLES OF OPERATION

This form of telemetry differs from the conventional by not having the rf carrier generated at the information sending unit. The backscatter of an externally supplied rf carrier beam from the sending unit antenna is modulated by the signal. The principle is like that of a heliograph sender, in which the reflection of sunlight off a mirror is controlled by a shutter.

As seen in Figure 1, a cw rf beam is sent from the interrogator to the transponder, which has a half-wave dipole antenna to intercept the incident rf. An electronic switch is located at the center of the antenna, allowing the dipole halves to be shorted or opened. When the switch is closed the dipole is

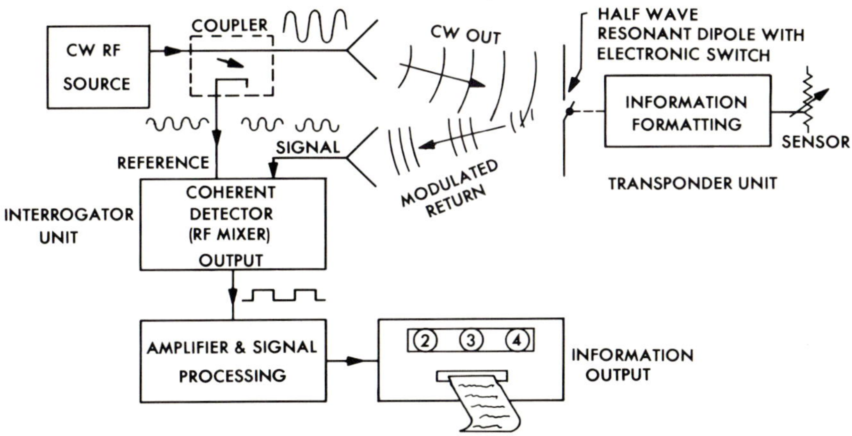

Fig. 1 BASIC BLOCK DIAGRAM

resonant at the incoming rf and relatively large rf currents will flow along the dipole, resulting in a relatively large reradiated or "scattered" rf field, a portion of which is returned to the interrogator.

If the switch is opened, the antenna is no longer resonant and the rf currents will be much lower, resulting in a less scattered rf field. If the switch is periodically opened and closed, the fluctuation in the rf signal returned to the interrogator can be sensed and information sent by varying the pattern of these fluctuations.

The amplitude modulated return signal is intercepted and coherently detected in a mixer, using the cw rf as the reference signal. The mixer output is the modulation envelope, which is then amplified and processed to recover the desired information.

DOPPLER EFFECT INTERFERENCE PROBLEMS

Note that this is similar to a homodyne Doppler radar system, the difference being that the telemetry system recovers the modulation envelope in the returned signal whereas the Doppler radar produces the Doppler shift in the rf reflected off a moving object. The relationship between carrier power, antenna gains, bandwidth, noise figure, and useful range is the same.

Since the telemetry system is so similar to Doppler radar, it will also respond to reflections from objects in the rf field and it is necessary to avoid interference from such responses. The response is dc from stationary objects and low frequency ac from moving ones. Fortunately, the Doppler signal frequency is limited because of the limited velocity of physically moving objects and the Doppler response can be discriminated against if the desired telemetry signal frequency content is placed sufficiently higher than the highest expected Doppler frequency. A good way to do this is to use a subcarrier of at least several kHz, which is then frequency modulated with the information signal.

Another problem, related to the Doppler effect, is caused by relative motion between the interrogator and transponder, which changes the round trip rf path length and therefore the relative phasing between the information and reference signals at the mixer. The mixer output amplitude is proportional to the cosine of this phase angle, and therefore has a variable amplitude, two polarities, and can go to zero. The consequence is that the mixer output will, as the rf phase changes with changing distance, go through zero once every quarter wavelength as the distance is changed.

This can be avoided by using two mixers, as shown in Figure 2. Both mixers have the same transponder return signal, but the reference signals differ in phase by 90° at the subcarrier frequency and are then summed with the other signal. The resulting signal is of constant amplitude.

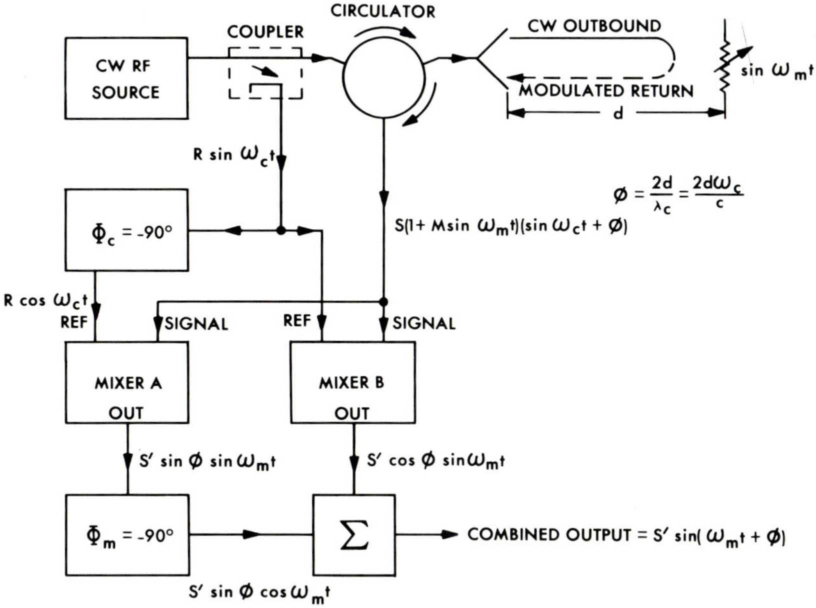

Fig. 2 ANTI-DROPOUT RF FRONT END CIRCUITRY

SYSTEM CHARACTERISTICS AS THEY AFFECT POTENTIAL APPLICATIONS

The biggest advantage of such a telemetry system is that the transponders can be very simple and low in cost. This is so since there is no rf circuitry other than the antenna loading switch. The power needed to modulate the antenna reflectance is very low, as is that for a subcarrier generator. If the power for the sensor and signal processing is also low, it may be possible to get all the required dc by rectifying rf from the antenna, thereby avoiding the need for a battery or other power source in the transponder.

The interrogator is necessarily more complex than a conventional telemetry receiver. This method is therefore most advantageous in situations requiring a multitude of low cost transponders and where it is permissible to interrogate the transponders one at a time by pointing the interrogator beam at them.

The range is relatively limited at practical power levels since the return signal varies as the fourth power of distance. If the transponder is powered by the rf beam, the range is further limited by the need to develop enough dc voltage for the circuits. Practical ranges over a line of sight air path are from several meters for an implanted rf powered transponder to several hundred meters with battery power. The range could be extended by using a directional antenna at the transponder and by reducing the system bandwidth.

APPLICATION

Los Alamos has built a system to monitor the temperature of cattle with subdermally implanted transponders. Figure 3 shows an unpotted transponder with hybrid circuitry, printed circuit antenna, and an rf rectifier that both makes dc for the circuitry and acts as the antenna shorting switch. A system with 6 transponders has been under test since December 1975 at the USDA's National Animal Disease Laboratory in Ames, Iowa to determine the usefulness of temperature data obtained this way.

Temperature data are sent by making the subcarrier frequency proportional to temperature. The frequency is 20 kHz at 40°C. Resolution is 0.1°C and the reading time is under 0.1 second. A range of 2 meters is obtained with 30 watts of carrier at 560 MHz.

A demonstration system with combined temperature and 3-digit identification code should be completed by the summer of 1976. This system, like the temperature-only one, will use hybrid circuitry in the transponder. Future plans include extension to a 15-digit code, which will require a custom integrated circuit.

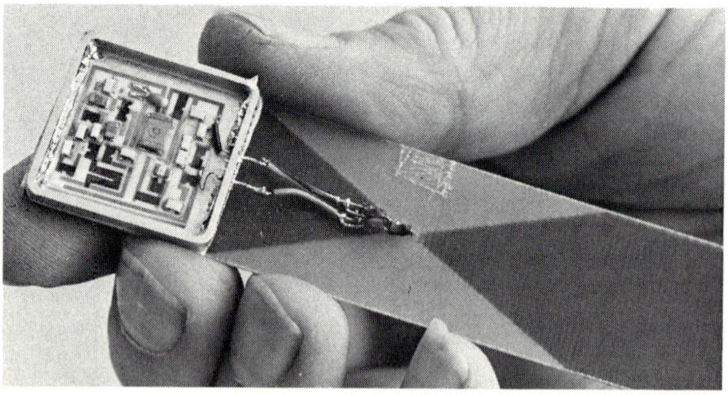

Figure 3 Transponder before encapsulation.

Acknowledgements

This work was supported by the Animal and Plant Health Inspection Services of the U.S. Department of Agriculture and by the Energy Research and Development Administration.

References

1. Koelle AR, Depp SW, Freyman RW: Short-Range Radio-Telemetry for Electronic Identification, Using Modulated RF Backscatter. Proc IEEE 63: 1260, 1975

2. Koelle AR, Bobbett RE, Landt JA, et al: Passive Electronic Identification and Temperature Monitoring System, July 1, 1974 - June 30, 1975. LA-6088-PR, Sept 1975

THE ROLE OF SURFACE ACOUSTIC WAVES AND INFRA-RED COMMUNICATIONS IN BIOTELEMETRY

C. Weller[1] and M.F. Lewis[2]

1. Medical Research Council, Clinical Research Centre, Harrow, Middlesex, U.K. 2. Royal Signals and Radar Establishment, St. Andrews Road, Great Malvern, Worsc., U.K.

INTRODUCTION

Biotelemetry is used to maximum advantage, if, as well as allowing observation of physiological or behavioural states, it enables the observer a measure of control over these factors. When two-way or interactive systems are required, simultaneous radio and optical telemetry obviates some of the problems associated with the use of radiofrequencies alone. Very stable FM transmitters may be fabricated from surface acoustic wave (SAW) devices that are more suitable for biotelemetry than conventional quartz crystal stabilized telemeters.

SURFACE ACOUSTIC WAVES AND FM

Frequency modulation is generally considered the most appropriate telecommunications method for biotelemetry, due to the ease with which the signal may be made independent of the carrier level and to the satisfactory signal-to-noise ratio. As far as small research groups designing telemetry systems for their own use are concerned, there is another important consideration that makes FM attractive, namely the ease with which very simple oscillators may be frequency modulated. Even a cursory search of the literature reveals that there are nearly as many simple FM circuits as there are workers in the field. These elementary telemeters, especially when used in conjunction with subcarrier oscillators, are adequate for many research projects - particularly if the designer is available to carry out minor adjustments. Large deviation ratios and AFC compensate for poor stability. Crystal control is an acceptable alternative, in that narrowband systems that comply with licensing regulations are realizable; unfortunately, transmitter size, power consumption, and cost all increase. A linear frequency deviation entails operating the crystal in the fundamental mode and employing several stages of frequency multiplication, especially if a wideband or multichannel system is required. Alternative methods, such as mixing a high-frequency crystal oscillator with a comparatively low-frequency voltage-controlled oscillator (VCO), involve additional stages as well as sometimes effectively degrading the properties of the crystal. Phase modulation is sometimes used but this does not preserve DC or low-frequency information. A low-cost compact alternative to the conventional VCXO is therefore desirable; SAW devices meet these criteria adequately. Surface acoustic waves were first described by Rayleigh in the 1880's but it was not until piezoelectric crystals and photolithographic techniques were available that practical devices based on surface waves became realizable (1). Acoustic waves are normally generated on the surface of a piezoelectric material by interdigitated transducer (IDT) electrodes whose finger spacing defines the wavelength. There are several applications for SAW devices, which may be relevant to biotelemetry but the most significant is probably the SAW delay line. A delay line consists of two thin film aluminum IDT's formed by photolithography on the surface of rotated Y cut quartz. Propagation takes place along the X axis and the separation of the IDT's determines the delay time. If the delay line is used in the feedback loop of an amplifier, oscillation will occur when the phase shift is an integral multiple of 2 π radians. The frequency of oscillation may be fixed by making the length of one IDT equal to the centre to centre separation of the transducers, or by more complicated designs (see Figure 2 in Ref. 2). For FM it is only necessary to modify the phase shift by a varactor diode, for instance. Stability depends on the cut of the quartz; the Q and the degree of FM (up to 1%) are dependent on the design and layout of the IDT's. The delay line designed at the Royal Radar Establishment for use at the Clinical Research Centre has been designed for high-stability, low-deviation applications (102.36 MHz $\Delta f = \pm$ 10 kHz) and measures 15 x 10 x 1.5 mm. The long-term stability of SAW oscillators is not yet quite as good as bulk quartz, nevertheless it has not been necessary to adjust the frequency of this device during the past two years despite some rough handling. A two-channel transmitter (BP and ECG or bloodflow) complete with 575 mAh battery, voltage regulator, and signal conditioning, using standard components, is housed in a 60x70x30 mm package. The TX will operate continuously for more than 60 hours and has a range of 200m. The production process for SAW devices is very simple; once the mask has been produced, quantity production is assured. Since it is possible to fabricate multiple IDT's on one substrate the method should appeal to manufacturers who wish to produce one unit that is compatible with the telemetry regulations of several countries.

I R COMMUNICATIONS

For some time now infrared (IR) radiation has been used in the biomedical field for applications such as plethysmography and optical isolators. However, recent improvements in photodetectors allow large areas to be flooded with IR light as a practical alternative to radiofrequency radiation. Many would-be users of modulated light as a communications medium have a scepticism stemming back to their schooldays--light travels in straight lines and is therefore only suitable when the transmitter and receiver are fixed line-of-sight installations. Light does of

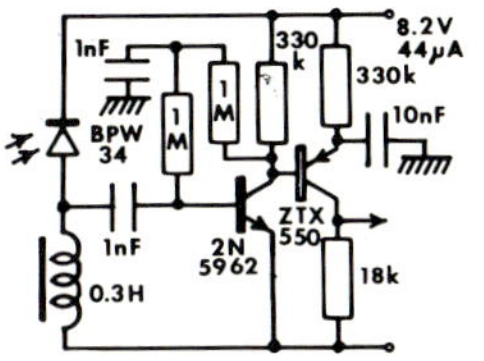

Figure 1. Shows a suitable receiver circuit consisting of one or more tiny photodetectors followed by a simple two-stage amplifier (A=1500) drawing only 44μA. Various encoding techniques may be used in conjunction with such a receiver for multifunction or selective control.

course travel in straight lines but whereas the reflections that occur at the radiofrequency end of the electromagnetic spectrum cause interference and nulls, optical reflections can be used to advantage as the interference patterns are small in comparison with the dimensions of the sensor. One has only to stand in a room illuminated by a low-level point source to appreciate this; there are very few areas of the room that cannot be seen, i.e. light from the lamp reaches the remotest corner and returns to the observer's retina, even though there may be obstacles between the lamp and the observed area. If this is true with visible light, it is even more so with IR radiation, because surfaces that do not reflect much visible light, such as matt black, act as good reflectors for IR. Clearly the IR radiation will not pass through opaque objects and therefore IR telemetry is restricted by the size of the room. There are situations where this restriction is a positive advantage, but if a wider or selected area coverage is required multiple emitter/receiver stations may be strategically sited. IR methods are ideal for telestimulation or control purposes especially when simultaneous radiotelemetry is required - "blocking" of preamplifiers and receivers by the stimulating transmitter is circumvented.

Even receivers intended for model aircraft control are bulky, more prone to spurious operation, and consume more power than their optical counterparts; the range of RF systems is much greater, but where experiments are confined to laboratories the advantage is negated (figure 1).

IR transmitters are also very straightforward, consisting of an array of GaAs luminescent diodes usually operated in a pulsed mode from suitable encoders. The low efficiency of emitters (5%) suggests IR communications are more appropriate when the transmitter is the fixed station and the receiver is mobile. Nevertheless there are exceptions; short-range communications from within diving bells where standing waves cause signal drop-out or in electrically-noisy environments where the photodetectors do not respond to RFI, are typical examples. Such a system has been developed for ECG monitoring from within a whole body calorimeter (basically a thermally-insulated metallic enclosure). The transmitter consists essentially of a preamplifier, a pulse position modulator, and two arrays of LED's (LD 241). The average pulse separation is 1.0 mS and the emitters are activated for only 10μS giving a duty cycle of 100:1. The diodes are arranged in two small clusters of three and attached to the patient's clothing (one on each shoulder). The peak diode current is 400 mA and a 575 mAh battery (8.2v) powers the device (60x65x20mm) for at least 48 hrs. The receiver consists of three BPW 34 photodetectors mounted on three faces of a tetrahedron and an inexpensive IC amplifier (LM 382N) together with a time-to-voltage converter. With the detectors mounted in the centre of a room (6x6m) the signal is independent of the patient's position. If a room of these dimensions can be flooded with IR when there are severe restrictions on the radiated power it will be appreciated that without such limitations much larger areas may be readily illuminated.

Where small animals are involved several photodetectors may be required to form a diverse receiving system. Simply combining the outputs of three or more diodes equally spaced around a collar prevents null positions when the animal rolls, etc. - very much less complex than a RF diverse system, as the diode outputs are scalar quantities. Fluorescent lighting can cause interference; the BPW 34 photodetector peaks at 850 nm but has a broad spectral response. Many fluorescent tubes generate noise that falls within this band so we have incorporated a simple gelatin filter (Kodak No. 87C) that virtually eliminates this problem. An inductive load prevents strong sunlight or filament lamps from causing saturation.

Wideband information may be handled without infringing licensing regulations; an important consideration where telemetry is to be used by the general public for applications such as the control of domestic appliances by the disabled.

REFERENCES

1. Smith WR, et al: Design of surface wave delay lines with interdigital transducers. IEEE Transc on Microwave Theory and Techniques. MTT-17, 1969 pp 865-873

2. Lewis MF: SAW devices and applications No. 6. Oscillators - the next successful SAW device? Ultrasonics, May 1974, pp 115-123

BODY-MOUNTED TRANSMITTING ANTENNAS:
RADIATION PATTERNS AND DESIGN OF HELICAL DIPOLE ANTENNAS

Peter A. Neukomm

Laboratory of Biomechanics (Head: Prof.Dr.J.Wartenweiler)
Swiss Federal Institute of Technology, Zürich

INTRODUCTION

Continuous data transmission is required in monitoring athletes and patients. In order to minimize the encumbrance of the test subject TS, a small antenna has to be mounted close to the trunk of the TS. The radiation pattern should be omnidirectional with only a minimum attenuation of the far field strength in the equatorial plane. Depending on the frequency f and the chest-antenna distance d, the human body acts like a reflector, an absorber and a shielding. The aim of this investigation was to study these effects with an unchanged telemetry dummy, varying $25 < f < 1000$ MHz and $35\,mm < d < 4$ m.
Small telemetry devices often have an insufficient earthing plane. Combined with a small antenna with a low radiation resistance, the VSWR is excellent, but the radiation is weak. A helical dipole antenna is small in size, needs no counterpoise and can be easily matched on a $50\,\Omega$ coaxial cable. The helical dipole is elliptically polarized and is an efficient radiator. Mounted on the TS, both vertical and horizontal polarizations are strong and allow the TS to rotate on more than one axis using a single vertical dipole receiving antenna.

THE HUMAN BODY AS A REFLECTOR AND ABSORBER

It is known (1) that the body has a negative influence on the azimuthal radiation pattern, if the antenna is close to the trunk and if f exceeds 100 MHz. Data obtained from diathermy researchers (2,3) allow the following predictions: a body with a radius $a = 15$ cm, irradiated with a *plane* wave, will absorb a large amount of the incident power, if $130 < f < 500$ MHz, because the Relative Absorption Cross Section RACS, reaches its maximum (2,4). The second prediction is that horizontal polarized waves will be reflected from biological tissues. The reflection coefficient R amounts to $0.4 < R < 0.9$ and the phase-shift ϕ to $169 < \phi < 177^0$ in the regarded frequency range (3).

Method of Measurement

A wooden transmitting antenna holder (1.16 m high) and a cross LPD receiving antenna (6.5 m high) were installed on a large lawn. Ground reflections could be reduced by using vertical polarized antennas and choosing a distance (31 m), where the Brewster angle (14^0) occurs. Six short transmitting antennas (T_{ant}) with the following resonant frequencies were applied:helicals of 180,300,360 MHZ; and whips of 500,630, 1060 MHz. The telemetry dummy was a copper can with a diameter of 70 mm and a height of 100 mm. The antenna was fed by a long coaxial cable. Waves propagating from the cable surface could be eliminated by forming the cable under the can in a shape of a matched HF-chock coil. Load variations for the generator could be prevented by a 20 dB attenuator,built into the can. The whole feed line,the choke included,was imbedded in absorber tubes. The vertical polarized field strength was recorded after checking the free-space values with a 50 Ω terminal,different antennas and chokes,and in both polarization axes,in order to obtain a signal/ noise ratio of better than 20 dB. The test subject TS (173 cm tall 94 cm chest measurement) moved slowly towards the Tant on the axis 0° (TS between the antennas) and on the axis 180° (TS behind the antennas). The distance d (chest surface to the Tant) was recorded from 35 mm up to 4 m by means of the rubber band goniometry (5) with an accuracy of 2 cm.

Results

The reflection maxima (see Fig.1 on the top) occurred in the range $50 < f < 1000$ MHz at $0.7\ \lambda/4 < d < 1.1\ \lambda/4$. The reflection minima occurred at the double distance. Starting with 50 MHz, the reflection amplitude

BODY-MOUNTED ANTENNAS: RADIATION PATTERN

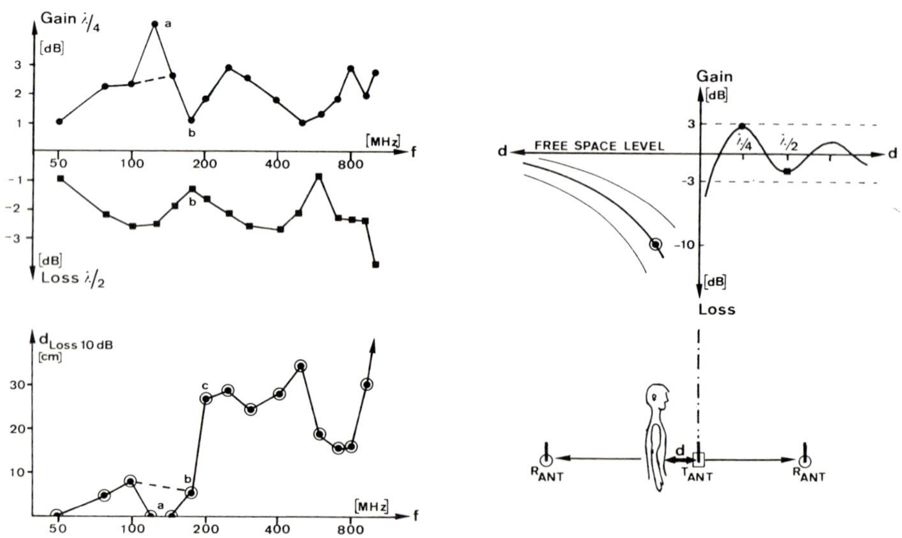

Fig.1. Influence of the distance d (chest to the transmitting antenna) on the far field strength at vertical polarization.
Top left: Reflection amplitudes, test subject behind the antennas.
Below: Distance d required for a maximum 10 dB attenuation, test subject between transmitting and receiving antenna.

increased up to 125 MHz. The body acted as a capacitive coupled resonant antenna, if $125 < f < 150$ MHz (especially with small d). In the region of 175 MHz, the echo was small, evidently due to the high RACS. The reflection increased between 175 and 300 MHz, decreased at 400 to 600 MHz, and increased again above 600 MHz. It seems that the body is like a thick reflector rod below 300 MHz, changes to a rough plane (reflections from the chest and the arms with phase differences) at 400 to 600 MHz, and becomes a large plane above 600 MHz. The field attenuation behind the subject should be treated by field computation. The following hypothesis can be made: Below 100 MHz, the body is transparent due to the small body dimension/wavelength ratio (RACS< 0.1). Between 200 and 500 MHz, the body is an efficient absorber (RACS is approx. 1). In the range 600 to 800 MHz, resonances of E- and H-mode may occur, and above 800 MHz, the body is a shielding plate.

AZIMUTHAL RADIATION PATTERNS IN THE EQUATORIAL PLANE

The same transmitting antenna equipment was mounted on a revolving stage and rotated together with the TS. The minimum vertical field strengths E_{vmin}, normalized at 0 dB = free space antenna, are given in Table 1 for two different distances d. The minimum did not always occur at 0^0, but at approx. $\pm 45^0$ azimuthal angle at some frequencies. With d decreasing from 77 mm to 35 mm, the losses increased by about 1- 8 dB. In practice, small antennas with a narrow bandwidth will be detuned by the body material, causing high additional losses.

Table 1. Minimum normalized vertical polarized field strength

Frequency f (MHz)	25	75	100	150	200	300	400	500	600	700	800	900
E_{vmin}, d=35 mm (dB)	0	+2	-4	-18	-20	-23	-20	-22	-28	-23	-21	-29
E_{vmin}, d=77 mm (dB)	0	+4	-7	-14	-17	-21	-19	-18	-20	-21	-18	-24

HELICAL NORMAL MODE ANTENNAS

A helical monopole antenna consists of a helical wound wire. The main advantages are the small size and the elliptical polarization. The theory and the computer methods may be found in the references (6) to (11). The most important parameters of a helical antenna are:

λ	(cm)	: wave length in free space
h	(cm)	: physical length of the helix axis (vertical)
D	(cm)	: diameter of the helix coil (horizontal)
p	(cm)	: pitch of the helix
N	(1)	: total number of turns
n	(1/cm)	: turns per cm vertical length

Depending on the size of the helix, different radiation mode are possible. Maximum radiation in a direction *normal* to the helix axis (normal mode) will occur, if $D, p \ll \lambda$ and $h < \lambda$. Such a helix may be regarded as N loops and a whip of $N \cdot p = h$, producing an elliptical polarized E-field. The ratio of E_v/E_h in the far field at zero elevation angle is $2p\lambda/\P^2D^2$. Since the phase velocity along the helix axis is reduced, a resonant $\lambda/4$ antenna can be made with $h < \lambda/4$. For $h > 3D$, n and $N = n \cdot h$ may be calculated quite accurately with this formula:

$$\log n = 0.4 \left(\log\left(\frac{\lambda}{h} - 4\right) + \log\left(\frac{\lambda}{h} + 4\right) + 0.5 \log \lambda - 3 \log D \right) - 1$$

The radiation resistance of the resonant helical antenna above perfect ground amounts to $R_r = (25.3\ h/\lambda)^2$ and is relatively small. A helical dipole antenna for 231.5 MHz was calculated (9,10,11) with the following features: 2h = 22 cm, D = 2 cm, 2N = 18, wire diameter 2 mm. The calculated efficiency was 94 %, the E_v gain was +1.4 dB and the E_h gain - 17 dB. The helix was fed over the four center windings and matched to the 50 Ω coaxial cable by a $\lambda/4$ compensating stub. The measured VSWR was 2:1, the - 3 dB bandwidth 17 MHz and the free space values were E_v = 0 dB, E_h = - 10.5 to - 21.5 dB (0 dB = isotropic radiator). Mounted dorsally (resp. on the chest) on the TS with d = 5 cm, E_v varied from 0 to - 15 dB (resp. - 2.5 to - 18 dB) and E_h varied from - 4 to - 32 dB (resp. - 11.5 to - 20.5 dB). Better results were achieved with larger d. The helical dipole antenna is a good solution, if a small antenna must be mounted close to the trunk of the TS, if the counterpoise of the telemetry case is small, and if the TS rotates on more than one axis.

ACKNOWLEDGMENTS

The author wishes to thank W.Oehen and N.Balzarini for their computer analysis; R.Graner, M.Knaute and his team of the Division of Military Aerodromes for the measuring equipment and assistance; and all who contributed to these investigations.

REFERENCES

1. Buchanan H, Moore WF, Richter CR: Human body effect on signal patterns of personal telemetry transmitters. SAM-TR-70-4, USAF School of Aerospace Medicine, Brooks Air Force Base, Texas, 1970
2. Schwan HP: Microwave radiation: Biophysical considerations and standards criteria. IEEE Trans Biomed Eng BME 19: 304, 1972
3. Johnson CC, Guy AW: Nonionizing electromagnetic wave effects in biological materials and systems. Proc IEEE 60: 692, 1972
4. Neukomm PA: Health hazards from telemetry RF exposure ? Proc 3rd Int Symp on Biotelemetry, Asilomar, Ca. May 17-20, 1976
5. Neukomm PA: The rubber band goniometry. J Biotelemetry 1:12, 1974
6. Kraus JD: Antennas. New York, McGraw-Hill, 1950, chapt 7, pp 441
7. Kandoian AG, Sichak W: Wide frequency range tuned helical antennas and circuits. IRE Nat Conv Rec, part 2:42, 1953
8. Tong DA: The normal-mode helical aerial. Radio Commun: 432, 1974
9. Swift WD: Numerical analysis of normal mode helical dipole antennas. Diss Jowa St Univ. 1971, Mic film Ann Arbor H 01:72-12 600
10. MBA Information service: Antenna Modeling Program, System Manual IS-R-72 110, 1973. Nat Tech Inf Serv, Springfield, AD 767 420
11. Oehen W, Balzarini N: Helix Biotelemetrie Antennen. Study work at the Lab Microwave Tech and Lab Biomechanis, ETH Zürich, 1975

TIME-SEQUENCED TELEMETRY FOR THE TRANSMISSION OF BIOMEDICAL DATA FROM DIFFERENT STATIONS OPERATING ON THE SAME CARRIER FREQUENCY

Hans-Joachim Klewe

DFVLR, Institute for Flight Mechanics
Braunschweig/Germany

For telemetering data from different persons or places to a central receiving station several methods are known. For example, one telemetry transmitter operating on a different carrier frequency must be attached to each of the different persons involved, and at the receiving station a corresponding number of receivers are provided with each tuned to the corresponding frequency. This method requires as many transmitters and receivers as there are persons. The number of receivers can be reduced by tuning one single receiver in periodical sequence to the different transmitters. The continuous operation of the transmitters in both mentioned cases brings up problems concerning energy supply of the generally battery-operated transmitters. There exists further the possibility, to switch on and off a number of transmitters operating on the same frequency by telecommand given by the central receiving station. This permits the operation of all transmitters on one carrier frequency and furthermore to lower the power consumption of the transmitters. However, there must be attached a telecommand receiver to each telemetry transmitter, which adds weight and bulkiness to the entire arrangement. Moreover, this telecommand-receiver must be continuously operative and adds therefore to power consumption. Since telemetry plays, in general, an essential role in solving problems arising with rescue and recovery systems, which is a concern of the Institute for Flight Mechanics, the following mentioned problems were to be solved (1).

There were two problems concerning the telemetering of data from different persons to one receiving station and a solution which fits to both problems. Within a distinct flight-medical investigation the rectal temperature of a number (about 10) of persons had to be measured in short intervals over a three week period and telemetered to a central receiving station. The following requirements were given: telemetry coverage about 50-100 m inside a building, e.g. a hotel; low power consumption, and, most important, expenses should be as low as possible. A similar problem was to telemeter the heart rate of several persons in sequence periodically to a central receiving station. Since the first mentioned state of the art for telemetering data from several persons to one receiving station was not to be accepted due to high expenses, a time sequenced telemetry for telemetering data from different persons using transmitters operating on the same carrier frequency has been proposed. The principal idea was, that all transmitting stations were to be switched on at the beginning of the experiment with a required time-displacement and thereafter switched on and off in the periodic time sequence which has been given. For example, in telemetering rectal temperature from 10 persons a period between two measurements of 20 min. and a duration of transmission of 30 s was prescribed. At first, the experiment station (transmitter) 1 is switched on. It transmits data for 30 s and will be switched off automatically by the timer circuit. 1 min. 30 s later. i.e. 2 min. after starting the operation of station 1, station 2 will be switched on, which also transmits 30 s and then stops transmitting by automatic time control. This operation continues until all 10 stations are switched on in the proper sequence. At the end of the first 20 min. cycle, station 1 will be automatically switched on again and transmits data for 30 s. After 22 min station 2 will be switched on automatically and transmits also 30 s etc., until the 2nd 20-min. cycle has ended. This operation continues as long as required or until the batteries are emptied.

TIMER CIRCUIT

The requirement for a timer circuit as described, is that long-time-constancy is sufficient and the power consumption of the circuit is low enough that it will not add substantially to the power consumption of the entire arrangement. A timer circuit such as described became possible when modern integrated circuits for quartz-controlled clocks existed in CMOS-technique. So as an oscillator for the timer circuit a CMOS-IC of the Type SAJ 300 S (ITT) was selected (2). This IC contains an oscillator circuit together with a 21-step programmable frequency divider. The divider-ratio can be adjusted to the required value with an accuracy of 1×10^{-6}. At an oscillator-frequency of 4.194812 MHz, the output of the circuit supplies a symmetrical rectangular signal with a duty cycle of 0,5 and a frequency of 0,5 Hz. The closest adjustment-step is 1,9 ppm, whereby an accuracy of 10^{-6} becomes possible.

So a time-accuracy of $\pm$ 2,5 s/month (i.e. 30 s p.a.) can be achieved. The output-pulses of the SAJ 300 S will be supplied to a further CMOS programmable divider by "N" Counter. The divider-ratio N of this IC CD 4059 AD (RCA) (3) is between 3 and 15.999, programmable by outside wiring. With the beat-frequency of 0,5 Hz given, programming of the divider output-control pulses with time sequences of 6s up to 8 h 53 min can be achieved. The control-pulse from the divide by "N" counter is fed to a CMOS low-power monostable multivibrator CD 4047 AE (3), which operates a Reed-Relay PRME 15.005 that serves to switch the telemetry equipment on and off.

TELEMETRY EQUIPMENT

As already mentioned, the entire arrangement should be established at low cost. Therefore microminiature telemetry modules are

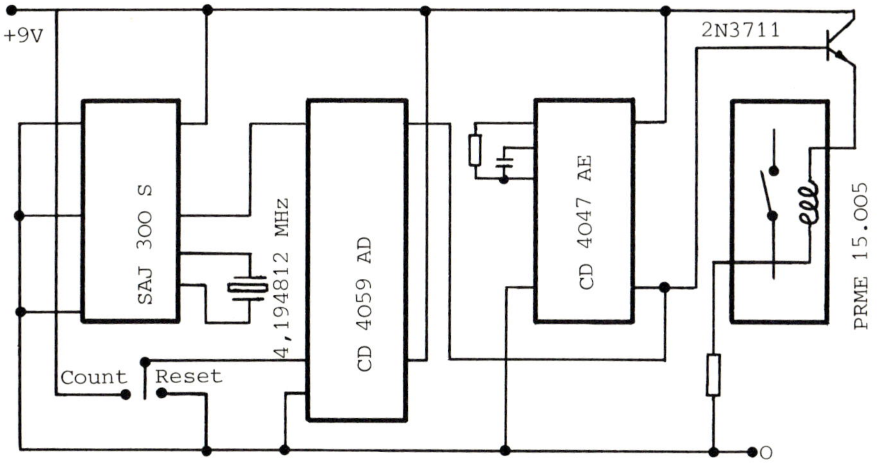

Figure 1. Block-diagram of the timer circuit

used which are prepared for the use in ballistic missiles (4,5). With those modules a one-channel FM/FM-telemetry station was built to operate on a subcarrier frequency of 7,35 kHz (IRIG-Proportional Channel 11) and a VHF-carrier-frequency of 105 MHz. A stabilizer stage KS 416, a VCO KS 413-a,a modulator circuit KS 415, an RF-oscillator KS 358,and an RF-buffer-amplifier KS 435 were used.

Figure 2 shows an embodiment of a time-sequenced telemetry station for telemetering heart rate. In the lower part of the picture is the timer circuit,and on the right-hand is the proper telemetry equipment. In the upper left-hand part, the power supply can be seen. This consists of two alkali-manganese batteries (Mallory MN 1604) with a voltage of 9,0 V and a capacity of 525 mAh. The apparatus has the dimensions 100 x 80 x 30 mm and weighs 300 gr. At its frontplate there is an On-Off-switch and a further switch,which in position T provides constant operation, e.g. to enable tuning or adjustment,and which in position R resets the timer circuit to zero.

From the two embodiments of time-sequenced telemetry described,models have been built for laboratory purposes. The embodiment for measuring rectal temperature has been operated under laboratory conditions for a longer time-period. By periodic operation of 30 s duration with 20 min intervals, an operating time of about 17 days can be achieved while discharging the batteries down to 5,6 V.

The achievable telemetry coverage of the transmitter, which transmits 30 mW RF power, corresponds to the requirements mentioned. The embodiment shown in Fig.2 for telemetering heart rate has a different period and duration of operation. Heart rate is telemetered 15 s with intervals of 2 min. Thus, an operating time of about 80 h can be achieved for heart-rate telemetry.

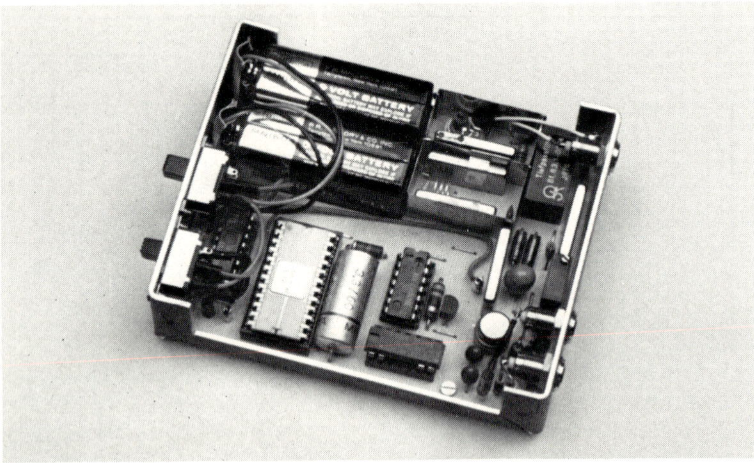

Fig. 2 Time sequenced telemetry for telemetering heart rate

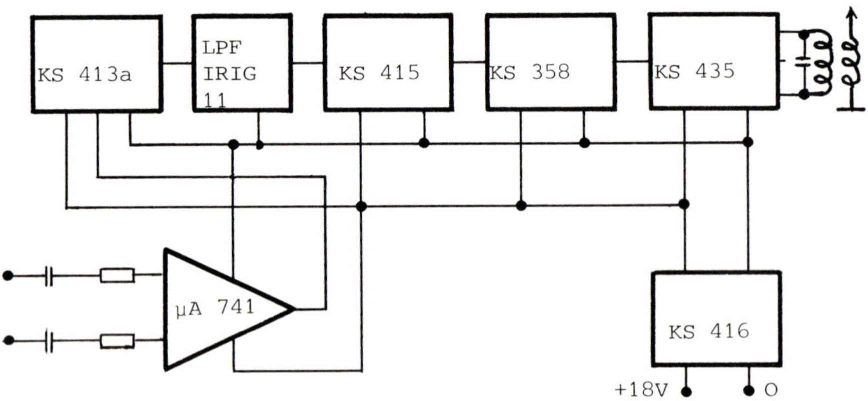

Fig. 3 Block-diagram of the telemetry equipment for telemetering heart rate

REFERENCES

1. Klewe, H.-J.: Der Einsatz von Telemetrie- und Telekommandoanlagen in der Meßtechnik bei Rettungs- und Bergungssystemen DLR-Mitt. 74-36, 1974
2. Intermetall Semiconductors ITT: Integrierte Schaltungen und Einzelhalbleiter 1975/76
3. RCA: Solid State 74 Datenbuch Serie SSD-203 B
4. AEG-Telefunken: Industrie-Telemetrie, Bausteine in Präzisions-Dickfilmtechnik
5. Schulze, G., Rateau, Ph.: Deutsch-Französisches Forschungsinstitut Saint Louis, Technischer Bericht RT 9/70

PREPROCESSING OF PULSE CODE MODULATION TELEMETRY DATA

M. L. Hull
C. D. Mote, Jr.
L. W. Lamoreux

Department of Mechanical Engineering
University of California, Berkeley, California

In field applications involving telemetry, it is often not feasible to analyze data in real time at the receiving site. A bulk storage device is required so the data can be subsequently reduced. Pulse code modulation (PCM) data analysis is only realistic on a digital computer and data must be transferred from the field storage device to the computer. This paper discusses the system in Fig. 1 which transferred serial PCM data recorded on analog tape in the field to a CDC 6400 computer compatible tape.

The PCM data, transmitted and recorded in the bi-phase (Bi-ϕ) L format, were collected from field experiments in snow skiing and consisted of a number of approximately 30 second runs sequentially recorded on a 7 track, 250 kHz Sanborn instrumentation recorder. The 16-word data frame shown in Fig. 2 was comprised of 13 data channels, 2 frame-counter words, and a sync word (word with a constant value for synchronization of data frames). Each 12-bit word was sampled 520 times/s so that the data transmission rate was ~ 100 kbits/s.

Serial bi-phase data, reproduced from the analog recorder, were played into a custom built PCM decoder. The decoder design is discussed in detail in Refs. (2,3). As the data were decoded, each word was output along 12 parallel (bit) lines to the data interface in the Nova minicomputer. The PCM decoder provided status and control commands along three additional lines to drive the minicomputer interrupt logic. On one control line, a 1 μsec <u>frame pulse</u> denoted the beginning of each new data frame; on another line, a <u>word pulse</u> signified a new word. On the third <u>sync level</u> line, a transition from +4.0 V, or logical true, signified a synchronization loss arising from an information dropout. A circuit in the decoder monitored the sync words and controlled the level transition. If three consecutive sync words were 16

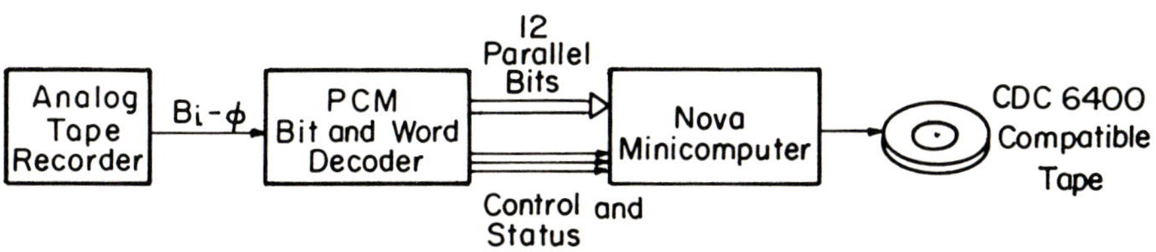

Figure 1: Data Handling System

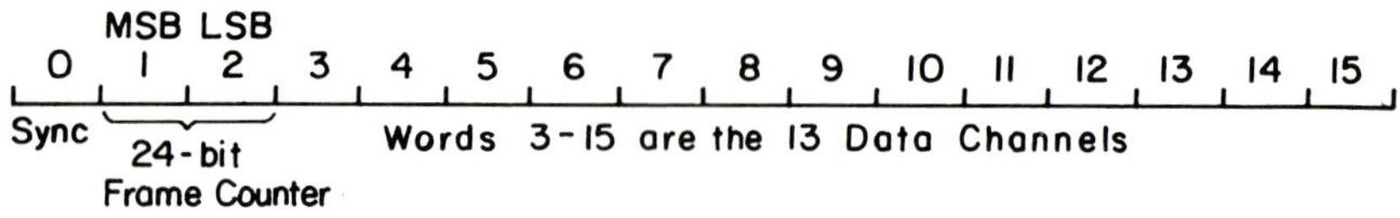

Figure 2: 16 Word Data Frame

words apart, then the circuit was synchronized (true); if three consecutive sync words failed to appear or were improperly timed, then a lock-loss (false) was registered. Therefore, the minimum information loss was six data frames.

A Nova (Data General Corp.) minicomputer was used to buffer the data, check the timing of the control commands, and write CDC 6400 compatible tapes. Word pulses were counted and checked against arriving frame pulses. Any word count different from 16 was a fatal error. As data were accepted, they were stored in one of two 1500-word buffers which were alternately being loaded with data and written onto CDC 6400 tape. The sync word (word 0) at the beginning of each frame was not stored. The CDC 6400 utilizes a 60-bit word so that the fifteen 12-bit words comprising a data frame were written onto digital tape in three 60-bit groups. Using software developed for the CDC 6400, data tapes were ultimately rewritten so that data frames consisted of fifteen 60-bit words each with 12-bits right justified. When synchronization was lost, storage of the data ceased while the machine waited for a sync level true. During this period, the computer backed up three frames and began storing from that point since those frames contained faulty information. The data transfer rate was 50 kbits/s.

Several program features were added to facilitate the data transfer. The program was controlled through two teletype commands which initiated or terminated data storage. Any two data channels could be displayed whether or not the initiate-data-storage command had been entered. In this way data were scanned so that only those sections of interest were written onto digital tape. Computer costs were reduced because the volume of data was minimized. A final software package was written to handle the tape. The data transfer was checked by playing back the data tapes and displaying channels.

Information losses were occasional and stemmed from marginally adequate equipment and nonoptimal telemetering techniques. One loss was due to the receiver becoming detuned when RF reflection patterns changed as the skier moved down the mountain. The duration of these losses typically ranged from .1 to .3 seconds. Another loss was accountable to the Sanborn instrumentation recorder. The bi-phase waveform (serial PCM data recorded @ 100 kHz) was degraded by the recorder frequency response (3 db down @ 250 kHz) and synchronization was lost during data playback when the degradation became too severe. These data losses were usually ~ .02 s so that only 6 to 10 data frames were lost.

Recall that the minicomputer ceased storing data when information was lost and began when data again became valid. A data discontinuity or time base shift at the loss point resulted (Fig. 3b). If the data sampling interval is Δt, then for continuous data

$$t_{i+1} = t_i + \Delta t \tag{1}$$

where the i represents data points on the digital tape and t is the absolute time. In the case of an information loss of say 50 frames (~ .1 second)

$$t_{i+1} = t_i + 50 \cdot \Delta t \tag{2}$$

This discontinuity is displayed in the 2 Hz sine wave of Fig. 3.

Fourier coefficients computed over data segments with discontinuities do not accurately give the frequency content of the original data. Fig. 3 compares the power spectral density estimates for the signal waveforms shown and illustrates the inaccuracy. These power spectrum estimates were obtained following a procedure outlined in Ref. (1). Fourier coefficients were computed via the Cooley-Tukey Fast Fourier transform (FFT) method with a data sampling rate of 1024 points/s. For the discontinuous waveform, zeros were added at the end so that the record lengths for analysis were equal to .5 s. The dropout spectrum exhibits a 25% reduction in power at the 2 Hz coefficient and a

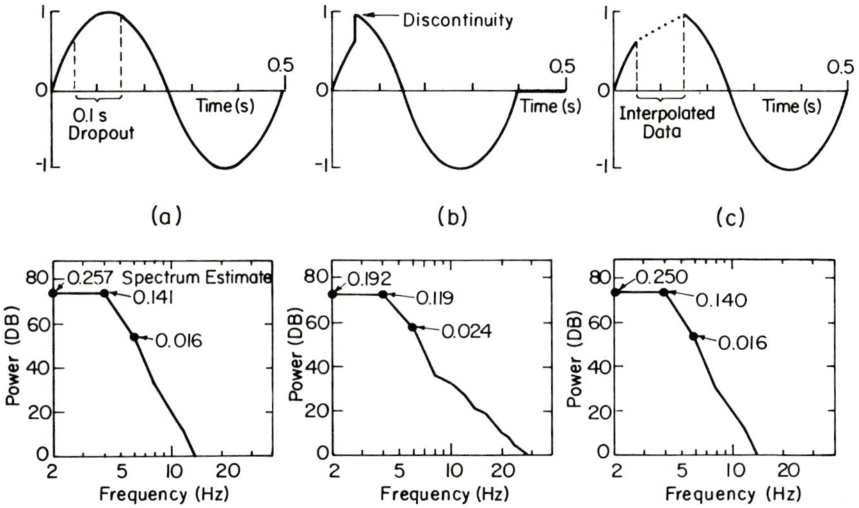

Figure 3: Signal Enhancement for 2 Hz Sine Wave Cycle with .1 s Dropout and Comparison of Power Spectra. a) actual data; b) minicomputer processed data; c) reconstructed data.

bandwidth increase from ~ 15 Hz to ~ 25 Hz. The bandwidth increases due to high frequency components in the discontinuity. Clearly, interpreting such results may lead to erroneous conclusions and it is desirable to minimize errors.

Errors were minimized by reconstructing the original time base via a software technique coded in Fortran. After data had been read by the CDC 6400, dropouts were monitored by checking for sequential counts in the least significant frame counter word. A result of +1 caused storage of a single frame and the next count was checked, etc. Counts differing from +1, except for a counter turnover, originated an information loss. The number of lost frames was computed from the word count difference. Lost frames were recovered by storing data word values linearly interpolated from the data bounding the loss. Fig. 3 shows the reconstructed 2 Hz waveform and the improved accuracy in the power spectrum estimates. Note that if a dropout is longer than half the period of any frequency component, then interpolating cannot enhance the accuracy of the wave form and the error of the corresponding spectrum estimate may be large. Data intervals with either a high dropout density or a long duration (> .3 s) dropout were not analyzed.

The data preprocessing system features ease of operation, versatility, and flexibility. Much higher telemetered bit rates can be handled by using an analog recorder with compatible frequency response and reducing the data playback speed. The data transfer capability is dependent on the minicomputer cycle time rather than the writing capabilities of digital tape machines. Minicomputers such as the PDP-11 T55 are available with cycle times 6 times greater than the Nova, so that the data transfer rate could be faster than 50 kbits/s. Different data frame lengths and word sizes can also be accommodated through various degrees of software modification depending on the computers.

ACKNOWLEDGEMENTS

We are indebted to the United States National Science Foundation and the Veterans' Administration for financial support. We also thank Professor Forrest S. Mozer and Mr. Henry Heetderks of the Space Sciences Laboratory, University of California, Berkeley, for designing the PCM encoder and decoder.

REFERENCES

1. Bendat, J. S., and Piersol, A. G., Random Data: Analysis and Measurement Procedures, Wiley-Interscience, New York, 1971, pp. 286-330.
2. Hull, M. L., and Mote, C. D., Jr., "PCM Telemetry in Ski Injury Research: I. Instrumentation," Biotelemetry, Vol. 1, No. 4, 1974, pp. 182-191.
3. Hull, M. L. and Mote, C. D., Jr., "Skiing Injuries: Field Loading and Analysis," Dept. of Mech. Engr., Univ. of Calif., Berkeley, Nov. 1975, 116 pp.

PACE-O-TELL

A TELEPHONE-TELEMETRY SYSTEM FOR THE FOLLOW-UP OF PACEMAKER PATIENTS BY TELEPHONE. MEASURING THE PACEMAKER ACTIVITY AND INTERVAL TIME, AND THE STIMULUS RESULTS AND HEARTRATE OF THE PATIENT'S HEART

A. Vliegenthart[1], H. Bakema[2], and K.H. Go[2]

1. Research Department, Aryans Electronics, DeBilt, Netherlands
2. Medical Electronics Department of St. Antonius Hospital, Utrecht, Netherlands

ABSTRACT

The system described in this paper is the result of more and more coming wishes of pacemaker centers to decentralize the routine controls to the patient's home,and using the public telephone as a direct connection to the pacemaker center to overcome over-population at the pacemaker center and to minimize traveling discomfort and costs for the patients.

The whole system is based on a small hand-held transmitter placed on the patient's thorax, and a receiving unit located at the pacemaker center. The patient is in direct voice contact with the operator at the center. The transmitter generates a toneburst of 2000 Hz for each detected pacemaker spike and a toneburst of 1600 Hz by each detected ventricular ECG. If the transmitter detects a pacemaker spike directly followed by an ECG complex within 200 mS, a double tone is generated. The transmitter is acoustically coupled with the telephone handset. The receiver detects the transmitted tonebursts after filtering, gain control, and noise blanking in a digital circuit into what we call "Traffic light Philosophical Displays" which means: RED is danger, ORANGE is take care, GREEN is safe. Translated to our system this philosophy gives us the following display modes: a) RED only = no stimulation, b) ORANGE only = pacemaker in on demand, patient's own rhythm is displayed, c) GREEN only = patient's heart is normal, triggered by the pacemaker, d) RED is followed by ORANGE at undefined times = failure of pacemaker sensing circuit, bad wire contact, or competition between pacemaker and patient's own ventricular stimulation.

The SS interval in mS of the pacemaker and the RR interval in P/min. of the ventricular rhythm are displayed with a digital tachometer. The received pacemaker spikes and ECG's can also be recorded on a pen recorder as event marks.

Keywords: a) Pacemaker spike--shape and amplitude from implanted pacemaker on patient's thorax, b) PCG--stimulus results after stimulation by the implanted pacemaker, c) ECG--patient's own rhythm (ventricular stimulation).

DESCRIPTION OF PACE-O-TELL SYSTEM

What kind of signals can be expected from the patient?

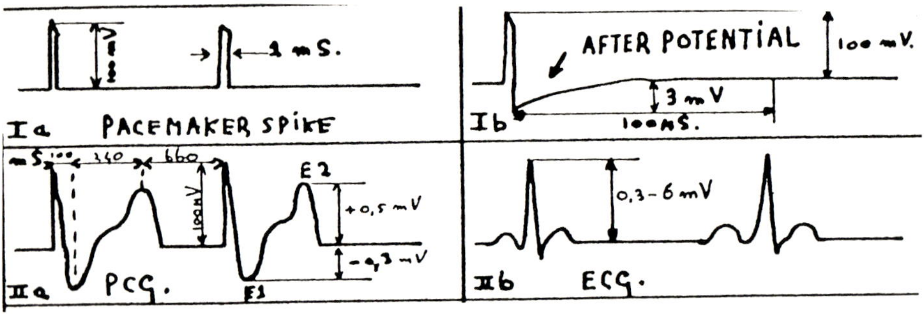

1. The amplitude of the pacemaker spikes are 3mV - 200mV (Fig Ia). 2. The PCG's and ECG's on the patient's thorax can have amplitudes between 0.3mV and 6mV (figure IIa and b). 3. There are great differences between the fall and rise times of the pacemaker spike, the PCG, and the ECG (figure I and II). 4. The pacemaker impulse has rise and fall times of 1ns and a width of ± 1 mS, but also a repolarization effect with an amplitude of ± 2-3mV and a time constant of 100 mS-unity with about the same fall time as can be found in the PCG's and ECG's (figure I b). This effect is called "after potential".

SHORT DESCRIPTION OF THE TRANSMITTER (Figure III).

After amplification the signal is rectified and split into PCG/ECG and pacemaker spikes. The differentiated and amplified spikes trigger one shot OSI, which generates a pulse of 20 mS. OSI triggers OSII, which activates the 2000 Hz osc. during 200 mS and, via the mixing amplifier, the loudspeaker.

During the 20 mS of OSI, S1, S2 and S3 are blocking the input amplifier and the bandfilter, to avoid triggering of the PCG/ECG detection system through the "after potential", and through overdriving. The PCG/ECG detection takes place via the clipper, band and lowpass-qrs filter, schmittrigger, and oneshot OSIII, which activates the 1600 Hz osc. during 200 mS, and,via the mixing amplifier, the loudspeaker. Oneshot OSIII, closes during its on-period of 200 mS S1, S2, S3 and S4, after triggering by a detected PCG/ECG. The time of 200 mS will be lengthened to 500 mS by means of S4, R22, C11, which forms an extra oneshot.

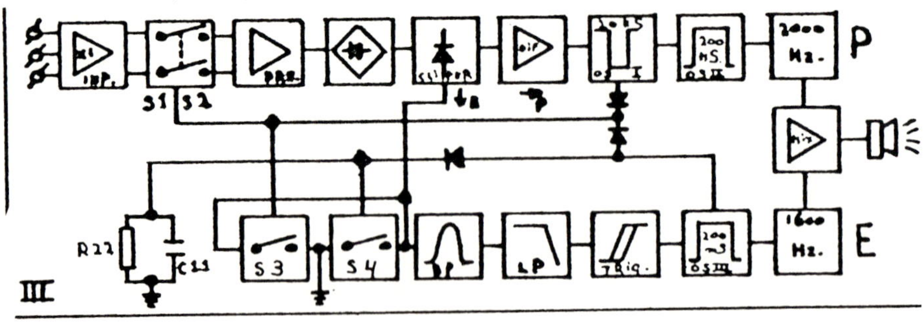

The total block time of 500 mS after triggering gives a better assurance against muscle and motion artifacts and double triggering through the E2 pulse of the PCG. During pacemaker-stimulated heart muscle contraction, the transmitter is generating a frequency mixture of 2000 and 1600 Hz at the same time. Some preamplification has been used for the 1600 Hz output level to compensate for the frequency character of the acoustic coupling with the telephone handset. The distance between transmitter and telephone handset may be up to 40 cm.

DESCRIPTION OF THE RECEIVER (Figure IV)

The transmitted PCG's/ECG's are picked up at the receiving station by means of a magnetic coupler which is placed under the telephone apparatus at the pacemaker clinic to hold the telephone handset free for voice communication with the patient.

The two magnetic pickups under the telephone drive two identical tone decoders of special construction. Each channel has a narrow-band-pass filter of 45 Hz width, a log. amplifier, and an avc amplifier with a dynamic of 80 dB. The output of the last amplifier is constant for 80 dB of output signal variations. After rectifying a schmittrigger converts the signal into logic "ones" and "zeros" to drive the noise-spike detection circuit which controls the incoming pulses on their length. If longer than 50 mS, the output will be a logic "one", if not, the output remains zero. The output of the above-mentioned circuit drives the lamp display decoder, the tachometer, and the recorder decoder. The lamp display decoder has a special awaiting oneshot in its circuit to decide if a PCG/ECG received pulse is the stimulus result of the pacemaker, or a ventricular stimulation. In the first case, green is displayed; this means stimulation has been effected within 200 mS after pacing. In all other cases red, followed by orange, will be displayed. The recording output generates a negative spike of 40 mS for each pacemaker spike and a positive pulse of 120 mS for each PCG/ECG complex.

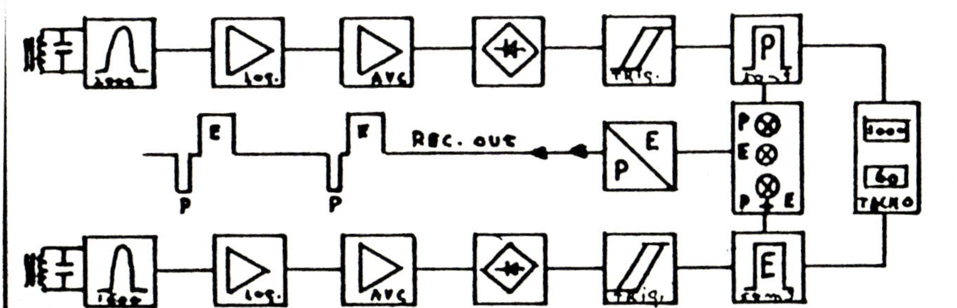

IV

CONCLUSION

A very useful system has been described which at this moment is under practical evaluation with a group of pacemaker patients and it is providing very satisfactory results.

REFERENCES

1. Netherlands PTT Regulations for Data Transmission and Telephone Lines. Ontwerp eisen lijst HtF 02-37, 1 July 1970

2. Vliegenthart A: A one channel telephone telemetry system using acoustic couplers. Proc 1st Int Symp on Biotelem, 1972, p 89

3. Furman S: Transtelephone observation of implanted cardiac pacemakers. Medical Instr 7:3, May-August 1973, pp 196-202

4. Funke HD: Ein einfaches Gerat zur Funktionskontrolle von implantierter Herzschrittmacher. Deutsche Medizinische Wochenschrift, April 1973, pp 889-890

5. Thull R, Schaldach M: Moderne Methoden der Schrittmacher-Uberwachung. Diagnostik 6, 1973, pp 517-520

6. Green GD, Thomas DL: Improvements in the technique of assessing implanted cardiac pacemakers. MBE, 1974, pp 589-592

7. Serge Barold MB, Gaiduala JJ: Evaluation of normal and abnormal sensing functions of demand pacemakers. Amer J Cardiol August 1971, pp 201-210

8. Furman, Escher, Parker: The pacemaker follow-up clinic. Prog in Cardiovasc Dis, March 1972, pp 515-530

9. Doris JW, Escher MD, Furman S: Modern Methods of follow-up of the patients with an implanted cardiac pacemaker. Amer J Cardiol, September 1971, pp 359-362

10. Furman S, Parker, Bryan et al: Transtelephone pacemaker clinic. Jour of Thoracic and Cardiovascular Surgery, May 1970, pp 827-834

A DETECTION AND MEASUREMENT OF BIOMECHANICAL STRESSES ON A WHEELCHAIR

G. Catalano[1], W. Smith, C. Winget, H. Sandler, B. Ragent[2], and L. Vistnes[3]

1. Div. of Engineering, San Francisco State Univ., San Francisco, CA., 2. NASA-Ames Research Center, Moffett Field, CA., 3. V.A. Hospital, Palo Alto, CA.

INTRODUCTION

Serious skin problems occur at pelvic pressure points in wheelchair-confined patients after varying periods of time. This condition is compounded in paraplegics and quadraplegics due to lack of skin sensation in affected areas. These occurrences vary greatly with different chair designs and for varying lengths of chair use. It is the purpose of this study to determine the multidimensional forces acting on the seat of a wheelchair. These were analyzed with respect to their shearing and compressive forces and associated moments and torsion. These data are presently used in the investigation of ulcers over the buttocks. They can also be used to predict efficiency of design of wheelchairs.

The detection of these forces was accomplished by a multiple-component strain-gauge transducer that detected static and dynamic forces in the wheelchair seat. The generated signals were amplified, telemetered, and recorded. The biotelemetry unit was an 8-channel FM-PWM multiplexed system with a frequency response of 100 Hz per channel. Two of the eight channels were used for synchronization and a reference signal.

METHOD

A block diagram of the telemetry and associated instrumentation is shown in Figure 1. Six analog signals from the strain-gauge transducers are coupled to an extracorporeal multichannel telemetry transmitter (1), amplified, time multiplexed, and transmitted on a FM carrier using pulse with modulation. The RF signal is transmitted to an FM receiver and demodulated. The received analog signals are continuously recorded for dynamic motion or printed every minute for static forces on the seat. A direct

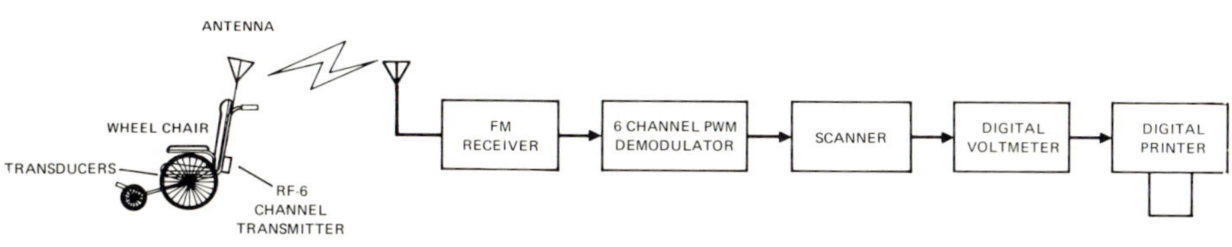

Figure 1. Wheelchair telemetry and associated instrumentation.

interface to a computer is available for data reduction and graphic displays.

A wheelchair was modified so that the seat was entirely supported by a six-component force transducer. Three components of force and three components of moment were measured using a rectangular coordinate system as shown in Figure 2. The transducer was a solid block of aluminum with material cut away leaving four small columns to support the seat. Size and spacing of the columns separated and isolated stresses due to the force input for a maximum signal output and minimum crosstalk between channels. Stress analysis of the four supporting columns revealed the locations for mounting the four 120 ohm strain gauges to produce bridge circuits which respond uniquely to each of the six components of force or moment. DC voltage to power the strain-gauge bridges was provided by the same batteries which powered the transmitter.

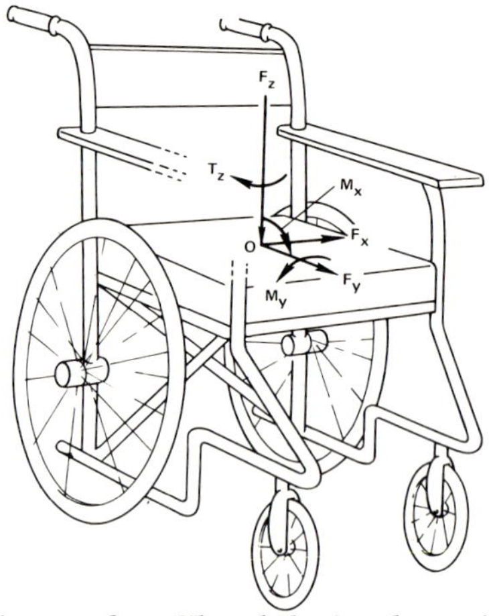

Figure 2. Wheelchair dynamics and coordinate system.

A basic 1 MHz clock pulse is time-multiplexed to provide a frequency response of 100 Hz per channel. The basic telemetry system was designed for use with implantable transducers for physiological monitoring (2). A range of approximately 100 feet is obtainable by the use of a few turns of wire to couple from the extra-corporeal antenna, a buffer amplifier, and a ¼-wave dipole antenna. Each telemetry frame consists of eight pulses with six of the pulses varying in width, in proportion to the analog signal of the transducers. The sampling rate per channel is 1.25 kHz. One of the remaining sampling channels is used for synchronization and the last is used for reference zero voltage at the input. Pulse width modulation with the addition of an FM carrier is an optimum system choice due to improved signal to noise ratios with the small signal bandwidth requirement (3). The FM transmission bandwidth is 150 kHz and provides compatibility with commercial FM receivers. The PWM subcarrier is coupled directly to the demodulator, as shown in Figure 1. The output of the demodulator is the demultiplexed analog signals which are connected to a scanner. The signals from six channels are printed, recorded, and processed by computer. A zero-reference channel is incorporated so that any drifts in the telemetry link can be compensated for in the DC signal. This channel is also used to verify the adequacy of signal reception.

RESULTS AND DISCUSSION

A sample of the recorded dynamic forces is shown in Figure 3. The motion of the wheelchair is for a normal subject propelling himself in a straight path for a distance of fifty feet and

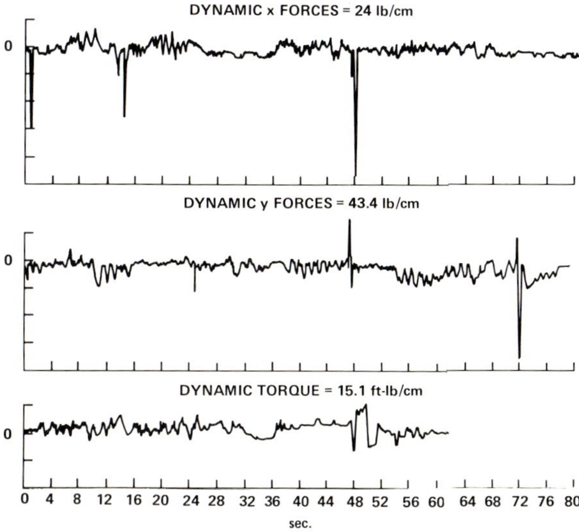

Figure 3. Dynamic forces and torque as a function of time.

returning. The correlation of these time axes are approximate due to a separate recording technique used. Static force distributions are shown in Figure 4, as a function of time. This shows the mean force on the wheelchair seat for a normal subject. For longer time durations these data are printed for better utilization. These data are in the form of three forces, namely F_x, F_y, and F_z, and three moments, M_x, M_y, and T_z. Quantization of these components is determined by the transducer sensitivity and telemetry gain. If we consider that the z-force acts only in the downward direction, the effective center of force is found by the following set of equations:

$$M_x = F_z\ y; \quad M_y = F_z\ x; \quad S_x = Fx/A; \quad S_y = F_y/A \quad (1)$$

where S = shearing stress, and A = effective buttocks area. The center of force is determined by the magnitude and sign of the measured forces and moments. An estimate of pressure and shearing stress is made by using an average area under the buttocks (4), or by planimeter methods. This procedure will result in determining the location and magnitude of the center of pressure. The first two equations of (1) are used to determine the center of pressure (or force) in compression and the latter two are required to determine shearing stress. The area used in this procedure is of importance due to the mechanical pressures exceeding mean capillary pressure (5).

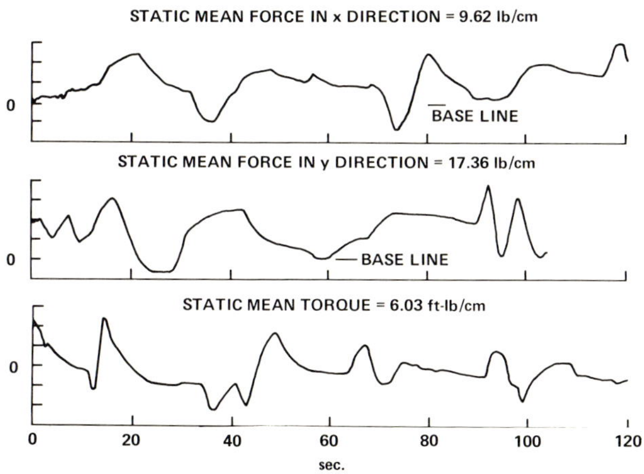

Figure 4. Static force and torque vs time.

An approximate measure of efficiency of operation of a wheelchair is accomplished by measuring the three components of reactive forces transmitted from the buttocks to the seat in order for the patient to propel the wheelchair in a forward path. The efficiency and work done by the patient to travel a known distance is based on the magnitude of the force in the direction of travel (y-direction in Figure 2) divided by the sum of the forces in the x, y, and z directions.

CONCLUSION

Accurate results of forces and moments are obtained with the use of the described telemetry system. Six high-speed channels that yield a 100 Hz bandwidth are transmitted to a nearby FM receiver for data processing. The reception distance is about 100 feet, which is adequate for most applications in which the physical, psychological, and physiological effects on wheelchair-confined patients need evaluation.

A detailed force and pressure profile can be achieved over short (for dynamic studies) or long periods of time (for static studies), and correlated by the researcher or clinician with the subjects's dynamic motion, motor activity, and heart rate. Long-term measurements may be used to relate the generation of skin ulcers to the subject's size and weight as well as the time needed for their formation. Such research can lead to the development of wheelchairs that tend to prevent tissue damage.

REFERENCES

1. Miranda R, McEwen G, et al: 3rd International Symposium on Biotelemetry, Asilomar, CA. May 17, 1976.

2. Fryer T, Sandler H, et al: J. Appl. Physiol. 39: 318, 1975.

3. Carlson A: Communication Systems. N.Y., McGraw-Hill, 1975, p 309.

4. Lindan O, et al: Arch. Phys. Med. Reh. 46:378, 1965.

5. Siegel R, Vistnes L, et al: Plast. Reconst. Surg. 51:31, 1973.

A Six-Channel Extracorporeal Telemetry System

B. Fletcher, E. Konigsberg

Konigsberg Instruments, Inc., Pasadena, CA., USA

INTRODUCTION

A six-channel extracorporeal telemetry transmitter with a compatible demodulator has been developed. The system is suitable for transmitting data in a "backpack" configuration from either chronically implanted or extracorporeal sensors. Parameters which can be transmitted are pressure, force, strain, EKG, EMG, EOG, EEG, acceleration and up to two channels of electromagnetic flow. The system's compatibility with such a wide variety of parameters makes it useful for both animal and human response studies.

DESCRIPTION

The transmitter signal conditioning package (Fig. 1) is approximately 8 x 6 x 2.5 cm and weighs 180 gm. The transmitter power pack is 1.9 x 5.4 x 5.7 cm and weighs 100 gm including batteries. With electromagnetic flow, an additional battery pack is used to supply ±1.25 V to the field coil of the flow transducer. The battery pack also contains a command switching circuit which is used to select the operating mode of the flowmeter. There are three flowmeter states -- OFF, ON, and CALIBRATE. The signal conditioner and transmitter require less than 5 mA of current, excluding strain gage sensors. The power pack uses two alkaline 9 volt batteries, with a total capacity of 380 mAH, giving the system 76 hours of continuous operation. Fig. 2 shows a block diagram of the signal conditioner/transmitter portion of the system.

There are six available data channels, each with its own GAIN and ZERO controls. The amplifiers are instrumentation types, and therefore exhibit a high CMRR (Common Mode Rejection Ratio) and input impedance. The gain is adjustable from 15 to 150X, and zero can be varied ±50%.

The outputs of the six data channels are fed to an eight-channel CMOS multiplexer where time division multiplexing takes place at a 1250 sample per second rate. This easily allows the system to achieve its D.C. to 100 Hz frequency response requirement for each channel. The first and last channels of the multiplexer are used for a system reference and a blank sync. word. The reference channel is set to give a continuous half scale output. The analogue voltage derived from it is inverted and then summed into the other six data channels in the demodulator. By doing this, compensation for a transmitter clock frequency shift is maintained.

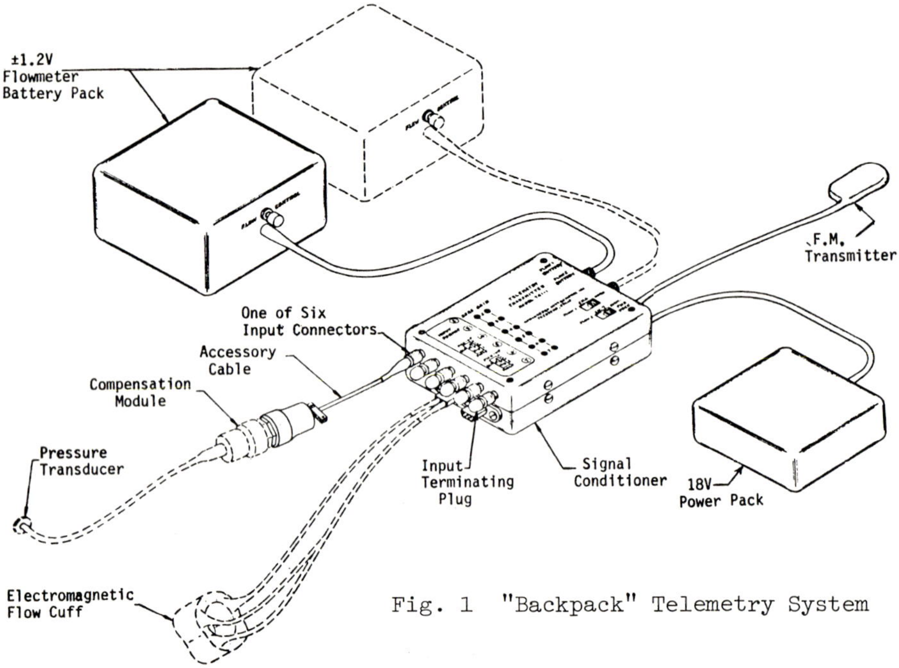

Fig. 1 "Backpack" Telemetry System

The common signal line output from the multiplexer is an eight word pulse amplitude modulated (PAM) wavetrain of periodic pulses, each 100 μsec in duration. Each PAM pulse is then converted to a pulse width modulated (PWM) wavetrain in a linear ramp/comparator stage. The signal is then ready for R.F. transmission and since it is a PWM signal, it is not susceptible to the errors caused by changes in amplitude that a PAM signal would be. The clock generator also supplies a three line binary address required for the electromagnetic flowmeter electronics. It is used for field coil drive synchronization and sample gating.

The transmitter is a hybrid circuit mounted on a cable some distance from the main package and provides up to 25 meters range under ideal conditions. Its deviation is set to ±75 KHz making it compatible with commercial F.M. receivers while allowing adequate bandwidth to faithfully receive and reproduce the PWM wavetrain.

The voltage regulator circuitry in the signal conditioning package provides three dual polarity voltages. The six input amplifiers use ±1.87 V, ±2.2 V for transducer excitation and ±3.73 V for all other analogue and digital circuits. The regulator is constructed on one P.C. board with discrete components. Also on this board are the two completely hybrid electromagnetic flowmeters. A second P.C. board contains all six hybrid instrumentation amplifiers and the hybrid multiplexer and clock generator package.

Fig. 3 shows a block diagram of the demodulator. The PWM signal from the receiver is reconstructed in the input signal conditioner to provide a noise free PWM wavetrain. The missing pulse

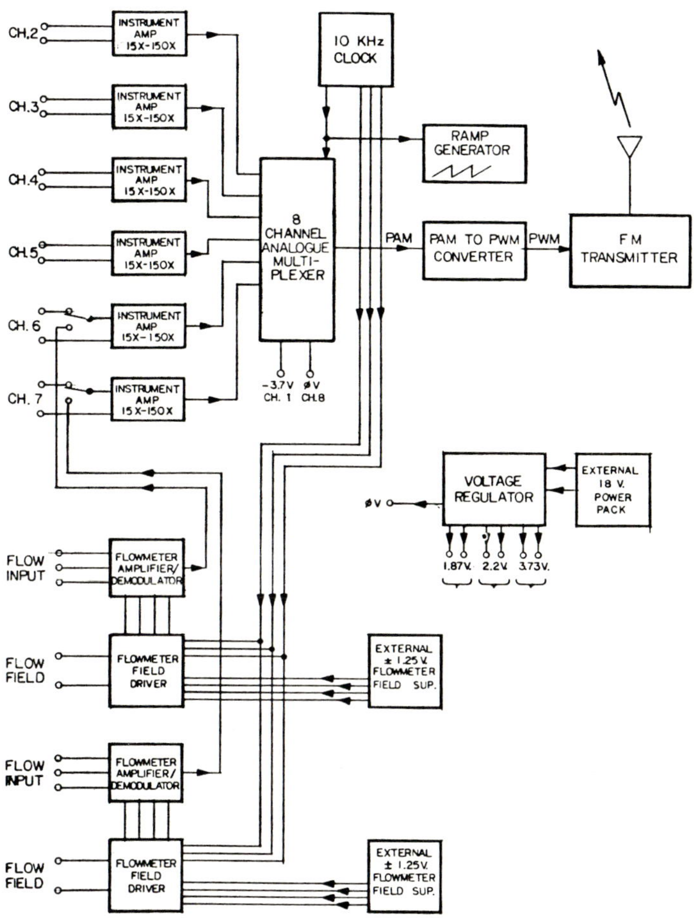

Fig. 2 "Backpack" Transmitter Block Diagram

detector senses the blank word at the beginning of each frame, and generates a sync. pulse used to reset a Johnson counter. The synchronous outputs from the counter are used to sequentially gate each of the seven PWM-to-voltage converters at an appropriate time to receive its incoming PWM pulse. The outputs of these converters, the analogue signals, are then filtered to remove the higher frequency sampling artifacts. Each channel has an adjustable zero control for initial offset compensation. The output from the reference channel, Ch. 1, is fed back prior to this stage in each of the six data channels to provide correction in output voltage due to transmitter clock variations.

All six data channels are simultaneously available for monitoring or recording.

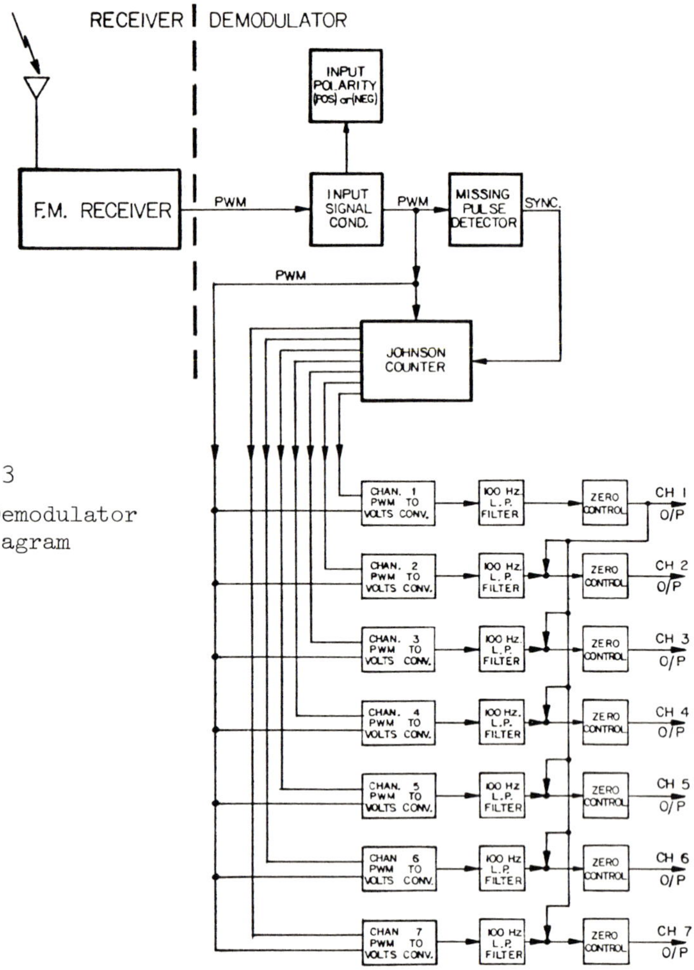

Fig. 3
Six Channel Demodulator Block Diagram

SUMMARY

Six systems such as reported on herein have been fabricated. Use of one such system is reported on elsewhere in this symposium (1). The availability of a multi-purpose miniature telemetry system which can accept a wide variety of inputs with minimal sensor interface problems should prove valuable in physiological research.

ACKNOWLEDGEMENT

This system was developed under the sponsorship of the NASA-Ames Research Center, Moffett Field, CA under contract NAS2-8690.

REFERENCES

1. Catalano et al: Detection and Measurement of Biomechanical Stresses on a Wheelchair. 3rd International Symposium on Biotelemetry, Asilomar, California, May 17-20, 1976

NOISE STABILITY IN DYNAMIC BIORADIOTELEMETRY

Jacob V. Fraden

Sverdlovsk Research Institute of Industrial Hygiene and Occupational Diseases, USSR

INTRODUCTION

Every engineer faces the problem of noise stability if he constructs a bioradiotelemetry system (BRTS) for a motion patient. We can divide all disturbances into 3 parts - equipmental, physiological, and methodical. In this article some principles of minimizing of methodical artifacts will be shown.

ELECTRODE ARTIFACTS

All electrodes, which are being used for EEG, ECG, etc. form electrochemical contact between skin and metal (1). Even though electrodes are top-quality, there is always different EMF, which comes up to 10-20 mV. During the patient motion that EMF (galvanic and polarization) is changed and the changes may reach 1.0 mV. We can say the same about static electricity, which may have changes of 40 mV to 100 mV (2). Indeed it is these changes that account for most large artifacts. Other ones are not so significant. The problem of reducing electrode artifacts to a minimum is not simple. It goes along the path of the improvement of electrode construction and attachment.

The dynamic range of BRTS is not as large as the voltage ranges of electrodes and static electricity (Figure 1). These circumstances do not allow to use a direct-current channel through whole BRTS. It is convenient to use a simple formula for computation of direct-current amplification factor in the first stage of the pre-amplifier:

$$K < \frac{AE_s}{2\Delta E} \qquad (1)$$

Where A - Voltage utilization index, ΔE - peak EMF, E_s - Voltage of the power source. We use K=20.

More perspective is the use of dry electrodes (3,4), that do not form electrochemical contact and act on the basis of capacitative transition between the skin and the pre-amplifier. Dry electrodes require pre-amplifiers with great input impedance (up to 10^9-10^{10} Ohms). The use of these pre-amplifiers is conditioned by the influence of HF-power on non-linear elements.

HIGH FREQUENCY INFLUENCE

All the elements of BRTS, including electrodes, in the strict meaning are non-linear. They always occur in the electromagnetic HF-field which is produced by the proper transmitter. The square of noise amplitude U^2 in the non-linear input impedance Z, denoted by the influence factor W(t) and the transmitter power P(t) is:

$$U^2(t) = P(t)W(t)Z \tag{2}$$

The HF-power P(t) and W(t) continuously change synchronously with the patient motion. The influence of HF-power, corrected for non-linear input is:

$$U_i \approx \frac{1}{a_1}\left[U_o + 1/2 a_2 U^2(t) + a_1 U(t) \cos\omega_o t + 1/2 a_2 U^2 \cos 2\omega_o t\right] \tag{3}$$

Third and fourth summands may be filtered out because the frequency of the transmitter (ω_o) is high. The first summand is a constant bias, but the second one is an artifact because it contains a variable-amplitude HF-voltage, U (a_1-linear and a_2-non-linear amplification factors). Finally the voltage of the artifact can be presented as

$$U_a \approx \frac{a_2 U^2(t)}{2} = \frac{a_2 W(t) Z}{2 a_1} P(t) \tag{4}$$

The artifact is directly proportional to a non-linear factor a_2, power of transmitter and input impedance.

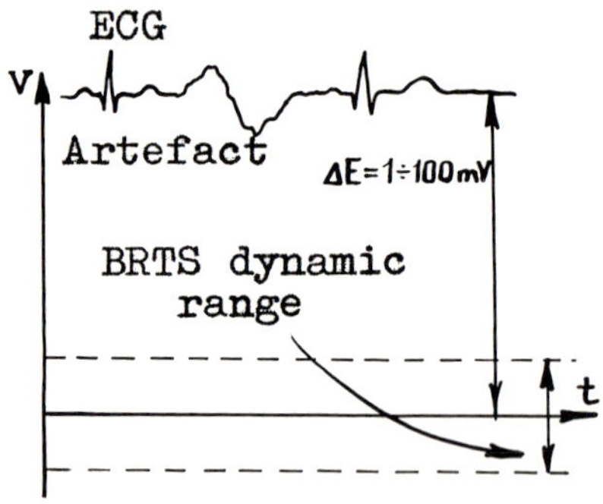

Figure 1. Voltages and dynamic range of a whole BRTS

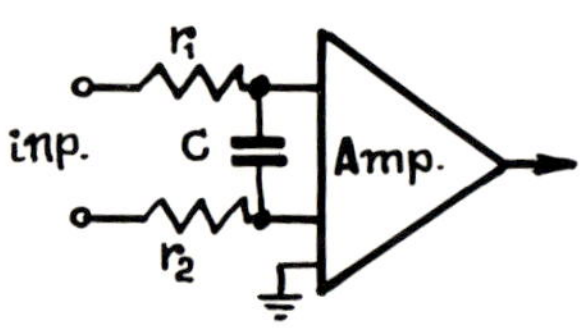

Figure 2. The optimal interlocking by the RC-filter

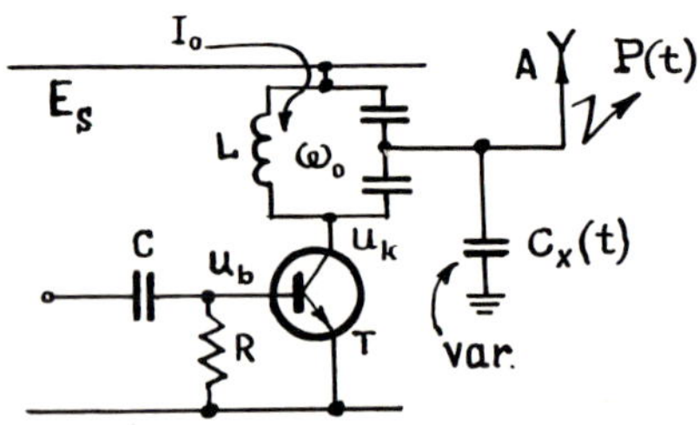

Figure 3. The output stage of the transmitter

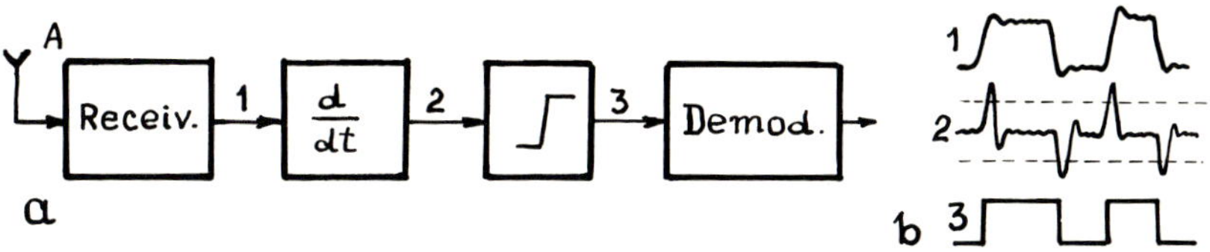

Figure 4. Simplified diagram of DPS (a) and oscillograms in the same points (b)

The most simple method to reduce these artifacts is the use of the parallel RC-filter (figure 2).

Why does the HF-power change synchronously with the patient motion? The antenna capacity C_x (see figure 3) depends on such factors as the patient overall dimensions, his motions, surrounding medium, etc. The capacity C_x is variable because of the patient motions and it changes the output-stage tuning. That tuning causes the change in d.c.I_0 and power P(t). The direct current I_0, taken from a power source, on one hand influences the pre-amplifier, and on the other hand, P(t) influences the non-linear elements and the receiver (as a fading).

PULSE SHAPING (PS)

The shaping of the received pulses is of a great significance in motion artifacts. The HF-signal is changed in the receiver input during a patient motion and a wave interference exists. In BRTS with pulse modulation the fading may reach 80%. It causes the changes of the pulse width in the threshold output and forms artifacts. The differentiated PS(DPS) reduces fading disturbances up to 40-60 dB (5). The signal is differentiated before the threshold and then the rectangular pulses are formed (figure 4).

RESULTS

On the basis of these researches, a new 4-channel BRTS "Almaz-2m" was designed. It has a double pulse-width modulation (2-PWM-FM) with non-constant period. The utilization of a 45 MHz 30 mW transmitter permits a service range of hundreds of meters. The patient equipment weighs 250 g. Interchannel interferences are less than 0.5%; and total error is no more than 3%. The portable tape-cassette recorder is used for an intermediate recording of EEG, ECG, etc.

REFERENCES

1. Pacela AF: Collecting the body's signals. Electronics 40:14, 1967, pp 103-112

2. Tam HW, Webster JC: Motion artifacts in biopotential measurements. Proc 27th ACEMB, 1974, p 320

3. Ferris CD, Hakes SD: A Dry-Electrode Electrocardiograph system. ISA BM 75304, pp 15-20

4. Rich R, Neuman M & Ko WH: Evaluation of dry surface electrodes and their attachment. 27th ACEMB, 1974, p 319

5. Fraden JV: Magnitnaja zapis telemetricheskoy informaciy. In: Otbor y peredacha informaciy. Naukova dumka, Kiev 47, 1975, pp 101-109

DISCUSSION

WOLTRING: Mr. Westbrook, could you comment on the directional uniformity of emitted radiation?

WESTBROOK: The antenna is a 1 1/4" loop around a printed-circuit board. In free air it is very directional. When placed in proximity to the subject's leg it is coupled into the body, thereby completely changing the radiation pattern. In the 40' x 20' room where we have been working, we have not found nulls to cause any dropouts, though there is probably a 20 to 30 db change in RF signal strength. As long as we maintain 5 uv or better signal level at the receiver, that is adequate for clear reception.

THE STATUS OF BIOTELEMETRY

- The Editors' View -

The Symposium proceedings, including formal and informal discussion, indicate that biotelemetry has advanced to the state where it is contributing to patient diagnosis and care, and to research in medicine and the biology and behavior of humans and animals. However, biotelemetry is still in the early, rapid stages of development; it is still low on the learning curve, so that large, rapid advances might be expected. This publication can be considered a status report at this point in the rapid development of biotelemetry.

The first link in the chain of a biotelemetry system (except in the case of biostimulation) is a sensor to sense the magnitude of the parameter of interest. Except for a simple voltage sensing electrode[1], most sensors used for biotelemetry combine the transducer function with the sensor function. A few sensor transducers (generally referred to as "transducers") are available commercially, but most of the projects described at the Symposium involve the development of special purpose transducers just for the project at hand. During the present stage, biomedical researchers generally develop their own biotelemetry systems. Several variety of physical reactions are used to sense changes of pressure, temperature, acceleration, displacement, or some other biological function into analog voltages or coded pulses suitable for teletransmission. The principal problem with transducers is in achieving long-term stability in both baseline and calibration slope. These are both items of critical importance for implanted transducers. The area of transducer stability and ease of calibration is in a state of flux and requires continued research and development.

Neither electronic components nor circuits seem to be limiting factors, due to the availability of a variety of transistors and other miniature, low-power, solid-state components and to the ingenuity of electronic engineers. Development of improved technology for economical production of special purpose integrated circuits in small quantities[2] would be a great boon.

---

1 Miller HA, Harrison DC: Biomedical Electrode Technology: Theory and Practice. Academic Press, New York and London, 1974

2 Combs SR, Meindl JD: KITCHIP-A Generalized Integrated Circuit for Biomedical System Design.

Modern techniques have made possible amazingly compact implantable transducers and short range transmitters. Of course, efforts to this end should continue as long as significant gains are possible. An addage commonly heard from implant researchers on the size of implantable hardware is "if it's big enough to see, it's too big".

There appears to be available an adequate choice of biologically compatible materials for construction of implantable units.

Transmitting range requirements vary from a few centimeters in clinics or laboratories, to several miles for roaming wild animals, to hundreds of thousands of miles for astronauts. These requirements can be met with commercially available or special transmitters provided sufficient power is available. In some communities, however, congestion of the air waves poses a real problem. Resort to telephone lines to alleviate this problem is being tried in at least one large metropolitan area.[3] Coordination of many interests is needed to achieve the best use of the limited radio frequencies.

Receivers are seldom a problem. Commercially available receivers, including FM broadcast receivers used as is or with minor modifications, generally meet the need. Standard types of antennas are adapted to the various needs.

Off-the-shelf recorders and display devices can generally be adapted to meet the needs of biotelemetry. Likewise, commercial computers can generally be used for storing and analyzing the data. In some monitoring and research projects such masses of data are recorded that analysis is a monumental task. For such cases,thought is being given to including some "pre-computer" capacity in the transducer-transmitter unit so as to transmit only the most useful information.

Batteries to power implanted telemetry units for long periods of time are always a problem despite the considerable effort that has been devoted to improving them and to reducing power requirements. Such efforts should be encouraged, but until batteries reach zero size and unlimited life, the users of implanted devices, particularly the patients, will continue to long for smaller batteries of greater capacity with complete reliability. An attractive alternative to batteries, supplying power to implanted instruments by induction from external sources, is reported in some papers. The demands on batteries are minimized in some cases by ingenious switching schemes which activate the telemeter system to transmit only on command.

Reliability of biotelemetry will probably never achieve the utopian 100%, although some users have found radiotelemetry to be

---

3 Nagel EL: Emergency Medical Telemetry, 3rd International Symposium on Biotelemetry, Asilomar, California, May 2976

more reliable than wired systems. Meticulous design, fabrication, and use have made many biotelemetry systems reliable enough to win general acceptance for use with humans even in critical situations. Efforts for further improvements must be unrelenting to make systems completely safe and reliable in the hands of physicians, researchers, nurses and paramedics without the need for constant attendance of electronics experts.

The safe limits of exposure to RF radiation from biotelemetry instruments and in our daily environment are not adequately known.[4] Even the ways in which radiation affects biological processes are not known. This area deserves early, intensive study.

Cooperation between physicians and biologists on the one hand and engineers and physical and chemical scientists on the other has been increasing. The results have been encouraging. Already, biolotelemetry is playing a part in reducing sudden death from heart failure, the greatest single cause of sudden death in this country through its application to monitoring both in- and outpatients. The multidisciplinary cooperation should be extended and enlarged with apparent benefit to all mankind.

4 Neukomm P: Health Hazards from Telemetry RF Exposure: A Review of the Interactions Between Electromagnetic Fields and Biosystems. 3rd International Symposium on Biotelemetry, Asilomar, California, May 1976

DESIGN STANDARDS FOR BIOTELEMETRY EQUIPMENT - ARE THEY COMING?

- The Editors' View -

INTRODUCTION AND BACKGROUND

During 1975 and early 1976 the United States Federal Communications Commission (FCC) considered the questions of necessity and appropriateness of having design standards for one type of emergency biotelemetry equipment (1). The type of equipment under consideration was that used in emergency medical, e.g., voice, ECG, and other physiological data transmitted by radio or telephone between remote or mobile field equipment and a hospital or other health care facility. The FCC also considered whether such a standard should, if adopted, be one proposed by the United States Interdepartmental Radio Advisory Committee (IRAC).

The impetus for the IRAC proposed standard and for the FCC consideration of its adoption was the diversity of signal characteristics among the equipment being produced by different companies for biomedical telemetry purposes. This diversity has led to incompatibility between transmitters and receivers of different manufacturers. Thus a transmitter which is part of one biomedical telemetry system might be unable to communicate with another hospital system because of the signal incompatibilities.

PROPOSED STANDARD

The IRAC standard, as proposed, required that the data be transmitted on a 1.4 kHz subcarrier within a voice channel in the 460 MHz band. The standard specified the subcarrier frequency and deviation and the characteristics of the input voice and physiological signals, the composite output, the voice output, the maximum levels of distortion and noise, the received signal processing characteristics, and the equipment calibration procedures. IRAC also incorporated in its standard, by reference, the Electronic Industries Association (EIA) Land Mobile Standards.

FCC INQUIRY

The FCC followed the standard process of publishing a "notice of inquiry and proposed rule making" in the Federal Register and soliciting comments from all interested sources. As reported in (1), comments were received from 14 parties including Federal and State agencies, professional associations, manufacturers, and one individual. Some respondents supported the proposed rule making, but the majority opposed it. Supporters argued that uniformity in equipment design was required to assure intersystem compatibility and that the rule was required to provide a legal basis for assuring compliance to the standards. Opponents argued that the field of biotelemetry technology was young and changing, and that the imposition of rigid design standards at this time would hinder innovation and delay needed advances in the state-of-the-art in an area that is vital to life saving in emergency situations. There was general agreement that the IRAC standards would promote compatibility among biomedical telemetry systems and that they were the best that could be developed at the present time. However, the majority believed that the disadvantages noted above were sufficient to make adoption of standards inappropriate at this time.

FCC CONCLUSIONS

As a result of its deliberations, the FCC concluded that incorporating a biomedical telemetry equipment compatibility standard into its Rules at this time would not serve the public interest. Nevertheless, the FCC expressed support for efforts to achieve compatibility and strongly urged the development of area wide medical radio systems with central dispatch flexibility. The area wide concept could achieve compatibility by regional planning and coordination of medical response systems.

INDICATIONS FOR THE FUTURE

One may draw several inferences from the events and opinions reported in the Federal Register (1) and paraphrased herein. Among these inferences are:

(a) The suggested "regional planning and coordination" may well result in several regional sets of standards.

(b) Aggressive efforts continually to improve biotelemetry systems and the natural time lapses between the establishment of various regional systems can result in parallel advances in both equipment and regional standards.

(c) Unless major changes occur in the tendencies toward ever-increasing regulation by nearly all governments, design standards for biotelemetry equipment are coming.

(d) It would behoove the biotelemetry equipment developers, therefore, to exercise a maximum of aggressiveness in seeking solutions to technical problems and devising improved equipment in order that various regional health-care planning groups may have available demonstrably better state-of-the-art equipment upon which to base their "regional standards". In this way, it may be possible for regulation to encourage rather than deter innovation and advancement.

REFERENCE

1. Federal Communications Commission: Biomedical telemetry equipment standards for design. Federal Register 41:107, p 22280, June 2, 1976